Cities, Heritage and Transformation

Series Editor

Fatemeh Farnaz Arefian, Bartlett Development Planning Unit
Silk Cities, University College London
London, UK

This book series addresses contextual and global urban challenges and opportunities that cities in historic regions face within the process of urban transformation, linked with their historic past. It publishes peer reviewed books on the relationships between the built environment, urban transformation and cultural heritage, whether to be tangible or intangible, overarched by the notion of resilience and sustainable development. Special attention is given on real-life under-explored topics. Challenging existing assumptions and disciplinary divides, the series takes an interdisciplinary position and brings together innovative researches from different areas within Geography, Social Sciences and Humanities.

The aim is to create a knowledge hub for academics and practitioners, researchers and research organisations, as well as managers and policy makers worldwide, presenting advances and case studies, which connect academia to practice.

The series benefits from the existing and growing Silk Cities' network, as well as ongoing collaborative activities Silk Cities undertake as part of its intellectual leadership on the subject matter.

The series is open for empirical, theoretical, and methodological high-quality contributions, which advance the global understanding on the subject matter. The areas covered in the series include but are not limited to the following subjects:

- Cities and intangible cultural heritage
- Community participation in managing urban heritage
- Urban history
- Urban heritage
- Cultural memory, society, and the built environment
- Development management in historic contexts
- Disaster management in historic cities
- Disaster risk reduction and resilience of historic cities and societies
- Disaster recovery and reconstruction
- Environment and historic cities
- Governing historic cities
- Geotourism and geoheritage
- Historic contexts and complex urban systems
- Post-crisis reconstruction and recovery
- Smart historic cities
- Sustainable heritage tourism
- Urban Silk Roads
- Urban economy in historic cities
- Sustainable urban heritage
- Urban conservation
- Urban design and public spaces in historic context
- Urban planning of historic cities
- Urban morphology
- Urban transformation of historic cities
- Urban connectivity and Silk Roads
- Historic cities, urban policies, and territorial planning

Eugenia Bitsani • Theodoros Tsekos •
Isidora Thymi • Georgia Marava

The Biocultural City as a Utopia or the Future Reality?

Models, Policies, and the City of Ixelles

Springer

Eugenia Bitsani
Business and Public Administration
University of the Peloponnese
Kalamata, Greece

Isidora Thymi
Business and Public Administration
University of the Peloponnese
Kalamata, Greece

Theodoros Tsekos
University of the Peloponnese
Kalamata, Greece

Georgia Marava
Business and Public Administration
University of the Peloponnese
Kalamata, Greece

ISSN 2731-5363 ISSN 2731-5371 (electronic)
Cities, Heritage and Transformation
ISBN 978-3-032-17583-0 ISBN 978-3-032-17584-7 (eBook)
https://doi.org/10.1007/978-3-032-17584-7

This Springer imprint is published by the registered company Springer Nature Switzerland AG
The registered company address is: Gewerbestrasse 11, 6330 Cham, Switzerland

Ethics Approval

We clearly state that informed consent was obtained from individual participants.

"A landscape is not simply a collection of land, plants, and water. It is the projection of a people's soul onto matter."
—Odysseas Elytis, Greek poet *Excerpt from "The Public and the Private", Ikaros Publications, 2011.*

Competing Interests The authors have no competing interests to declare that are relevant to the content of this manuscript.

Introductory Letter

Cities are not made of stones, but of relationships; and to build their future, we must first learn to imagine it—because sustainability is, ultimately, a cultural choice.

Dear Readers,

It is with great pleasure that we introduce this volume, the culmination of an extended research journey shaped by a shared commitment to rethinking the future of European cities. Our starting point is a critical observation: despite decades of strategies, frameworks, and declarations, the challenges cities face across Europe continue to be addressed in fragmented and technocratic ways. Urban policy remains largely embedded within neoliberal paradigms that prioritise operational, managerial, and economic tools while treating culture as an optional or decorative element rather than a foundational dimension of urban life.

This reductionist approach has produced top-down policies that, although couched in the language of participation, inclusion, and diversity, frequently ignore the relational subjects who shape the city from below. Neighbourhoods, communities, and grassroots networks—where bonding, bridging, and linking capital is formed—rarely serve as the starting point for designing the urban future. Cities are living cultural ecologies made of relationships, meanings, practices, and identities; yet contemporary governance often fails to recognise this basic truth.

The originality of this book lies precisely in overturning the dominant economic narrative and placing *culture at the heart of the paradigm.* Through a biocultural lens, we argue that genuine sustainability, resilience, and social justice cannot emerge without understanding the cultural grammars, social ties, and ecological sensibilities that shape our cities. Culture becomes the "umbrella" under which environmental, social, economic, and technological policies can be reconnected and given coherence.

Our contribution provides new scientific knowledge by bringing together theoretical inquiry, critical policy analysis, and empirical research. The case of the Municipality of Ixelles—located at the centre of Brussels and the European

Union—demonstrates how a cultural–ecological approach can be successfully implemented in practice. Ixelles shows that when neighbourhood communities develop strong internal bonds, when diverse groups build bridges between them, and when institutions create meaningful links, a city can evolve into a vibrant mosaic—a *biocultural city* shaped by its people.

Our Mediterranean origins deeply informed this perspective. Born and raised in the small places of the Mediterranean—among the winds and light of the Aegean, in landscapes touched by Homeric memory, and within a way of life scaled to the human measure—we carry with us a cultural ethos centred on coexistence, reciprocity, and the everyday rituals of community life. To our surprise and delight, we encountered elements of this Mediterranean spirit in Ixelles: in its neighbourhoods, its plurality of cultures, its rhythms of public life, and the coexistence of many worlds within one shared space. This resonance guided our work and reaffirmed our conviction that culture is not a peripheral embellishment, but the essence of urban being.

We sincerely hope that this book will contribute constructively to scientific and political dialogue on the future of European cities. Our aspiration is for it to serve as a valuable guide for students, scholars, policymakers, municipal authorities, and citizens who wish to imagine and shape cities that are more inclusive, sustainable, and culturally vibrant. Ultimately, we hope to help build societies in which everyone has a place, a voice, and the possibility to belong—a future in which the sustainability of our local worlds and of our planet is safeguarded.

We extend our warmest gratitude to our families, who supported us unwaveringly throughout the entire research process. Our heartfelt thanks also go to *Mr. Christos Doulkeridis*, former Mayor of Ixelles; *Mr. Jean-Michel Defays*, former Professor at HE2B; *Maria-Rosa Salvado*; *Stavro Klognomo*; and *Mr. Mohamed Mashla*, for their generous participation in the study and for their insightful comments, reflections, and critical observations.

Kalamata, Greece

Eugenia Bitsani
Theodore Tsekos
Isidora Thymi
Georgia Marava

Contents

List of Acronyms

CBAM	Carbon Border Adjustment Mechanism
CBD	Convention on Biological Diversity
CE	Circular Economy
CEAP	Circular Economy Action Plan
CFC	Chlorofluorocarbon
CHCfE	Cultural Heritage Counts for Europe
CLIM	Climate
CO_2	Carbon Dioxide
COVID	Coronavirus Disease (COVID-19)
EGD	European Green Deal
EU	European Union
EURESFO	European Urban Resilience Forum
GCAP	Greenest City Action Plan
GDIP	Green Deal Industrial Plan
GPURL	Global Practice for Urban, Disaster Risk Management, Resilience and Land
HUL	Historic Urban Landscape
ICOMOS	International Council on Monuments and Sites
IFLA	International Federation of Landscape Architects
IoT	Internet of Things
IPCC	Intergovernmental Panel on Climate Change
JRC	Joint Research Centre
LGBTQ+	Lesbian, Gay, Bisexual, Transgender, Queer and others
NASA	National Aeronautics and Space Administration
NBS	Nature Based Solutions
Nox	Nitrogen Oxides
NZIA	Net-Zero Industry ACT
ODS	Ozone Depleting Substances
OECD	Organisation for Economic Co-operation and Development
REITs	Real Estate Investment Trusts

ROCK	Regeneration and Optimisation of Cultural heritage in creative and Knowledge cities
SDG	Sustainable Development Goals
TURaS	Territorial and Urban Potentials Connected to Culture
UCCN	UNESCO Creative Cities Network
UN	United Nations
UNCED	United Nations Conference on Environment and Development
UN-Habitat	United Nations Human Settlements Programme
UNESCO	United Nations Educational, Scientific and Cultural Organization
USA	United States of America
WCED	World Commission on Environment and Development
WEF	World Economic Forum
WHO	World Health Organization

List of Figures

List of Graphs

List of Tables

Part I
Theorising the Biocultural City: Principles, and Conceptual Approaches

Chapter 1
Introduction

Despite the growing academic and policy interest in sustainability, the interconnection between culture and ecology has rarely been examined as a single, integrated system of thought and practice. Existing studies tend to address cultural policy, environmental management, and urban development as separate or only loosely related domains. This research responds to that gap by proposing a biocultural framework in which culture is recognized as a foundational and transversal axis that intersects and shapes all dimensions of sustainable development. Within this context, the study seeks—both theoretically and empirically—to reverse the prevailing paradigm that places the economy at the center of sustainability. It instead positions culture, within the ecology of life, as the organizing principle of a new model: the biocultural approach, which redefines sustainability as a culturally embedded process of ecological awareness, social inclusion, and collective meaning-making.

From antiquity through the Middle Ages and into the present day, cities have faced both internal upheavals and external threats—economic crises, poverty, social instability, crime, and severe natural phenomena such as floods and earthquakes (Bitsani, 2013). Yet many cities have managed to preserve essential characteristics—such as their cultural heritage—or to reconstruct themselves and flourish anew (Bitsani, 2016a).

The shared history of European cities has endowed them with comparable identities. Today, their principal challenge lies in redefining and reshaping developmental frameworks and objectives, which concern not only their content but also their scale (Bitsani, 2013, pp. 9–10).

Throughout this study, the term *European cities* refers specifically to cities within the European Union, as research was conducted within the framework of EU policies. Urban space embodies the synthesis of all elements necessary to meet the functional needs of societies developing within its boundaries. It also reminds us that urban planning and development are the results of strategic thinking and craftsmanship.

E. Bitsani et al., *The Biocultural City as a Utopia or the Future Reality?*, Cities, Heritage and Transformation, https://doi.org/10.1007/978-3-032-17584-7_1

Modern cities face increasingly complex pressures stemming from natural factors, human activity, and socio-economic conditions. Planning practices and strategies that promote resilience and sustainability constitute essential tools for addressing these risks and, consequently, for safeguarding and maintaining urban environments.

Buildings of significant aesthetic or historical values such as heritage structures, monuments, and cultural sites, along with traditional residential ensembles and historic city centers, represent invaluable cultural assets that must be protected (Rodwell, 2007). At the same time, these structures serve as repositories of knowledge and methodological frameworks for achieving urban resilience and sustainability (Meier et al., 2007).

Urbanization exerts multifaceted effects on settlement patterns, transforming cities into centers not only of employment but also of economic and cultural life (Stevenson, 2003). As early as the early twentieth century, Simmel (1971) examined how the sensory experiences of individuals within metropolizes could transform accidental encounters into sources of misunderstanding. His work, grounded in a sensory approach to social life, reveals the psychological implications of modern urban existence.

The Chicago School of Sociology used the city of Chicago as a living laboratory for exploring urbanization and its accompanying challenges (Bitsani, 2013). In the twenty-first century, the rise of megacities within an increasingly globalized society, combined with the transformative power of the Fourth Industrial Revolution, has generated unprecedented complexities. The global city has become a pivotal stage for political, economic, social, and cultural interactions (Pooch, 2016; Sassen, 2005).

Within this evolving landscape, fluctuating populations and technological advancements reshape material resources, redefine ideologies and values, and restructure institutional frameworks. These shifts demand new developmental paradigms capable of reconciling economic efficiency, cultural vitality, and environmental sustainability.

Significant challenges emerge during the process of reformulating developmental plans and objectives. These challenges are not only related to the context of these plans but also to their scope, which is influenced by the phenomenon of "globalization." In his 2003 argument, Habermas posits that the concept of globalization ought to be considered as a dynamic rather than a static phenomenon (Habermas, 2003). It facilitates the extent and intensity of movement, communication, and exchange beyond a country's geographical boundaries. In the contemporary era, it is no longer feasible for any city to adopt a legislative stance in complete isolation; as such, it is imperative that cities do not neglect their financial policies or vulnerabilities when confronting the monumental challenges arising within the context of the prevailing economic crisis, recession, and rampant globalization (Bitsani, 2013).

Bauman (2003) characterizes the contemporary era as a mutation in the process of transitioning from an ageing world that is fading away. The advent of globalization has precipitated the formation of a new world order. The fall of the Berlin Wall was a seminal moment in global history, precipitating a paradigm shift in which previously insular countries became open to external influences (Bauman, 2003).

Economic globalization and global colonialization remain blind toward democratic deficiencies, social exclusions (Stiglitz, 2003 in Ferri, 2003; Kovel, 2007), financial inequalities, and the environmental destruction (Beck, 1992) that it creates.

Utopia from the Greek ou-topos, meaning no place, or nowhere, the word utopia has become adopted in the English language to mean a place where everything is ideal or perfect. In addition, Utopia is an imaginary community or society that possesses highly desirable or nearly perfect qualities for its citizens. The term was coined by Sir Thomas More for his 1516 book Utopia, describing a fictional island society in the south Atlantic Ocean off the coast of South America.

The opposite of the concept of utopia is the concept of "dystopia," which etymologically means "bad place."

For a long time, it was believed that John Stuart Mill was the one who coined the term "dystopia," which he used in a speech to the British Parliament in 1868. But recent work has shown that this word appeared in 1747. It then had the meaning of "unfortunate country." But this term was not used in the current language until the 1950s, in a context marked by the Cold War, the fear of a nuclear conflict and the success of the novel "1984" by George Orwell.

These two concepts can help us see where we are, how we got here, and the path that lies before us. Utopias are intended to predict the distant future, offering predictions of what awaits us. In doing so, utopia offers us an alternative vision, allowing us to imagine a much better world that we could build.

Until now policies worldwide aim at a utopian ideal, that of universal prosperity. This is an unattainable goal, to the extent that the planet's resources are insufficient and seem to be depleted by over-consumption. A realistic diagnosis of our environmental situation demonstrates that since the middle of the twenty-first century, we have entered an unprecedented era of climate warming. The temperature will undoubtedly increase by 3 °C by 2050, which will cause the extinction of many forms of life on Earth, possibly including humanity. We are therefore facing the worst dystopian scenario ever predicted. It is, however, possible to imagine a utopian solution to this extreme ecological disaster, revolving around a set of ideals such as sustainable development, rational consumption, and peaceful intercultural coexistence.

This study aims to deepen the understanding of European cities' identities, but also seeks to develop a proposal, perhaps "utopian" for the future, based on harmonious equal coexistence in the context of holistic approaches.

At this point we consider it necessary to clarify that the term European cities refers to cities within the European Union and the scope of the research is within the European Union. Therefore, where the term "European Cities"is mentioned below, it refers to cities within the European Union.

Culture and ecology have a powerful role to play in regenerating the potential of diverse communities and contexts and connecting them on a global scale to benefit all and the planet. The fourth industrial revolution with the new emerging technologies contributes to this direction.

To this end, cultural and environmental policies on the one hand, must integrate emerging technologies and, on the other hand, must be integrated into the concept of sustainable development, both as an autonomous backbone of sustainable development and as a contributor to its many dimensions.

Specifically, in this book we argue that a cultural shift is essential to fulfill the sustainability goals as it requires a paradigm shift, which is first and foremost a change in thinking.

We propose that culture must be at the heart of all policy areas if sustainability goals are to be achieved, as we argue that biocultural lenses are the key tool for urban development. Our basic research question is whether cultural lenses in biocultural approach of sustainability help to address issues of inequalities like power, minorities, gender, etc. and at the same time address the pressing issues of climate change, adaptation, and sustainable development. To strengthen and substantiate our argument, the city of Ixelles-Brussel's is used as a case study, where we conducted large field research using mixed methods to explore, among other things, how residents and municipal authorities perceive their city daily life and to what extent it could be used as an asset for sustainability.

Most of the population on the European continent lives in cities. Cities constitute centers rich in history and traditions, industrial, commercial, and other activities. The features cities have form citizen's physiognomy and their way of life, as a result they function as a point of reference.

The major problem that cities face today is the re-formulation or the recreation of the developmental schemes and goals, which concern not only their context but their scale too.

The scale is determined by what is known as "globalization." This term is a dynamic process, a necessary tool of capitalist economy, which, however, produces continuous crises at multiple levels. No city is able to completely self-legislate today; thus, no city is able to ignore its financial policies or its weakness to deal with the huge problems which are created within the circumstances of the economic crisis, pandemic crisis and recession, and the absolute globalization that we are experiencing today. The ongoing crisis on a global scale started in the USA in 2008 with the collapse of the financial system and was transferred—in a "domino" effect—to Europe with the collapse of the European banking system respectively. These events resulted in the economic crisis of the European countries, recession, and unemployment. The burden of this crisis fell on the European industrial cities but also on the countries of the European south (Bitsani, 2013).[1]

---

[1]The economic crisis—especially the 2008 global financial crash and the subsequent Eurozone debt crisis—had deep and long-lasting consequences for Southern European countries and many industrial cities across Europe.

1. *Southern European Countries (Greece, Italy, Spain, Portugal).*

Public Debt and Austerity, Unemployment, Recession and GDP Contraction. Emigration Unemployment drove many young and educated people to emigrate, especially to Northern European countries. Political Consequences Growing Euroscepticism and mistrust toward EU institutions and policies

2. *Industrial Cities in Europe.*

Additionally, 2020 was a year of historic turning point for humanity, due to the outbreak of the COVID-19 pandemic, a fact that had not been recorded again for about a century. From the beginning of this year until today, millions of people have either been infected or died from the virus. The pandemic was the catalyst and accelerator of the recession and the prolonged economic crisis, as well as the imposition and prevalence of a series of practices, restrictions, and behaviors that were not unknown in the technologically advanced societies of late modernity. The pandemic, as a biopolitical event, in the philosophical sense of the word, runs through every form of life and transforms it anthropologically and socially, in a fast and corrosive way (Contiades & Fotiadou, 2020).

The pandemic has left an indelible mark in the panorama of the city and induces so urgently to rethink the essential development of cities in different dimensions: social, cultural, infrastructural, economic, environmental, and more. Cities were among the first areas of the globe to be affected, but cities, which have always been home to vibrant economic and cultural life, are also where creative solutions are devised.

Then, from 2022 onwards, the issue of climate change began to be raised with greater urgency. The issue of climate change is considered, nowadays, both an emergency and a threat to the entire planet (UNCTAD, 2022). The phenomenon of climate change is not a sign of the times, but a process that has been happening for centuries and continues to happen naturally. However, the global warming observed in recent years is the result of anthropogenic activities and not of any natural climate variability. According to the scientific community, the average global temperatures over the last 50 years are the highest in the last 1300 years, as a consequence of the

---

Traditionally industrial cities, particularly in Southern and some parts of Northern Europe, were also heavily affected by the crisis.

3. Job Losses in Manufacturing.

The crisis accelerated deindustrialization in cities such as Turin (Italy), Bilbao (Spain), and Thessaloniki (Greece).

Small and medium-sized enterprises (SMEs) struggled with falling demand and restricted access to credit.

4. *Urban Decline.*

Rising unemployment led to poverty, homelessness, and—in some cases—increased crime.

The housing market collapsed in overdeveloped cities, such as Valencia in Spain.

5. *Reduced Public Services.*

Austerity measures led to cuts in health care, education, and transportation.

Many municipalities were heavily indebted, limiting their ability to invest in infrastructure or social programs.

6. *Social Fragmentation.*

Inequality and social exclusion grew, creating tensions—especially in multicultural or economically uneven neighborhoods. Report - A7–0034/2010 European Parliament. https://www.europarl.europa.eu/doceo/document/A-7-2010-0034_EL.html

rapid increase in greenhouse gases, mainly carbon dioxide (CO2), while based on various climate models, it is estimated that the earth's temperature will continue to rise in the coming decades (Bitsani & Anselmi, 2021).

Climate change, which nowadays tends to be replaced by the term "climate crisis" due to the intensity and quality of the phenomenon, is caused mainly through two mechanisms: (a) the release of greenhouse gases, originating from human activities and (b) anthropogenic changes to the planet's surface that result in a reduction in its ability to absorb greenhouse gases by transforming solar energy into other forms of energy, besides heat. Indicatively, some of the negative impacts of the climate crisis are drought, disruptions in the availability and quality of water resources and in agricultural production, the increase in the frequency of extreme weather events, desertification, rising sea levels, and the spread of tropical diseases. Although the intensity of climate change will not be of equal magnitude in all regions of the planet and by extension the negative impacts will not affect all countries uniformly, cooperative measures are required at a global level (Bitsani & Anselmi, 2021).

However, climate change is not a purely environmental problem, but a complex issue, with political, scientific, technological, and economic implications, raising global concern. According to estimates, the wider Mediterranean region is one of the areas that will be significantly affected by climate change. Climate change (increasing average temperatures and decreasing precipitation) will have the effect of affecting the hydrological cycle and the availability of water, both in surface and groundwater. The issue of water is a critical issue regarding the impacts of climate change, since it is linked to a wide range of human activities, such as livestock farming, agriculture, tourism, ecosystems, but also survival itself (UNCTAD, 2021, 2022). According to estimates, global warming will continue to place additional pressure on water systems in countries around the world, which requires a reassessment of policies, technologies, and priorities in this critical sector (Perry & Praskievicz, 2017).

In this framework as the modern state enters the stage of its liquidation, it is apparent that public discussion regarding ethnic diversity and climate change or crisis dominates the social sphere. Diversity has become a myth ready for consumption in various cultural spaces: politics, literature, mass media, advertising, leisure activities, as Anselmi and Gouliamos have proved in their book "Elusive margins: Consuming media, ethnicity, and culture," since 1998. They revealed the patterns of exclusion, falsehood, and disorder constructed systematically by power elites in order to obscure diversity and quash the autonomy of subordinated communities (Anselmi & Gouliamos, 1998).

Today, democracy faces new threats; the UK and Europe are in a period of upheaval after Brexit; the US plots a tricky path out of the Trump administration; coronavirus ravages much of the world. The situation in Asia and Africa is much worse with authoritarian regimes, huge poverty, and barbarism. Inequality is rising, and divisions in society are widening, often fueled by inflammatory rhetoric from governments.

Today's problematic political-economic scene is basically the result of globalization. According to Habermas (2003), globalization is not a final stage but a process that facilitates the extent and intensity of any type of movement, communication, or exchange beyond the geographical boundaries of a country.

The most important parameter is unquestionably the fact that economic globalization, which is ungovernable, since there seems to be a lack of actual inter-state regulations, since it is argued that any kind of regulations would supposedly undermine balance around the globe (Global Governance Future 2030, 2019) .

In addition, the coercive dissemination of unequal economies undermines the very existence of sovereign states.

The reality that we are experiencing is described by Baumman as our mutation in the process of change from an old, "very old," "ageing" world that is disappearing. Through globalization a new world is taking shape. What we have seen from the fall of the Berlin Wall was an opening to the outside world for countries that were locked up in their own little worlds (Baumman, 2004). This is an entrance to the world open societies for billions of people. This opening leads to a redistribution of the global comparative advantages; it also leads to a world where new ideas in the market and in free enterprise prevail (Centeno & Cohen, 2012).

On the other hand, according to Atchison and Shames (2019), as they argue in their book "Survive and Resist: The Definitive Guide to Dystopian Politics," the definition of dystopia is political. Dystopia is not a real place; it is a warning, usually about something bad the government is doing or something good it is failing to do. Actual dystopias are fictional, but real-life governments can be "dystopian"—as in, looking a lot like the fiction.

Atchison and Shames (2019) demonstrate that dystopias, both real and imagined, help bring theories of governance, citizenship, and the state down to earth. They emphasize nonviolent resistance and change, exploring ways to challenge and overcome a dystopian-style government.

Industrial culture is substituted by the society of knowledge whose main symbol is the best internet, and it is guided by the advent of powerful technologies (Anselmi & Sheena, 2011).

This is what Alvin Toffler describes in the history of humanity using the term "the third wave" which leads to the "economy of knowledge" (Toffler, 1981). We are at the foot of a new Himalaya as far as scientific and technological progress is concerned and yet we have not gone further than a couple of millimeters. In this new world the most important thing is creativity, consequently individual initiative acquires a pivotal role that entails personal freedom and personal independence (Toffler, 1981, pp. 32–33).

The fourth Industrial Revolution, artificial intelligence and the digital transformation of the economy, the internet and its culture, teleworking and flexible forms of employment, surveillance societies, biomedical experiments, transformations of public space, digital inequalities, retirement of individual rights, and environmental risks have occupied the public debate since the late twentieth century (Contiades & Fotiadou, 2020).

On this horizon, it is to be expected that the welfare state has come to an end. Large networks of financial markets don't have the ability to print money indiscriminately. Markets impose a maximum limit regarding the accumulation of national debts. It is now not possible to increase taxes indefinitely. It is no surprise, especially for those who have heard of Milton Friedman, and are familiar with how fit he is that "there is no such thing as a free lunch" (Friedman, 2007 in Bitsani, 2013).

In this framework, the continuous increase of the internalization of the political, economic, and cultural activities contributes to the reinforcement of the cultural exchanges, to the broadening of the multi-ethnic activities, and to the even bigger signification of the supranational political institutions. In other words, structures are formed and strengthened beyond the narrow and limited confines of the nation state and simultaneously intercultural relations are established and organized between organizations and individuals of a national society with organizations and individuals in other societies (Bitsani, 2016a; Bitsani, 2016b).

The necessity to construct relations and structures based on solidarity, in times of prolonged economic recession and declining welfare systems, is evidenced in the significant widespread initiatives that have been taken, and the broad network of social services provided (Bitsani & Pantazopoulos, 2018; Thymi et al., 2020).

Finally, economic globalization and global colonialization remain blind toward democratic deficiencies, social exclusions, financial inequalities, and the environmental destruction that it creates.

Living in these times of uncertainty and emergency and in order to face challenges such as climate change, environmental degradation, digital transformation, energy demands, migration, social exclusion, mortality and population aging, pandemics, economic turmoil, growing inequalities, conflicts, etc., local governments must adopt innovative responses and new governance approaches, as traditional practices fail to respond to the above issues.

To effectively address these challenges, we need a variety of actors, resources, and tools that can unleash our collective imagination and unite communities in joint efforts to change the current status quo. Culture and ecology can play a powerful role in regenerating the potential of diverse communities and contexts and in their global interconnectedness for the benefit of all people and the planet. To this end, cultural and environmental policies must be integrated into the concept of sustainability, both as an autonomous backbone of sustainability and as a contributing factor to its many dimensions.

In the horizon of developments as presented above, urban areas in Europe are undergoing significant ecological and demographic transformations. These changes have intensified interest in the concept of biocultural diversity and biocultural approach—the dynamic, interconnected relationship between biological diversity and cultural diversity. Historically addressed in rural or Indigenous contexts, biocultural diversity is increasingly being recognized as highly relevant in urban settings. European cities, shaped by centuries of migration, colonial legacies, and environmental shifts, offer rich ground for studying how nature and culture co-evolve (UNESCO, 2018).

The core of the biocultural approach is that it understands humans as biological, social, and cultural beings in relation to the environment (McElroy, 1990). It views human biological variability as a function of response and adaptation to the environment, with particular emphasis on the role of sociocultural environments. The concept of the biocultural approach originates from the field of biological anthropology, where it has been used primarily to describe the effects of social environments on human health and culture (Akinwale & George, 2022). From anthropology, the concept has spread to other fields, and the definition of the concept has shifted from human biology to an emphasis on the close relationships between human societies, especially their cultural sphere, and the natural and biophysical environments in which they exist (Maffi & Woodley, 2010), including the anthropogenic environment in all its dimensions. The word "biocultural" is most often used in the fields of biocultural diversity, biocultural heritage, and biocultural approaches. New areas of research and application have emerged in the last decade, including biocultural heritage, biocultural approaches to conservation, biocultural landscapes, and biocultural rights, among others (Lukawiecki et al., 2022).

The biocultural approach provides a comprehensive theoretical framework for interpreting the relationship between culture, environment, and human development. It rejects the dichotomy between nature and culture, proposing instead that both dimensions coexist within a dynamic continuum. This paradigm views human societies as integral components of ecosystems, whose survival depends on balance, reciprocity, and the creative adaptation of cultural practices to environmental conditions (Maffi & Woodley, 2010).

In the context of urban sustainability, the biocultural approach highlights how cultural expressions—traditions, architectural forms, artistic production, and collective rituals—are embedded in ecological processes. These elements not only reflect a community's environmental awareness but also shape its capacity for resilience and innovation. As such, culture becomes a key driver of sustainable transformation rather than a peripheral or decorative aspect of urban planning.

Adopting a biocultural perspective encourages a holistic understanding of cities as living systems where cultural vitality and ecological health are mutually reinforcing. This approach aligns with contemporary sustainability discourses that call for integrating environmental, social, and cultural dimensions within development policies (UNESCO, 2017; UN, 2015).

Furthermore, it challenges the reductionist view that economic or technological solutions alone can ensure sustainable futures. Instead, it insists on the importance of cultural values, social imagination, and local knowledge as fundamental components of adaptive capacity. By embracing cultural diversity and creativity, cities can cultivate resilience that extends beyond infrastructure and governance—toward an ethos of coexistence, empathy, and respect for all forms of life.

The advantage of biocultural approaches is that they include values, knowledge, and needs, which, as Sterling et al. (2017) suggest, shape the local cultural perspective. In other words, this approach can bring indigenous knowledge to the research, which is defined as: "a cumulative body of knowledge, practice and beliefs that evolves through adaptive processes and is transmitted across generations through cultural transmission" (Oloriz & Parlee, 2020).

Using a theoretical lens such as the biocultural approach can be seen as a feedback loop in which cultural factors and biological factors interact (Akinwale & George, 2022). This perspective can "recognize the feedback between ecosystems and human well-being" and once this approach is used, researchers can identify problems, identify solutions, and facilitate knowledge transfer from local to national and international levels (Gavin et al., 2015).

The principal aim of this research is to develop and demonstrate a biocultural framework for understanding and advancing urban sustainability through the lens of culture. Specifically, the study seeks to explore how cultural values, practices, and policies interact with ecological and social systems to produce more inclusive and resilient urban environments. By situating culture as the transversal axis that informs governance, participation, and environmental awareness, the research aims to generate both theoretical insight and applied knowledge capable of informing future urban policy and planning. In doing so, it contributes to redefining sustainability as a culturally embedded process that reflects local identities while addressing global challenges.

To fulfil these aims, the research is guided by three central questions that collectively translate the theoretical framework into empirical investigation.

To what extent can culture-centered strategies advance the goals of biocultural sustainability within the urban context of Ixelles?

How does the prioritization of cultural values and practices influence local responses to environmental, social, and economic challenges?

In what ways do residents, community actors, and municipal authorities perceive culture as a resource for biocultural transformation and inclusive urban governance?

Together, these questions provide the analytical foundation for examining how cultural processes mediate the relationship between ecology, policy, and everyday urban life, thereby operationalizing the study's core premise that culture functions as the integrative medium of sustainability.

Today, more than 90 years after the Athens Charter, we are once again renegotiating the future of our cities. What everyone is asking for is a redefinition of the strategy toward sustainable urban regeneration through the preservation of historical and cultural values, the preservation of social fabric, the reduction of inequalities, and the improvement of the quality of life of citizens who understand the urban space both physically and culturally (Artmann et al., 2019; Guzman et al., 2014).

The 2030 Agenda recognizes the role that cities play in promoting sustainable lifestyles for all people. The implementation of the 2030 Agenda is a multi-stakeholder process that requires a renewed emphasis on open communication and coordinated efforts across governance sectors. This is confirmed by the Seville Commitment, which highlights the critical role of local governments in implementing the global agenda, in laying the foundations for the social transformations needed to achieve the ambitious goals of the 2030 Commitment (UN, 2025).

This study assesses biocultural diversity in European cities, focusing on key indicators, challenges, case studies, and recommendations for future urban sustainability.

Our research treats biocultural diversity as inherently synergistic, yet this can risk romanticizing the relationship between nature and culture, overlooking cases where cultural practices may harm biodiversity. As highlighted by Gavin et al. (2015), a critical biocultural perspective should also be scrutinized by cultural practices that conflict with conservation goals. Moreover, the study misses recent frameworks such as the "biocultural ethics" model proposed by Rozzi (2013), which stresses ethical obligations toward both ecological and cultural others.

So far, the literature on urban sustainability and urban sustainable development offers few examples of European cities in terms of biocultural diversity but generally lacks a critical lens on systemic inequalities:

While Berlin's integration of biodiversity with climate policy is commendable, research highlights that gentrification around green spaces often displaces lower-income, frequently immigrant communities (Bitsani, 2016b). Similarly, in Barcelona, the Huertos Urbanos initiative holds promise, yet the commodification of urban agriculture for tourism or aesthetic purposes risks eroding its socio-political foundations (Anguelovski et al., 2018). In Tallinn, the challenges of incorporating indigenous ecological knowledge are underplayed, particularly within a post-Soviet planning context where such knowledge was historically marginalized (Raudsepp et al., 2009). Meanwhile, London's diversity is recognized, but recent studies point to persistent racialized barriers to green space access and meaningful community participation in planning processes (McDonagh, 2022).

While urban areas are indeed "melting pots" of biodiversity and culture, the report could benefit from integrating concepts from urban political ecology, which views cities as contested terrains shaped by socio-economic power asymmetries (Heynen et al., 2006). For instance, who has access to green space and who defines its use are deeply political questions.

In recent decades, cultural policy has evolved from a sector-specific concern into a multidimensional field that intersects with economic, social, and environmental agendas. This transformation reflects a growing recognition of culture's pivotal role in shaping collective identities, influencing behavior, and fostering social cohesion (Throsby, 2010; UNESCO, 2013). Culture is no longer perceived merely as heritage or artistic production but as a dynamic force that drives innovation and contributes directly to sustainable development.

The inclusion of culture within the sustainability paradigm emerged through several key milestones. The Agenda 21 for Culture (2004) and UNESCO's subsequent reports (2015, 2017) positioned culture as the fourth pillar of sustainability—alongside the economic, social, and environmental dimensions. This perspective emphasizes that development cannot be sustainable if it disregards the symbolic, ethical, and creative dimensions of human life.

Integrating culture into sustainability policies entails recognizing its dual function: as both a means and an end of development. On one hand, cultural initiatives can stimulate participation, education, and community resilience; on the other, the safeguarding of cultural heritage and diversity constitutes a developmental objective in itself. As UCLG (2015) notes, sustainable cities are those that protect their cultural resources while promoting creativity as a source of renewal.

In the urban context, cultural policies can serve as catalysts for rethinking the use of public space, encouraging citizen participation, and fostering environmental awareness. The concept of the biocultural city builds upon this rationale, proposing that culture and ecology must be considered together when designing urban policies. This convergence challenges the fragmented logic of traditional planning and opens new pathways for inclusive, adaptive, and ethically grounded urban futures.

In the following pages we present the theories of utopia and dystopia, and we will explore Plato's first utopian approach to the coexistence of city and governance, and Foucault's views on space and more, in Chap. 2.

The third chapter introduces sustainability theories from the Brudland report to more contemporary approaches, followed an in-depth analysis of the concept of the city and the contemporary challenges it faces.

Chapter 4 discusses contemporary approaches to policymaking, focusing on the concepts of "polycrisis"and "wicked problems." It also presents both UN and EU policies aimed at sustainable development, as well as modern city models that have emerged in urban centers over recent decades.

The fifth chapter presents the methodological framework followed in conducting the research. The research spans from 2018 to 2025 and is designed as a longitudinal, multidimensional case study.

In the next Chap. 6 the Municipality of Ixelles is presented through a visual content analysis of its websites, using a multimodal and visual approach to content analysis and participatory observation too, during 2018–2025. Chapters 7 and 8 provide a thorough analysis of the quantitative and qualitative data collected during all phases of field research in Ixelles between 2023 and 2025. Drawing from a range of research methods, including interviews, focus groups, participatory workshops, and ethnographic observation, a comprehensive overview of how residents, community actors, and institutions perceive and experience sustainability, governance, cultural diversity, and social inequality is provided. Chapter 9 synthesizes and cross-references the findings from the visual quantitative and qualitative analyses. Through triangulation, it identifies key patterns and contradictions in the different data sources and analytical approaches. The findings are interpreted and linked with research questions and existing literature. Recognizing the inherent limitations of the present study proposes potential areas for future research.

The study offers a robust typology of indicators but stops short of assessing their applicability or measurability across different urban contexts. Ethnolinguistic diversity, for example, is complex to measure without disaggregated census data, and often underrepresents marginalized groups. Moreover, biocultural indicators such as community gardens are rich sites of contestation and negotiation—spaces where race, migration, and land use politics intersect (Certomà, 2020).

The study examines how cities have become crucibles of both vulnerability and resilience, where new forms of collective imagination and spatial organization emerge. Contemporary cities are re-enacting the tensions at the heart of utopian and dystopian thinking. The study argues that the city remains a privileged site for the projection, negotiation, and critique of utopian aspirations and dystopian anxieties.

Development multi-scalar, participatory indicators co-designed with local communities and supported by qualitative and quantitative data (Maffi & Woodley, 2010).

In this context the study introduces the concept of biocultural development and the model of biocultural city, which combines the concept of biopolitics with the cultural dimension of human society. In Foucault's interpretation, biopolitics refers to the way in which power uses science, knowledge, and institutions to manage and regulate the lives of populations. Biocultural development, in turn, incorporates the principles of cultural sustainability and viability, focusing on sustainable development that is intertwined with the protection of cultural and environmental values.

The challenges posed by this study concern focusing and distinguishing more sharply between structural constraints (e.g., neoliberal urbanism/Inequity and Exclusion) and institutional shortcomings (e.g., fragmented policy). Urbanization is not merely about space loss; it reflects deeper capitalist pressures on land use, which must be addressed if biocultural diversity is to be protected.

As the IPBES (2022) notes, the lack of integrative governance frameworks hampers the implementation of biocultural approaches. The report correctly identifies this challenge but could expand on how intersectional inequalities (race, class, gender) affect who benefits from and contributes to biocultural diversity.

While practical, the recommendations would benefit from a rights-based framing. The right to the city—as popularized by Lefebvre—should be invoked to argue for residents' participation in shaping ecocultural futures. Furthermore, policies should go beyond inclusion to actively redistribute power and resources in urban ecological governance (Bitsani, 2013; Purcell, 2014).

Community engagement must be sustained, not tokenistic. Long-term partnerships with marginalized groups, especially migrants and racial minorities, require capacity-building and institutional accountability (Kothari et al., 2019).

Urban governance has become a central concept in the analysis of contemporary cities. It encompasses the processes through which local authorities, private actors, civil society, and citizens interact to manage urban change and shape collective futures (UN-Habitat, 2020). Within this framework, governance is no longer viewed merely as top-down administration but as a participatory and multi-scalar system of negotiation, coordination, and shared responsibility.

In the twenty-first century, the cultural dimension of governance has gained increasing importance. Cities that integrate cultural strategies into their development agendas tend to demonstrate greater social cohesion, innovation capacity, and adaptability. Cultural governance involves not only the protection and promotion of heritage and creativity but also the cultivation of values that support inclusion, dialogue, and diversity.

The *biocultural city* model expands this understanding by embedding cultural governance within ecological and ethical frameworks. It calls for policies that recognizes the interdependence between cultural expression, environmental health, and social equity. In this sense, urban governance becomes an act of stewardship of caring, for both people and place, grounded in respect for the living systems that sustain human existence.

Implementing biocultural principles in governance demands cross-sectoral collaboration, integrating environmental planning, cultural management, education, and technology. It also requires a shift from growth-oriented paradigms toward value-based ones that prioritize well-being, empathy, and creativity. As the World Cities Culture Forum (2021) notes, sustainable urban governance depends on recognizing culture not as an accessory but as a central resource for imagining and building future cities.

The biocultural city thus represents a paradigm of governance rooted in ethics, imagination, and reciprocity. It envisions the city as a *living common* shared space of cultural exchange, ecological interdependence, and collective responsibility. Such an approach redefines what it means to be urban, challenging conventional dichotomies between human and natural systems, and between culture and sustainability.

To this end, cultural and environmental policies, on the one hand, must integrate emerging technologies and, on the other hand, must be integrated into the concept of sustainable development, both as an autonomous backbone of sustainable development and as a contributor to its many dimensions.

Specifically, in this book, we argue that a cultural shift is essential to fulfill the sustainability goals as it requires a paradigm shift, which is first and foremost a change in thinking.

We propose that culture must be at the heart of all policy areas if sustainability goals are to be achieved, as we argue that biocultural lenses are the key tool for urban development. Our basic research question is whether cultural lenses in biocultural approach of sustainability help to address issues of inequalities like power, minorities, gender, etc. and at the same time address the pressing issues of climate change, adaptation, and sustainable development. Our argument is supported by the findings from the case study of the City of Ixelles.

The study ends on an optimistic note, yet the realization of urban biocultural diversity requires confronting uncomfortable truths—about exclusion, commodification, and ecological injustice. A truly resilient and inclusive urban future will depend not only on conserving life forms but mainly on transforming urban governance structures to be more equitable and ecologically just.

1.1 Challenges for Today's European Cities

In this part, we argue in an analytical way the contemporary challenges of European Cities. New urban governance emphasizes collaborative, participatory, and networked approaches involving local government, local societies, and the private sector. It aims to increase multiple participation and promote social inclusion of different cultural groups, as a key demand of the 2030 Agenda for the Sustainable City. The benefits of urban green spaces are not only environmental but also social. Future priorities should recognize the role of power and politics in new urban governance, develop effective tools, and address challenges in implementing new governance.

1.1.1 Urbanization

European cities are undergoing rapid urbanization, which is resulting in increasing populations and strains on infrastructure and resources. As populations increasingly seek to establish themselves in new locations, the world is witnessing the largest wave of urban expansion in history. The global population exhibited an increase of 0.9% in 2021, reaching a total of over 8 billion by 2022. It is estimated that over 50% of the world's population is currently residing in urban areas, with this trend increasing at a rate of 1.5%. The United Nations (UN) has estimated the world population in mid-2023 to be 8,045,311,447, indicating an increase of 0.88% (70,206,291 individuals) from 2022 (UN, 2023). Rapid urbanization, when considered in conjunction with population growth, is effecting profound changes to the geographical distribution of human settlements and posing significant risks to quality of life, the environment, and development (UNSTAD, 2022).

A portion of this growth is attributable to elevated levels of urban fertility rates and the reclassification of rural land as urban; however, a substantial proportion of future urbanization is expected to be driven by migration from rural to urban areas (Pantazopoulos, 2024a, 2024b). In this regard, numerous researchers have examined the direct effects of urbanization on the loss and fragmentation of vegetation associated with urban expansion (Dawazhaxi et al., 2023). It is evident that a correlation exists between urbanization, economic growth, and environmental sustainability (Yang et al., 2023).

The effective management of population growth while ensuring adequate housing provision represents a significant challenge. Most European cities are experiencing a housing affordability crisis, whereby the cost of housing is becoming increasingly unaffordable for a significant proportion of the population, particularly younger generations (Aronson et al., 2016; Lee et al., 2022; Lux et al., 2023). A substantial body of research has attributed the recent affordability crisis to neoliberal housing policies, citing the "Italian paradox," the "Spanish paradox," and accusations against the property lobby as evidence of systemic housing market dysfunction (Colenutt et al., 2020; Gentili & Hoekstra, 2018; Hoekstra & Vakili-Zad, 2011). But, Lux et al. (2023) found in their research about young Czechs' perceptions that they are cautious about state intervention in housing, often favoring market-based solutions. The views expressed by these individuals are heavily influenced by familialism, post-socialist norms, and concerns regarding the fairness of redistribution. While many advocate for individualism and competition, their stance on policy is often incongruent. This finding suggests that the perspectives held by young people regarding housing are multifaceted and not uniformly supportive of the state.

In recent discourse, housing insecurity is affected also by climate change. Climate-induced displacement, predominantly characterized by migration from rural to urban areas, represents the most significant challenge confronting the developing world (Buhaug & Urdal, 2013). It is imperative for host communities to consider the provision of adequate and affordable housing for new residents as a fundamental aspect of their responsibilities. This is of particular significance for individuals who have been compelled to relocate due to housing insecurity resulting from the repercussions of climate change. The provision of low-cost housing, particularly in communities that provide shelter for migrants or refugees, poses an additional challenge for decision-makers (Li, 2021). It is an established fact that financial hardship can have a detrimental effect on an individual's mental health; as demonstrated by Dotsikas' findings, prolonged exposure to housing affordability issues over several years is associated with poorer mental health outcomes when compared to individuals who experience minimal or no affordability problems (Dotsikas et al., 2023).

Another aspect of rapid urbanization is the issue of aging infrastructure, a growing concern worldwide. These structures are unable to meet the demands of growing populations. Maintenance and upgrades are often delayed due to budget constraints, further exacerbating the problem. In the United States, for instance, water utilities face significant investment gaps (Gifford, 2024). The failure to undertake essential maintenance has been demonstrated to result in increased repair costs and to pose a threat to public health, as obsolete systems are unable to satisfy contemporary standards (Gifford, 2024). Moreover, declining populations in specific regions have been demonstrated to result in diminished utility revenues, thereby exerting further strain on budgets and engendering complications in infrastructure management (Gifford, 2024).

Skamnakis and Pantazopoulos (2015) and Kigadye (2014) all argue for the assertion that chronic underinvestment in infrastructure maintenance represents a key weakness in public financial management, a problem that is especially prevalent in developing countries. It is asserted that budgetary constraints, inconsistent reporting, and the absence of standardized maintenance planning contribute to the neglect of extant assets. The findings of both studies indicate a systemic capital bias toward new projects, frequently at the expense of maintaining current infrastructure. This negligence has been shown to result in inefficiencies, undisclosed expenses, and subsequent, costly remedial measures. In addressing this issue, both authors advocate for institutional reforms that prioritize maintenance as a strategic investment rather than a discretionary expense.

1.1.2 Environmental Sustainability

European cities are confronted with a range of environmental challenges, including air and water pollution, inadequate waste management, and the degradation of green spaces. The delicate equilibrium of urban development with the preservation and enhancement of the natural environment is of paramount importance for the promotion of sustainability.

As emphasized by Hamdani et al. (2021), land subsidence is a significant cause of environmental degradation in coastal cities. The importance of integrating monitoring, adaptation, and mitigation measures is also stressed. Rondinel-Oviedo and

Keena (2023) explore the contribution of urban construction and operation to negative externalities, including waste and pollution, which result in environmental degradation.

The quality and functionality of public spaces have been demonstrated to have a significant impact on the perception of a city. Consequently, the standardization of good planning and maintenance of these spaces is imperative. The Sustainable Development Goals (SDGs) established by the United Nations include the provision of safe, inclusive, green, and public spaces, particularly for women, children, older individuals, and persons with disabilities (United Nations, 2015). Consequently, the three stated objectives of safety, inclusivity, and accessibility must be given due consideration by responsible urban planners (Naya et al., 2023). Moreover, the presence of urban green spaces has been demonstrated to play a pivotal role in enhancing the quality of life within urban areas. Consequently, contemporary urban centers are confronted with the imperative to conserve and enhance their urban green spaces, which play a pivotal role in fostering sustainability and environmental enhancement. The World Health Organization (WHO) has set out recommendations for the provision of green spaces, suggesting a minimum area of 0.5 ha. These should be located within a linear distance of 300 m from each household. However, research indicates that 62% of the European population resides in areas with inadequate access to green spaces (Barboza et al., 2021).

European cities are becoming increasingly susceptible to the effects of climate change, including elevated temperatures, severe weather occurrences, and rising sea levels. Extreme weather events pose significant challenges to urban infrastructure and the health and well-being of its inhabitants (IPCC, 2021), thus highlighting the urgent need for cities to take action to mitigate and adapt to the impacts of climate change. A large number of nations around the world have agreed to implement measures to limit global warming to below 2 °C (De Graaf & Ovink, 2021). As the planet warms, ice melts, sea levels rise, and air quality declines. Concurrently, there is an increased demand for water and energy, which in turn has an impact on public health, especially among vulnerable social groups (De Graaf & Ovink, 2021; Schär & Jendritzky, 2004). It has been argued that these changes will result in the dissolution of the cultural bonds that settlers previously held with their environment and with each other, as well as with their customs, traditions, and material and immaterial heritage (Aktürk & Hauser, 2021). The phenomenon of cultural abandonment may emerge as a consequence (Bridgewater & Rotherham, 2019), as may the absence of social equality, including gender and environmental justice (Lee et al., 2022). In recent times, the concept of resilience has assumed a pivotal role in the realm of infrastructure and structural systems planning, evaluation, maintenance, and repair. This is particularly evident in the context of bridges and transport networks, which are susceptible to both natural and human-induced hazards. In recent decades, there has been a growing emphasis on the impact of ageing and environmental aggression on the performance and functionality of structures. Consequently, the resilience of the system is contingent upon the timing of extreme events (Capacci & Biondini, 2022).

The World Bank's report (2022) emphasizes that the contemporary global economy consumes in excess of 100 billion tons of raw materials on an annual basis, while generating an estimated 90 billion tons of discarded materials (World Bank, 2022). This indicates that the prevailing linear economic model of "take, make, waste" is not sustainable and exerts a considerable influence on environmental degradation. The transition toward a circular economy, predicated on the reduction of waste, the maximization of resource efficiency, and the promotion of recycling and reuse, poses considerable challenges for European cities. In response to environmental challenges and the pursuit of global sustainable development, the transition to a circular economy (CE) is increasingly favored (Van Lange & Huckelba, 2021). This is because the CE provides significant environmental benefits when compared to the linear economic framework in which resources are extracted, used, and disposed of Murray et al. (2017) and Atanasova and Eckhardt (2021) posit that nature-based solutions (NBS), when meticulously designed and adapted to the local context, primarily rely on the regeneration of natural capital and the environment. The transition from linear to circular management of natural resources is facilitated, externalities are reduced by minimizing energy and heat consumption, and recuperated materials are utilized as building materials for construction. In order to address the challenges posed by shifting consumption patterns, the improvement of waste management systems, and the fostering of circular business models, concerted efforts are required (Försterling et al., 2023). Coskun and Taspinar (2022) posit that cities are pivotal in this endeavor, given that they are home to the largest human population worldwide and are responsible for approximately 60–80% of global energy consumption and 75% of carbon emissions. Governments encounter numerous challenges in their efforts to achieve urban circularity (UCC) in the transition toward a circular resource management model. These challenges encompass a range of actions, including the restoration and maintenance of the water cycle (including stormwater management), the treatment of water and waste, the recovery and reuse of nutrients, the production of food and biomass, and the enhancement of energy efficiency in building systems, among others (Atanasova & Eckhardt, 2021). It is imperative to address these challenges to facilitate a successful transition toward circular resource management in urban areas.

1.1.3 Social Inequality

A significant number of European cities are encountering issues of social exclusion, particularly in areas characterized by high levels of poverty and unemployment. In all societies, certain demographic groups are subject to systemic disadvantage as a consequence of discrimination. Discrimination manifests in a variety of settings, including public institutions such as the legal system, educational facilities, healthcare services, as well as domestic and community settings. However, the extent of discrimination varies from one society to another, depending on the form that social exclusion takes and the groups affected (DFID, 2005). As Pohlan (2019) mentions,

"…the term 'social exclusion' has become increasingly prominent in policy debates on poverty and social inequality, often referring to disadvantages in basic living conditions that reduce opportunities to participate in society…" (Beer & Foley, 2003, European Commission, 2010). A considerable number of individuals with visual impairments have attested to having been subjected to discriminatory practices (Jackson et al., 2019).

Unemployment is widely regarded as one of the primary risk factors for social exclusion. Furthermore, the concept of "stigma effects" posits that employers may interpret unemployment as a negative signal, thereby systematically reducing the likelihood of unemployed individuals being hired (Pantazopoulos, 2024a, 2024b). Exclusion from the workforce can result in marginalization, which, in turn, can engender a heightened risk of long-term reliance on welfare benefits (see Bhuller et al., 2017; Pantazopoulos, 2024a, 2024b), suicide (see Sullivan & von Wachter, 2009), involvement in criminal activities or victimization (see Freeman, 1999), and even political backing for extremist factions (see Falk et al., 2011). Income levels and rising unemployment are likely to limit the ability of households to borrow (Aksoy Khurami & Özdemir Sarı, 2022; McQuinn et al., 2021). Furthermore, social exclusion may not only have an impact on the current generation, but it may also have an impact on the next generation (Pantazopoulos, 2024a, 2024b).

1.1.4 Transportation and Mobility

Modern cities and megacities are constantly faced with traffic problems that require the right solutions. A significant number of European cities are confronted with challenges pertaining to transportation, which include traffic congestion, air pollution, and limited accessibility. Urban congestion is increasing due to population growth, motorization, urbanization, and changes in population density (Lu et al., 2023). This has a detrimental effect on journey times, fuel consumption, other operating costs, and environmental pollution, as compared with uninterrupted traffic flow (Muneera & Karuppanagounder, 2018). The quality of life in urban areas is contingent on the efficiency of the transport network. As Stepanchuk et al. (2016) argue, the primary function of a city's transport network is to provide convenient and rapid access to employment opportunities, amenities, and cultural and educational facilities. As Remillard et al. (2022) point out, there are also issues such as a paucity of education for sustainable transition, an insufficiency of green spaces, and problems of accessibility.

This poses a significant challenge for city authorities, as they need to invest in upgrading and maintaining this infrastructure to ensure it remains safe and reliable, cost-effective, and resilient.

1.1.5 *Governance and Policy Challenges*

Effective governance and collaboration among different stakeholders, including local authorities, businesses, and engaging communities, are crucial for addressing complex urban challenges.

Iaione (2016) claims that there are major challenges in the realm of urban governance, such as the inefficiencies and fragmentation of conventional top-down management structures, the exclusion of citizen participation, and the imbalance between public and private control of resources. The author highlights issues such as resource depletion, infrastructure neglect, and outdated institutional frameworks that limit adaptive and participatory governance.

As Watson (2018) emphasizes, there is an increasing withdrawal of the state from public involvement and expenditure, which has a disproportionately adverse effect on vulnerable urban populations. Watson emphasizes that meaningful self-organization requires collaborative and democratic processes that actively involve marginalized voices. Furthermore, she observes that such involvement frequently gives rise to conflict, which, if managed constructively, can serve as a positive catalyst for recognizing differences and negotiating solutions. The paper puts forward the notion of inclusive, conflict-aware approaches to urban governance, which prioritize equity and responsiveness to local needs.

Meijer et al. (2019) argue that traditional urban governance models face significant challenges in addressing the complex, decentralized nature of modern city collaborations. It is highlighted that government roles in these collaborations often become less visible, which complicates accountability. Existing paradigms, such as New Public Governance, have been found to be inadequate in capturing the spontaneous, "pop-up" nature of these collaborations and the extensive individualized participation occurring in urban areas. In order to address this issue, the authors propose a novel open governance paradigm. This paradigm views governance as a platform-based strategy to managing urban ecosystems. This ecosystem is better suited to support and explain innovative, digitally enabled urban collaboration, emphasizing the significance of digitally interconnected actors and the requirement for smart technologies.

1.1.6 *Public Health and Safety*

The promotion of the health and well-being of urban residents has been identified as a priority, as evidenced by the recent pandemic. The year 2020 signified a pivotal moment in the annals of human history, precipitated by the emergence of the novel Coronavirus pandemic (henceforth referred to as "COVID-19"), a global health crisis of unprecedented proportions that has not been witnessed for over a century. The disease has affected millions of individuals, leading to contraction of the virus. The pandemic functioned as a catalyst and accelerator for recession and a prolonged economic crisis. It also imposed a range of practices, restrictions, and behaviors that are characteristic of technologically advanced societies in the late modern era. The pandemic is to be regarded as a biopolitical event in the philosophical sense of the term. The consequences of climate change are pervasive and deleterious, causing rapid and harmful anthropological and social changes (Contiades & Fotiadou, 2020 in Bitsani & Anselmi, 2021).

Another issue that governments must address for the well-being of their citizens is ensuring food security. Food security is defined as the ability of a nation or region to enhance the availability, accessibility, utility, and stability of its food supply (Putra et al., 2020). In urban areas, the storage, transportation, and distribution of food is often inadequate due to a lack of necessary infrastructure, resulting in a suboptimal food supply system. In the context of rapidly increasing urbanization, it is imperative to address this issue (Djan, 2023). To ensure the procurement of adequate sustenance, it is imperative to establish a value chain that is conducive to the sustainable supply of foodstuffs, the promotion of dietary diversity, the enhancement of economic development, the reduction of food wastage, and the encouragement of innovation with a view to improving the health and well-being of citizens (European Parliament, 2023).

Addressing issues such as air quality and noise pollution, as well as the promotion of active lifestyles, necessitates integrated approaches that prioritize public health. In contemporary urban environments, noise constitutes a significant environmental contaminant that is frequently disregarded, despite research indicating a correlation between noise exposure and an elevated risk of premature mortality. The World Health Organization (WHO) has identified environmental noise as a significant public health concern. In a similar vein, the European Environmental Agency (EEA) has reported that air pollution represents the most significant environmental health risk in Europe, with road traffic being a primary source of polluting gas emissions. The correlation between these pollutants has prompted the exploration of the development of shared strategies by various nations (Montes-González et al., 2018).

The issue of safety and security represents a fundamental challenge for European cities. It is a widely accepted proposition that urbanization is associated with an increase in crime and violence, with deleterious consequences for those who reside in urban areas. The escalating levels of violence and insecurity experienced by urban populations represent a pressing global concern. The safeguarding of public safety is considered a fundamental human right. It is therefore imperative that local governments and other relevant stakeholders collaborate in order to develop and implement strategies and actions with the objective of eradicating violence, crime, and insecurity (Un-Habitat). According to the findings of the Special Eurobarometer (2017), concerns regarding security, and more specifically, terrorism, are among the foremost issues of concern for Europeans. The issue of public perceptions of safety represents a significant challenge for contemporary cities, as they are primarily responsible for ensuring the security of their citizens, yet are confronted with an inadequate set of tools and resources at their disposal.

In summary, these challenges are complex and interconnected, requiring holistic and systemic approaches that involve collaboration and new—innovative—ways to address them.

As we discussed above, cities are fundamental reflections of the contemporary world, representing the current organization of the economy, the structure of social groups, lifestyles, habits, and the activities of their inhabitants. They are thus complex subjects of study—multifunctional, multidimensional, and dynamic—blending the built and cultural environments. Furthermore, cities are not isolated from their surrounding space or nature. Their structure, functions, and internal activities both influence and are influenced by the natural environment (Skamnakis & Pantazopoulos, 2015).

As cities are human achievements, ensuring their sustainable development through addressing environmental, social, and economic pressures is critical for their maintenance and prosperity. The most significant threats faced by cities are directly linked to the core dimensions of sustainability, while resilience emerges as a short-term strategy to confront these challenges and a long-term pathway toward sustainability. Since resilience and sustainability are interconnected and mutually reinforcing, it is essential that urban strategies and planning under these frameworks recognize cultural heritage as both a valuable source of insight and a core preservation target (Bitsani & Elefteriou, 2014).

1.2 Conclusion

The challenges faced by contemporary cities—climate change, social inequality, rapid technological transformation, and cultural homogenization—demand new frameworks for understanding and action. The biocultural approach offers one such framework, combining scientific knowledge with cultural insight and ethical reflection. It invites us to consider the city not merely as a physical or economic construct, but as a living organism whose vitality depends on the quality of its relationships—with nature, with its inhabitants, and with its own cultural memory.

In this light, sustainable urban development requires more than technical solutions or policy reforms. It calls for a renewed cultural imagination, a capacity to envision ways of living that reconcile human aspirations with ecological limits. The biocultural city embodies this vision by integrating ecological sensitivity, cultural creativity, and social justice into a cohesive model of coexistence.

This chapter has outlined the theoretical foundations and guiding principles of the biocultural paradigm, situating it within broader debates on urban sustainability and governance. The next chapter explores how these ideas translate into practice through empirical analysis and case studies, examining how cities, particularly within Europe, can harness their cultural and ecological assets to foster resilience, inclusion, and innovation.

Ultimately, the biocultural city represents both a conceptual framework and a moral imperative. It challenges us to move beyond fragmented understandings of urban life toward a holistic ethos of care, reciprocity, and shared belonging, an ethos capable of sustaining not only our cities, but the very fabric of life on which they depend.

Building upon this theoretical and empirical framework, the study formulates a set of research aims and questions designed to investigate and substantiate this biocultural shift in practice, examining how culture—as an axis of meaning and action—can operate as a catalyst for a more equitable, participatory, and ecologically conscious form of urban sustainability.

Conflicts of Interest The authors have no conflicts of interest to declare that are relevant to the content of this chapter.

References

Akinwale, O. E., & George, O. J. (2022). Personnel brain-drain syndrome and quality healthcare delivery among public healthcare workforce in Nigeria. *Arab Gulf Journal of Scientific Research, 41*(1), 18–39. https://doi.org/10.1108/AGJSR-11-2021-0075

Aksoy Khurami, E., & Özdemir Sarı, Ö. B. (2022). Trends in housing markets during the economic crisis and Covid-19 pandemic: Turkish case. *Asia-Pacific Journal of Regional Science, 6*(4), 1159–1175. https://doi.org/10.1007/s41685-022-00251-w

Aktürk, G., & Hauser, S. J. (2021). Detection of disaster-prone vernacular heritage sites at district scale: The case of Fındıklı in Rize, Turkey. *International Journal of Disaster Risk Reduction, 58*, 102238. https://doi.org/10.1016/j.ijdrr.2021.102238

Anguelovski, I., Connolly, J. J. T., Masip, L., & Pearsall, H. (2018). Assessing green gentrification in historically marginalized neighborhoods: A longitudinal and spatial analysis of Barcelona. *Urban Geography, 40*(3), 1–23. https://doi.org/10.1080/02723638.2017.1349987

Anselmi, W., & Gouliamos, K. (1998). *Elusive margins: Consuming media, ethnicity, and culture* (Essay series) (Vol. 29). Guernica editions.

Anselmi, W., & Sheena, W. (2011). Inception (2010): When we dream, do we accumulate capital? *Bright Lights Film Journal, 74*., http://www.brightlightsfilm.com/74/74inception_anselmiwilson.php. https://doi.org/10.7939/R39J2X

Aronson, J. C., Blatt, C. M., & Aronson, T. B. (2016). Restoring ecosystem health to improve human health and well-being: Physicians and restoration ecologists unite in a common cause. *Ecology and Society, 21*(4) http://www.jstor.org/stable/26270035

Artmann, M., Kohler, M., Meinel, G., Gan, J., & Ioja, I.-C. (2019). How smart growth and green infrastructure can mutually support each other—A conceptual framework for compact and green cities. *Ecological Indicators, 96*(Part 2), 10–22. https://doi.org/10.1016/j.ecolind.2017.07.001

Atanasova, A., & Eckhardt, G. M. (2021). The broadening boundaries of materialism. *Marketing Theory, 21*(4), 481–500. https://doi.org/10.1177/14705931211019077. (Original work published 2021).

Atchison, A., & Shames, S. (2019). *Survive and resist: The definitive guide to dystopian*. Politics Columbia University Press.

Barboza, E. P., Cirach, M., Khomenko, S., Iungman, T., Mueller, N., Barrera-Gómez, J., Rojas-Rueda, D., Kondo, M., & Nieuwenhuijsen, M. (2021). Green space and mortality in European cities: A health impact assessment study. *The Lancet. Planetary Health, 5*(10), e718–e730. https://doi.org/10.1016/S2542-5196(21)00229-1

Bauman, Z. (2003). *Wasted lives: Modernity and its outcasts*. Polity Press.

Beck, U. (1992). *Risk society: Towards a new modernity*. Sage.

Beer, A., & Foley, P. (2003). *Housing need and provision for recently arrived refugees in Australia* (Final report no. 48). Australian Housing and Urban Research Institute. https://www.ahuri.edu.au/research/final-reports/48

Bhuller, M., Mogstad, M., & Salvanes, K. G. (2017). Life-cycle earnings, education premiums, and internal rates of return. *Journal of Labor Economics, 35*(4), 993–1030. https://doi.org/10.1086/692509

Bitsani, E. (2013). *"Intercultural city identity" and "human intercultural cities" (H.I.C): An ontological/dynamic model for the social cohesion of modern and post-modern cities. The research example of Trieste Italy*. Nova Publishers.

Bitsani, E. (2016a). Migration memory, cultural heritage: A vehicle of the city' intercultural identity. The case study of the Eastern Greek community and the Greek museums in Trieste Italy. In *Cogents arts and humanities* (Vol. 3(1), pp. 2–15). Taylor & Francis.

Bitsani, E. (2016b). The cultural heritage and the historic memory as a main factor for sustainable development of modern cities in the framework of crisis: The case study of the museum cluster 'Museumsinsel' in Berlin. In *Cultural heritage and sustainable development*.

Bitsani, E., & Anselmi, W. (2021). From the today's dystopia to the future utopia reality: Recovering the human cultural identity of European City. In *Virtual 27th international conference of Europeanists, Europe's past, present, and future: Utopias and dystopias*. Columbia University.

Bitsani, E., & Elefteriou, A. (2014). The management of natural world heritage sites as an essential component of cultural tourism and sustainable development - The Tokaj-Hollókö case study in Northeastern Hungary: From a national/local past towards an international/global future. In N. Columbus (Ed.), *Tourism development and management*. NOVA.

Bitsani, E., & Pantazopoulos, S. (2018). Cultural policies to combat social exclusion and strengthen social cohesion in times of crisis. In E. Bitsani, K. Mavreas, D. Sotiropoulos, & T. Tsekos (Eds.), *Institutions and policies in times of crisis* (pp. 161–194). Papazisis.

Bridgewater, P., & Rotherham, I. D. (2019). A critical perspective on the concept of biocultural diversity and its emerging role in nature and heritage conservation. *People and Nature, 1*(3), 291–304. https://doi.org/10.1002/pan3.10040

Buhaug, H., & Urdal, H. (2013). An urbanization bomb? Population growth and social disorder in cities. *Global Environmental Change, 23*(1), 1–10. https://doi.org/10.1016/j.gloenvcha.2012.10.016

Capacci, L., & Biondini, F. (2022). Resilience of aging structures and infrastructure systems with emphasis on seismic resilience of bridges and road networks: Review. *Resilient Cities and Structures, 1*(2), 23–41. https://doi.org/10.1016/j.rcns.2022.05.001

Centeno, M. A., & Cohen, J. N. (2012). The arc of neoliberalism. *Annual Review of Sociology, 38*, 317–340. https://doi.org/10.1146/annurev-soc-081309-150235

Certomà, C. (2020). The politics of mapping urban agriculture: The case of participatory GIS. *Urban Studies, 57*(4), 837–855.

Colenutt, C., Brown, E., Nelson, N., Paton, D. J., Eblé, P., Dekker, A., Gonzales, J. L., & Gubbins, S. (2020). Quantifying the transmission of foot-and-mouth disease virus in cattle via a contaminated environment. *mBio, 11*(4), e00381–e00320. https://doi.org/10.1128/mBio.00381-20

Contiades, X., & Fotiadou, A. (2020). Comparative constitutional change: A new academic field (with Dr. A. Fotiadou). In X. Contiades & A. Fotiadou (Eds.), *Handbook of comparative constitutional change* (Routledge handbook) (pp. 1–21).

Coskun, E., & Taspinar, N. (2022). Volatility spillovers between Turkish energy stocks and fossil fuel energy commodities based on time and frequency domain approaches. *Resources Policy, 79*, 102968. https://doi.org/10.1016/j.resourpol.2022.102968

Dawazhaxi, Zhou, W., Yu, W., Yang, Y., & Jing, C. (2023). Understanding the indirect impacts of urbanization on vegetation growth using the Continuum of Urbanity framework. *Science of the Total Environment, 899*, 165693. https://doi.org/10.1016/j.scitotenv.2023.165693

De Graaf, R., & Ovink, H. (2021). The five pillars of climate resilience. In *Climate resilient urban areas* (pp. 1–19). Springer. https://doi.org/10.1007/978-3-030-57537-3_1

DFID. (2005). Social exclusion review. https://odi.org/en/publications/dfid-social-exclusion-review-2005/

Djan, M. A. (2023). Urban food security: Examining the unique challenges and opportunities associated with ensuring food security in urban areas. *European Journal of Nutrition & Food Safety, 15*(9), 42–52. https://doi.org/10.9734/EJNFS/2023/v15i91335

Dotsikas, K., Osborn, D., Walters, K., et al. (2023). Trajectories of housing affordability and mental health problems: A population-based cohort study. *Social Psychiatry and Psychiatric Epidemiology, 58*, 769–778. https://doi.org/10.1007/s00127-022-02314-x

Ferri, G. (2003). Review of the book Globalization and its discontents, by J. E. Stiglitz. *Philosophy & Social Criticism, 29*(5), 473–487. https://doi.org/10.1046/j.0391-5026.2003.00107.x

Försterling, G., Orth, R., & Gellert, B. (2023). Transition to a circular economy in Europe through new business models: Barriers, drivers, and policy making. *Sustainability, 15*(10), 8212. https://doi.org/10.3390/su15108212

Freeman, R. E. (1999). Response: Divergent stakeholder theory. *The Academy of Management Review, 24*(2), 233–236. https://doi.org/10.2307/259078

Friedman, M. (2007). *The social responsibility of business is to increase its profits* (pp. 173–178). Springer.

Gavin, M. C., McCarter, J., Mead, A., Berkes, F., Stepp, J. R., Peterson, D., & Tang, R. (2015). Defining biocultural approaches to conservation. *Trends in Ecology & Evolution, 30*(3), 140–145. https://doi.org/10.1016/j.tree.2014.12.005

Gentili, M., & Hoekstra, J. (2018). Houses without people and people without houses: A cultural and institutional exploration of an Italian paradox. *Housing Studies, 34*(3), 425–447. https://doi.org/10.1080/02673037.2018.1447093

Gifford, P. (2024, March 8). Aging infrastructure and how to address it. *Water Utility Management Magazine*. Retrieved from https://www.waterworld.com/water-utility-management/article/14310075/aging-infrastructure-and-how-to-address-it

Global Governance Futures 2030 Fellows. (2019). *Make or break: How will cities shape future global challenges?* Robert Bosch Stiftung. https://www.bosch-stiftung.de/en/publication/make-or-break-how-will-cities-shape-future-global-challenges

Guzman, P., Pereira Roders, A., & Colenbrander, B. (2014, April). Bridging the gap between urban development and cultural heritage protection. In *Impact assessment for social and economic development 2014*. Viña del Mar. https://doi.org/10.13140/2.1.4633.7923

Habermas, J. (2003). *The future of human nature* (W. Rehg, Trans). Polity Press.

Hamdani, R. S., Hadi, S. P., & Rudiarto, I. (2021). Progress or regress? A systematic review on two decades of monitoring and addressing land subsidence hazards in Semarang City. *Sustainability, 13*(24), 13755. https://doi.org/10.3390/su132413755

Heynen, N., Kaika, M., & Swyngedouw, E. (Eds.). (2006). *In the nature of cities: Urban political ecology and the politics of urban metabolism*. Routledge.

Hoekstra, J., & Vakili-Zad, C. (2011). High vacancy rates and rising house prices: The Spanish paradox. *International Journal of Urban and Regional Research, 35*(1), 142–165. https://doi.org/10.1111/j.1467-9663.2009.00582.x

Iaione, C. (2016). The CO-city: Sharing, collaborating, cooperating, and commoning in the city. *American Journal of Economics and Sociology, 75*(2), 415–455. https://doi.org/10.1111/ajes.12145

IPBES. (2022). *Summary for policymakers of the methodological assessment of the diverse values and valuation of nature*. IPBES Secretariat. https://www.unep.org/resources/report/assessment-report-diverse-values-and-valuation-nature

Jackson, S. E., Hackett, R. A., Pardhan, S., Smith, L., & Steptoe, A. (2019). Association of perceived discrimination with emotional well-being in older adults with visual impairment. *JAMA Ophthalmology, 137*(7), 825–832. https://doi.org/10.1001/jamaophthalmol.2019.1230

Kigadye, F. (2014). Architectural conservation in rapidly urbanising cities: The case of Dar es Salaam. *Tanzania. Les Cahiers d'Afrique de l'Est, 47*, 29–54. https://doi.org/10.4000/eastafrica.342

Kothari, A., Salleh, A., Escobar, A., Demaria, F., & Acosta, A. (Eds.). (2019). *Pluriverse: A post-development dictionary*. Tulika Books.
Kovel, J. (2007). *The enemy of nature: The end of capitalism or the end of the world?* Zed Books.
Lee, C., Lee, J., & Park, S. (2022). Forecasting the urbanization dynamics in the Seoul metropolitan area using a long short-term memory–based model. *Environment and Planning B: Urban Analytics and City Science, 50*(2), 453–468. https://doi.org/10.1177/23998083221118002
Li, X. (2021). Housing insecurity and its impact on health outcomes: A comprehensive review. *Journal of Housing and the Built Environment, 36*(2), 345–367.
Lukawiecki, J., Wall, J., Young, R., Gonet, J., Azhdari, G., & Moola, F. (2022). Operationalizing the biocultural perspective in conservation practice: A systematic review of the literature. *Environmental Science & Policy, 136*, 369–376. https://doi.org/10.1016/j.envsci.2022.06.016
Lux, B., Schneck, N., Pfluger, B., Männer, W., & Sensfuß, F. (2023). Potentials of direct air capture and storage in a greenhouse gas-neutral European energy system. *Energy Strategy Reviews, 45*, 101012. https://doi.org/10.1016/j.esr.2022.101012
Maffi, L., & Woodley, E. (2010). *Biocultural diversity conservation: A global sourcebook* (1st ed.). Routledge. https://doi.org/10.4324/9781849774697
McDonagh, R. (2022). Inequities in access to urban greenspace during Covid-19 national lockdowns: The case of Islington residents. *LSE Journal of Geography and Environment, 1*, 180–207.
McElroy, M. B. (1990). The empirical content of Nash-bargained household behavior. *Journal of Human Resources, 25*(4), 559–583. https://doi.org/10.2307/145667
McQuinn, K., O'Toole, C., Coffey, C., & Disch, W. (2021). *Quarterly economic commentary: Autumn 2021*. The Economic and Social Research Institute. https://doi.org/10.26504/qec2021aut
Meier, H.-R., Petzet, M., & Will, T. (Eds.). (2007). *Cultural heritage and natural disasters: Risk preparedness and the limits of prevention/Kulturerbe und Naturkatastrophen: Möglichkeiten und Grenzen der Prävention* (Heritage at risk special edition 2007). ICOMOS Germany & Technische Universität Dresden. https://openarchive.icomos.org/id/eprint/3025
Meijer, M., Rehbach, K., Brunner, J. W., Classen, J. A., Lammertse, H. C. A., van Linge, L. A., Schut, D., Krutenko, T., Hebisch, M., Cornelisse, L. N., Sullivan, P. F., Peitz, M., Toonen, R. F., Brüstle, O., & Verhage, M. (2019). A single-cell model for synaptic transmission and plasticity in human iPSC-derived neurons. *Cell Reports, 27*(7), 2199–2211.e6.
Montes-González, D., Vílchez-Gómez, R., Barrigón-Morillas, J. M., Atanasio-Moraga, P., Rey-Gozalo, G., & Trujillo-Carmona, J. (2018). Noise and air pollution related to health in urban environments. *Proceedings, 2*(20), 1311. https://doi.org/10.3390/proceedings2201311
Muneera, C. P., & Karuppanagounder, K. (2018). *Economic impact of traffic congestion—Estimation and challenges*. University of Trieste, IST Studio Trasporti Integrazione Economica Europea (ISTIEE).
Murray, A., Skene, K., & Haynes, K. (2017). The circular economy: An interdisciplinary exploration of the concept and application in a global context. *Journal of Business Ethics, 140*(3), 369–380. https://doi.org/10.1007/s10551-015-2693-2
Oloriz, C., & Parlee, B. (2020). Towards biocultural conservation: Local and indigenous knowledge, cultural values and governance of the white sturgeon (Canada). *Sustainability, 12*(18), 7320. https://doi.org/10.3390/su12187320
Pantazopoulos, S. (2024a). Social and political change in post-war Greece: Demographic transformations in the in the 1970s. *Social Cohesion and Development, 18*(1), 69–88. https://doi.org/10.12681/scad.37336
Pantazopoulos, S. (2024b). *Social transformation and welfare: From the local to the supreme*. Dionikos. ISBN: 978-618-5665-11-1.
Perry, D. M., & Praskievicz, S. J. (2017). A new era of big infrastructure? (Re)developing water storage in the U.S. West in the context of climate change and environmental regulation. *Water Alternatives, 10*(2), 437–454.

Pohlan, L. (2019). Unemployment and social exclusion. *Journal of Economic Behavior & Organization, 164*, 273–299. https://doi.org/10.1016/j.jebo.2019.06.006

Pooch, M. U. (2016). *DiverCity—Global cities as a literary phenomenon: Toronto, New York, and Los Angeles in a globalizing age*. Transcript Verlag. https://doi.org/10.1515/9783839435410

Purcell, M. (2014). Possible worlds: Henri Lefebvre and the right to the city. *Journal of Urban Affairs, 36*(1), 141–154. https://doi.org/10.1111/juaf.12034

Putra, A. S., Tong, G., & Pribadi, D. O. (2020). Food security challenges in rapidly urbanizing developing countries: Insight from Indonesia. *Sustainability, 12*(22), 9550. https://doi.org/10.3390/su12229550

Raudsepp, M., Heidmets, M., & Kruusvall, J. (2009). Environmental justice and sustainability in post-Soviet Estonia. In I. J. Agyeman & Y. Ogneva-Himmelberger (Eds.), *Environmental justice and sustainability in the former Soviet Union*. (Cambridge, MA, 2009; online edn, MIT Press Scholarship Online, 22 Aug. 2013). https://doi.org/10.7551/mitpress/9780262012669.003.0010

Remillard, E. T., Campbell, M. L., Koon, L. M., & Rogers, W. A. (2022). Transportation challenges for persons aging with mobility disability: Qualitative insights and policy implications. *Disability and Health Journal, 15*(1), 101209. https://doi.org/10.1016/j.dhjo.2021.101209

Rodwell, D. (2007). *Conservation and sustainability in historic cities*. Wiley. https://doi.org/10.1002/9780470759547

Rondinel-Oviedo, D. R., & Keena, N. (2023). Entropy and cities: A bibliographic analysis towards more circular and sustainable urban environments. *Entropy, 25*(3), 532. https://doi.org/10.3390/e25030532

Rozzi, R. (2013). Biocultural ethics: From biocultural homogenization toward biocultural conservation. In R. Rozzi, S. Pickett, C. Palmer, J. Armesto, & J. Callicott (Eds.), *Linking ecology and ethics for a changing world* (Ecology and ethics) (Vol. 1). Springer. https://doi.org/10.1007/978-94-007-7470-4_2

Sassen, S. (2005). The global city: Introducing a concept. In D. Held & A. McGrew (Eds.), *The global transformations reader: An introduction to the globalization debate* (3rd ed., pp. 252–263). Polity Press.

Schär, C., & Jendritzky, G. (2004). Climate change: Hot news from summer 2003. *Nature, 432*(7017), 559–560. https://doi.org/10.1038/432559a

Simmel, G. (1971). The metropolis and mental life. In D. N. Levine (Ed.), *Georg Simmel on individuality and social forms* (pp. 324–339). University of Chicago Press.

Skamnakis, C., & Pantazopoulos, S. (2015). Social protection and local goverment: The evolution of a twin deficit. *Region and Periphery, 5-6*, 89–116. https://doi.org/10.12681/rp.18509

Stepanchuk, O., Bieliatynskyi, A., Pylypenko, O., & Stepanchuk, S. (2016). Peculiarities of city street-road network modelling. *Procedia Engineering, 134*, 276–283. https://doi.org/10.1016/j.proeng.2016.01.008

Sterling, E. J., Filardi, C., Toomey, A., et al. (2017). Biocultural approaches to well-being and sustainability indicators across scales. *Nature Ecology & Evolution, 1*, 1798–1806. https://doi.org/10.1038/s41559-017-0349-6

Stevenson, D. (2003). *Cities and urban cultures*. Open University Press.

Stiglitz, J. E. (2003, July 1). Ethics, market and government failure, and globalization. In *The governance of globalization* (Acta: Proceedings of the ninth plenary session of the pontifical academy of social sciences, 2–6 May 2003). Casina Pio IV.

Thymi, I., Bitsani, E., & Pantazopoulos, S. (2020). The role of strategic human resource management in cultural entities: The research example of Association "Lyceum of Greek Women" in Kalamata Greece. *International Journal of Cultural Management*. ISSN 1756-5677 paper accepted.

Toffler, A. (1981). *The third wave* (pp. 1–12). Bantam Books.

UNESCO. (2018). *Annual report 2017*. UNESCO. https://www.gcedclearinghouse.org/resources/unesco-2017-annual-report-2017?language=en

United Nations Conference on Trade and Development. (2022). *Rising to the challenge—The transformative power of trade in times of crisis*. United Nations. https://unctad.org/publication/unctad-annual-report-2022

United Nations Conference on Trade and Development (UNCTAD). (2021). *UNCTAD annual report 2020*. United Nations. https://unctad.org/webflyer/unctad-annual-report-2020

United Nations, Department of Economic and Social Affairs. (2015). *Transforming our world: The 2030 agenda for sustainable development*. United Nations. https://sdgs.un.org/2030agenda

Van Lange, P. A. M., & Huckelba, A. L. (2021). Psychological distance: How to make climate change less abstract and closer to the self. *Current Opinion in Psychology, 42*, 49–53. https://doi.org/10.1016/j.copsyc.2021.03.011

Watson, S. (2018). The challenges of collaboration and democratic participation in turbulent and unsettled times. *Tracce Urbane. Rivista Italiana Transdisciplinare Di Studi Urbani, 2*(3). https://doi.org/10.13133/2532-6562_2.3.14302

World Bank. (2022). World Development Report 2022: Finance for an equitable recovery. https://www.worldbank.org/en/publication/wdr2022

Yang, M., Chen, L., Wang, J., et al. (2023). Circular economy strategies for combating climate change and other environmental issues. *Environmental Chemistry Letters, 21*(1), 55–80. https://doi.org/10.1007/s10311-022-01499-6

Chapter 2
Utopia-Dystopia Theoretical Clarification in the Context of the City

The evolution of utopian and dystopian theory from the Renaissance to the twenty-first century is explored in this chapter, with particular attention being paid to the spatial and socio-political dimensions of urban environments in times of global crisis. Starting with the humanist ideals and speculative urban visions of the Renaissance, as exemplified by Thomas More's Utopia, the study then traces the changing nature of utopian thought through Enlightenment rationalism and nineteenth-century socialist idealism to the dystopian turn of the twentieth century. The analysis then integrates Michel Foucault's concept of heterotopias as a critical lens to understand real and imagined spaces that juxtapose, reflect, or invert normative societal structures. Foucault's concept of heterotopias provides a theoretical framework through which the city can be understood not only as a physical place, but also as a site of contestation, possibility, and resistance. Using this framework, the chapter examines how utopian ideas are expressed in urban planning, architecture, and policy, contrasting idealized cityscapes with the often-dystopian realities of socioeconomic exclusion, surveillance, and crisis governance. Urban spaces thus become laboratories and battlegrounds for competing visions of order, freedom, and community. It examines how cities have become crucibles of both vulnerability and resilience, where new forms of collective imagination and spatial organization emerge. Contemporary cities are re-enacting the tensions at the heart of utopian and dystopian thinking. The study argues that the city remains a privileged site for the projection, negotiation, and critique of utopian aspirations and dystopian anxieties.

The first utopian approach to coexistence in the city and governance is found in Plato. Plato was a creative disciple and a loyal student of Socrates. He consistently prioritized ethics in his philosophical inquiries. In all of his works, Plato speaks on behalf of Socrates. Nonetheless, he diverged from strict Socratic theory in one significant aspect: with a pronounced poetic sensibility, he elevated philosophy "from the earth to the heavens," as one might say. He exhibited profound intellectual audacity, was distinguished for his wisdom, and sought to grant ethics primary importance. He studied several of his philosophical predecessors and re-examined

E. Bitsani et al., *The Biocultural City as a Utopia or the Future Reality?*, Cities, Heritage and Transformation, https://doi.org/10.1007/978-3-032-17584-7_2

each philosophical issue in his own manner, formulating original viewpoints. Building upon the life and thought of Socrates, Plato developed a deep and multi-faceted philosophical system.

In the Platonic (or Socratic) ideal state, as previously mentioned, certain foundational organizational principles prevail, which Plato regarded as necessary and sufficient conditions for a just and well-governed polity. These principles, as articulated in his seminal work The Republic—composed when he was around 50 years old—include the following:

(a) The division of the polity into three fundamental "classes" or categories of citizens, corresponding to distinct parts of the human soul: the rational (philosopher-kings), the spirited (guardians), and the appetitive (producers).
(b) The assignment of all economic activity exclusively to the third class—the producers or creators—according to the principle of specialization. This class alone is permitted to engage in marriage, possess private property, and exercise economic freedom.
(c) The absence of private property and the institution of communal living among the first class, the political leaders and lifelong thinkers ("to make the things of friends common," Republic 424a1), who own nothing beyond their own bodies ("all else being held in common," 464d9). They require little beyond leisure (otium) for the pursuit of public service and philosophical contemplation, free from other distractions (negotium). It is worth noting, from a comparative perspective, that Hegel also identifies three "estates" or social categories in his Philosophy of Right (Part III, §§201 ff.) within the emerging "civil society": the substantial or productive estate (engaged in primary economic production and the maintenance of the nuclear family), the commercial and industrial estate (urban-centered, fostering liberty, property, and entrepreneurship), and the universal estate (state officials serving the common interest through devotion to public institutions).
(d) The continuous and rigorous intellectual education of prospective rulers up to the age of 50, after which they assume roles in governance and administration, contingent on demanding evaluations.
(e) The abolition of marriage and family life for those in public authority, so they remain undistracted and wholly devoted to serving public affairs and the common good—living akin to "monks" for the sake of a political life infused with philosophy.
(f) The equal treatment of women and men, including the acknowledgment of women's capacity to hold public office and undertake political duties ("to share equally in the governance of the city," Republic 455b2; "if women are to share in everything with men," 540c8), under the same conditions and prerequisites.
(g) The softening of the institution of slavery, which, in the Socratic spirit (more socratico), was reinterpreted as rational labor and service provided in support of higher state functions for the common good ("to be ruled by the divine and the wise… so that we may all be similar in power and friends, governed by the same principle," 590d1–5).

As for the use of utopia as a didactic tool, both Plato and Marx employed the imaginary construction of a perfect state to instill virtue and awareness among the members of existing societies. Within this framework belongs the notion of the "noble lie" from Book III of The Republic, where Plato openly admits that the use of fictional narrative (myth) is not only acceptable but necessary to minimize dissent in the establishment of the state. Plato reiterates this view in the Laws (665a ff), emphasizing the need to create new myths that guide citizens toward virtue, replacing older ones that failed to achieve this goal.

Utopia from the Greek ou-topos, meaning no place, or nowhere, the word utopia has become adopted in the English language to mean a place where everything is ideal or perfect. In addition, Utopia is an imaginary community or society that possesses highly desirable or nearly perfect qualities for its citizens. The term was coined by Sir Thomas More for his 1516 book Utopia, describing a fictional island society in the south Atlantic Ocean off the coast of South America (Claeys, 2020; Giroux, 2003; Shadurski, 2016).

The discussion about the concepts of utopia and dystopia is basically a cultural discourse in the context of the search for an ideal state, which will ensure the happiness and well-being of its inhabitants.

Thomas More, a humanist, political thinker (McConica, 2011), and Lord Chancellor of England (1529–1532), was a close friend of Erasmus and a collaborator of King Henry VIII until he refused to submit his conscience to royal authority, denouncing the king as head of the Church of England. As a result, he was imprisoned and later executed. Throughout his life, more engaged with the political and social issues of the lower classes and produced influential writings (Burkard, 1999).

In 1515, he began composing one of his most famous works, Utopia, which reflected the hardships of his era. It was completed and published in Latin, in two volumes (Parker, 2002). Utopia was largely influenced by classical antiquity, particularly Plato's Republic, and by descriptions of the New World brought by early exploration. Many scholars argue that More drew from Americo Vespucci's four voyages (1507) (More, 2003). The term "Utopia" is derived from the Greek words "eutopia" (a good place) and "outopia" (no place), indicating an ideal yet non-existent location.

The study of utopia and dystopia can no longer disregard their cultural and ecological frameworks. In the realm of cultural theory and spatial studies, these imaginary geographies gain crucial significance for understanding the human relationship with the environment, the self, and social structures.

Cultural theory regards utopias and dystopias as narratives of identity, memory, and otherness. As Ruth Levitas (2013) argues, utopia is not merely a depiction of the ideal but a methodology of social reconstruction—a way to imagine transformative futures. Utopian imagination functions as a vehicle for cultural memory, projecting hopes and traumas of the past into imagined futures (Goodwin & Taylor, 2009).

Conversely, dystopia reveals the failures of such projects, exposing mechanisms of oppression and marginalization. Cultural dystopias—from literature to film—offer a mirror to the pathologies of late capitalism, globalization, and technological alienation (Booker, 1994; Jameson, 2005). Dystopian landscapes often serve as zones of intensified memory and identity erasure, as seen in Orwell's 1984 or Atwood's The Handmaid's Tale.

The ecological dimension of dystopia has become increasingly prominent in recent decades, especially in the context of climate crisis. Eco-dystopian narratives portray environmental collapse as a symptom of cultural decay and technological hubris. Works such as Cormac McCarthy's The Road (2006) and series like Black Mirror depict landscapes of ruin and dehumanization. Eco-dystopia not only imagines future disasters but also reveals how ecology, capitalist development, and surveillance technologies permeate everyday life (Morton, 2013).

Spatial theory, from Lefebvre (1991) to Soja (1996), stresses that space is socially produced and ideologically charged. Utopias and dystopias are spatial models embodying ideals or fears. In utopias like More's or those of Fourier, space is conceived as ordered and symmetrical. In contrast, dystopia—from 1984 to The Handmaid's Tale—reveals space as a matrix of control, oppression, and biopolitics (Foucault, 1976). Contemporary theorists like Massey (2005) advocate for a dynamic conception of space, open to multiple narratives, in contrast to the essentialist or authoritarian utopian planning. Massey emphasizes that utopias must account for the heterochronic and heterotopic realities that compose contemporary social geography.

The cultural and ecological discourse of utopia and dystopia illuminates the tension between imagination and materiality, between design and lived experience, between ideal and historical reality. These imagined spaces are not merely representations but frameworks of thought and action that shape political, aesthetic, and ethical orientations. Through their engagement with cultural memory and identity conflicts, utopia and dystopia emerge as dynamic tools of political imagination inscribed in the cultural spatial map.

2.1 The Evolution of Utopian Theory in the Quest for an Ideal City

2.1.1 The Utopian Theory in the Middle Ages

The medieval conceptualization of utopia was primarily ecclesiastical in nature. The city was not envisioned as a rational, secular construct but rather as a *reflection of divine harmony*, deeply influenced by Augustinian and Thomistic metaphysics. Recent scholarship expands upon this theological vision, noting how religious cosmology shaped spatial hierarchies and power relations within medieval urbanity (Kostof, 1991).

The early medieval period is marked by social and territorial upheavals resulting from intense conflict and warfare. From the eleventh century onward, European medieval society began to take a more coherent shape. In this period, characterized by theological obscurantism, ecclesiastical authority intensified, while the feudal system emerged as a dominant model of socio-economic organization, consolidating hierarchical and fixed social structures. By the thirteenth century, urbanization began to accelerate. Cities developed around essential institutions and functions, reflecting both defensive and economic imperatives. In addition to fortified urban centers, either by enclosing ancient Roman urban cores or through the aggregation of smaller settlements (Morris, 1994), three new urban forms emerged: monastic cities, feudal stronghold cities, and market-based cities (Ragon, 1975). These cities were encircled by defensive walls, delineating the natural from the built environment. Even as population density increased, urban expansion was managed through successive layers of fortifications (Benevolo, 1980). This urban form revealed a fundamental discontinuity between city and countryside. While feudal lords controlled the rural domains, urban societies began forming independently—often composed of artisans, craftsmen, and merchants, who organized themselves into self-governing guilds. Amid this medieval climate, ecclesiastical thinkers such as Thomas Aquinas played a vital role in sowing the seeds of philosophical reflection on ideal social and urban organization. Theologically grounded visions of the ideal city began to emerge—not as mere architectural constructs, but as *cosmological and spiritual blueprints* where divine order was mirrored in the spatial layout of the polis.

Notably, the ideal city of this period rarely existed as a blueprint for implementation. Rather, it functioned as a *discursive space*, a normative ideal within which the moral order of Christian society could be symbolically embedded (Blumenfeld-Kosinski, 1998). Modern interpretations also link these symbolic frameworks with Michel Foucault's notion of *heterotopia*—spaces that are simultaneously real and ideal, functional and symbolic (Foucault & Miskowiec, 1986).

Furthermore, contemporary urban theorists observe that early visions of utopia laid the groundwork for later secular and modernist models of spatial planning. Concepts such as enclosure, centralization, and institutional governance became recurrent themes in later urban design ideologies (Harvey, 2000).

2.1.2 Renaissance, the Ideal Social and Spatial Organization

As European society transitioned into the Renaissance, it began to gradually emancipate itself from the theocratic worldview that had dominated the Middle Ages. Cities no longer developed in an organic, unplanned fashion, but were now subject to rational design and systematic urban planning (Ragon, 1975). This shift gave rise to a new urban aesthetic, grounded in symmetry, proportionality, and geometric clarity—elements central to Renaissance artistic sensibility.

Urban layouts increasingly embraced radial or axial forms, with streets converging upon a central square that housed religious, administrative, and commercial functions (Nikolaidou, 1993). These designs were not merely technical; they reflected deeper ideological currents—order, balance, and humanist values—that permeated Renaissance thought.

One of the foundational inspirations for the ideal Renaissance city came from Vitruvius, an architect of Augustan Rome. Although his work De Architectura lacked specific urban diagrams, its principles deeply influenced Renaissance architects. The Vitruvian city was often envisioned as an octagonal form, composed of radial streets extending from fortified edges toward a central plaza. Concentric ring roads interconnected the urban segments, each containing public squares, temples, and civic buildings. Further theoretical grounding came from Leon Battista Alberti (1404–1472), whose seminal text De Re Aedificatoria (1452) offered a comprehensive treatise on architecture and urban form, albeit without visual illustrations. It was Antonio Filarete, through his Trattato d' Architettura (1457–64), who produced some of the earliest visual representations of a planned ideal city. His concept, Sforzinda, envisioned an ideal, fortified city in the shape of an eight-pointed star, integrating both geometric harmony and hierarchical civic order. The ideal Renaissance city was the embodiment of classical revival and humanist ideology. Spatial organization reflected an anthropocentric worldview, emphasizing human reason, balance, and civic virtue. According to Benevolo (1980), this period marked the beginning of the city as a consciously designed political and aesthetic object—one that mirrored the rationality and centrality of the human subject.

Recent scholarship emphasizes that Renaissance utopias were not simply architectural exercises but ideological instruments, expressing a desire for order amid the socio-political uncertainties of post-medieval Europe (Alsayyad, 2000; Kostof, 1991). The ideal city served as a metaphor for the ideal polity: structured, centralized, and harmonious.

Additionally, urban theorists like Françoise Choay highlight how these early utopian plans set precedents for later urbanist ideologies, particularly during modernism, where form and function were likewise fused under rationalist principles. Renaissance urbanism becomes, in retrospect, a proto-modernist project, combining aesthetics, ethics, and governance through spatial form.

2.1.3 Utopian Thought from the Renaissance to the Twenty-First Century

Utopias from this period exhibit certain common features, such as the proper organization and functioning of an ideal society—social structure, labor, property ownership, and the relationship between the city and the countryside—as well as walls that protect the city from external threats. As Vercelloni points out, "the European thinkers of the time developed their utopian models in perfect coordination and with mutual knowledge of each other's works." However, there are also differences, resulting from the individual characters and experiences of each author. In general, the ideal cities of this period are situated on distant islands and serve as critiques of the sociopolitical conditions of their time.

As Berneri notes, the authors of these works, especially More, recognized the importance of harmonizing urban and rural environments. Their works embody the humanist spirit of the Renaissance, opposing extreme individualism and seeking a new unity among peoples. It is noteworthy that in these utopias, money is abolished (with the exception of Bacon's utopia), everyone participates in labor and contributes to the community. This practice fosters greater cohesion and solidarity (Berneri, 1950).

In the utopias of Bacon and Campanella, significant references are made to scientific research and study. They strongly advocate for its importance in improving human life and inaugurate a new era in which material prosperity is seen as a prerequisite for human happiness (Mumford, 1962). This perspective resonates undeniably with our own time. Furthermore, in Bacon's utopia, scientific knowledge is held and controlled by a specialized group, whereas in the works of Andreae and Campanella, knowledge is widely disseminated throughout the community.

The utopias of the Renaissance aim to address the issues of their era. They are, in general, conceptual frameworks that seek optimal models and living standards beyond further improvement—spaces in which divine harmony is realized through societal restructuring and scientific advancement.

In conclusion, during the Renaissance, there are shared features in the organization of ideal cities, inherited from Antiquity and the Middle Ages—such as walls and temples—economic reliance on agricultural production, and in most utopian theories, the abolition of money. Regarding social organization, it becomes evident that the majority of Renaissance utopias, in their pursuit of grand visions and in their neglect of human's natural aggressive instincts, often result in oppression. The Renaissance essentially connects the values of classical antiquity with the main current of European culture.

2.1.3.1 The Utopian City in the Enlightenment: Eighteenth Century—Pre-Industrial Age

The eighteenth century, known as the Age of Enlightenment, was a time of reform during which traditional intellectual values were challenged and replaced by progressive thought promoted by philosophers such as Jean-Jacques Rousseau. These thinkers advocated freedom of expression, justice, equality before the law, and democracy. During this period, utopia often took the form of distant narratives, political romances, or imaginative dreamscapes. Most authors used their works to critique the monarchy and express dissatisfaction with the political state of the time (Berneri, 1950; Nikolaidou, 1993).

At the same time, new mechanical innovations appeared, leading to the creation of early industrial units in the countryside, distant from residential areas (Servier, 1967). Labor was now organized into productive units that would later evolve into industrial cities. Within this framework, Nicolas Ledoux attempted to realize the first organized industrial city—a model of agricultural industry.

2.1.3.2 Nineteenth Century: The Golden Age of Utopia

The nineteenth century, often referred to as the golden age of utopia, is associated with three major events: the Industrial Revolution, the French Revolution, and the emergence of socialism (Muravchik, 2003). As Berneri observes, the history of utopias in this century is closely tied to the birth of the socialist movement.

From the late eighteenth century, a significant expansion of production units had already begun, typically located near energy sources, transportation routes, and raw materials. This marked the gradual onset of industrialization, which started in Great Britain (Gospodini, 2002) and spread across Europe. Profound changes followed, such as rapid population growth, urbanization, the development of transportation networks, and an increased demand for goods and services (Benevolo, 1980). This culminated in the emergence of large-scale industrial installations and the rise of a capitalist development system. The new industrial regime transformed labor dynamics, gradually replacing human effort with mechanical power.

Amid these upheavals, rural populations formed the backbone of the industrial workforce and were forced to live and work in dire conditions. The urban-rural relationship was dramatically altered. The massive population influx into urban centers led to expansion beyond natural boundaries and at the expense of the environment. Inadequate infrastructure, speculative exploitation of land, overcrowding, and the rise of epidemics prompted urgent reforms in urban planning and public hygiene (Gospodini, 2002; Nikolaidou, 1993).

Two primary approaches emerged for restoring the urban environment (Benevolo, 1980). The first involved predetermined interventions in the urban fabric for gradual environmental improvement—Haussmann's redesign of central Paris is a prime example. The second advocated for idealized industrial settlements, utopian communities designed as social models of organization. Within this climate of reconfiguration, utopian idealism flourished. As Jacoby (2000) argues, utopian thinking does not reject practical reforms but rather fosters fundamental changes. The French Revolution, indeed, served as the catalyst for these transformations and the development of liberal ideology, which fueled popular aspirations and hopes (Wallerstein, 2006).

This period gave rise to the major figures of utopian thought: Robert Owen, Henri de Saint-Simon, Charles Fourier, Jean-Baptiste Andre Godin, James Silk Buckingham, William Morris, and Etienne Cabet. Their utopias moved beyond abstract theory, challenging the status quo and engaging in real-world experiments.

Simultaneously, socialist political theory emerged, with Karl Marx (1818–1883) and Friedrich Engels (1820–1895) as central figures. They classified the ideas of the aforementioned thinkers as utopian socialism. For Engels and Marx, the city was a crucial issue that had to be understood within a broader political, economic, and social framework. Consequently, they rejected the proposal of specific spatial models for addressing urban problems. As scientific socialists, their theories are not extensively analyzed in this study.

2.1.3.3 Robert Owen's Utopias: New Lanark and New Harmony—Communities of Collectivity

Robert Owen, a British utopian socialist and social reformer, is closely associated with the concept of cooperative ownership. According to Gospodini, Owen's ideas centered on creating small ideal communities in which residents collectively managed the means of production, including industrial facilities and arable land.

In 1798, Owen became manager of the New Lanark textile factory in Scotland and implemented innovative reforms to improve workers' quality of life. He believed that in order to improve individuals' lives, the environment itself had to be redesigned. In 1815, he initiated several improvements for the 3000 New Lanark workers, raising wages and reducing working hours to a maximum of ten per day. He also implemented strict labor division to enhance productivity and prohibited child labor.

Owen prioritized children's education over labor, aiming to shape them into better citizens and future workers. In 1816, he founded the "Institution for the Formation of Character," which catered to children aged 3–25, offering both general and technical education. His views on education influenced subsequent utopian thinkers and reshaped pedagogical practices.

2.1.3.4 Utopia in the Twentieth Century

In the latter half of the nineteenth century, the dense cores of industrial cities, and pollution from unchecked factory expansion, low living standards, and unregulated urban growth severed humanity's connection with nature. Efforts to reorganize society began with the ideas of utopian socialists and were furthered by early twentieth-century architects such as Ebenezer Howard, Tony Garnier, and Le Corbusier (Gospodini, 2002; Silva, 2003).

This era saw the emergence of modernism, which, along with functionalist theory, laid the foundations for the modern city. Modernism marked a radical break from history and tradition, discarding inherited social practices. Urban space was now functionally segregated, surpassing conventional forms such as the square, the street, and the urban block. Yet, the rejection of the past and the pursuit of innovation led urban planners to believe that the ideal could now be realized—sometimes through extreme designs such as Takis Zenetos' "Hanging City."

Two major schools of modern architecture arose in the twentieth century. The first, associated with Mies van der Rohe, emphasized minimalist design, while the second, tied to Le Corbusier (Charles-Edouard Jeanneret), became known as New Brutalism (Gospodini, 2002; Silva, 2003; Stevenson, 2012).

The link between modernism and social utopias endured throughout the twentieth century and proved critical in shaping modernism as both art and ideology. Whether in utopian city designs or efforts to build new towns and societies, modernism sought to clear the slate and begin anew, in strict accordance with its principles (Silva, 2003).

2.1.3.5 Utopia in Twenty-First Century

In the twenty-first century, discussions around utopia for some authors include post-scarcity economics, late capitalism, and universal basic income; for example, the "human capitalism" utopia envisioned in Utopia for Realists includes a universal basic income and a 15-h workweek, along with open borders (Ostalska & Fisiak, 2021; Shadurski, 2019).

Sargent discusses the role of utopianism in literature, and in the development of colonies, and in immigration. The idea of utopia has become commonplace in social and political thought, both negatively and positively. Some thinkers see a trajectory from utopia to totalitarianism with violence as an inevitable part of the mix. Others see utopia directly connected to freedom and as a necessary element in the fight against totalitarianism.

"None of the abstract concepts comes closer to fulfilled utopia than that of eternal peace" (Adorno, 2005), our point of view in this study agrees with Adorno and proposes a model of equal co-existence in the ecological communities, based on biocultural approach.

A city or a place exists only if it has an identity, which originates from and is constructed by the people who live there, have lived there, or want to live there. The identity can convey a certain image, the image of the specific place, or the specific city.

However, the city or the place itself has a huge impact on its identity due to the interaction between the natural environment and anthropogenic factors. These dynamic components also have repercussions on the identity (Anselmi & Wilson, 2011; Bitsani, 2014).

2.2 Space/Place: City and Foucault's Heterotopias

Here the study revisits Michel Foucault's theory of heterotopias and its critical value for analyzing contemporary urban space. Moving beyond the classical framework, it incorporates perspectives from postcolonial, feminist, digital, and critical urban theory. Through new case studies and an interrogation of power, identity, and spatiality, this work proposes a renewed understanding of heterotopias for the twenty-first-century city.

Cities are no longer purely material constructs but hybrid constellations of digital flows, spatial policies, and symbolic orders. Michel Foucault's notion of heterotopia spaces that mirror, contest, and invert dominant spatial logic offers a powerful lens for rethinking the fractured, multipolar urban experience of today. In this reappraisal, heterotopias are not merely marginal or exceptional; they constitute active terrains where biopolitics, exclusion, and subjectivity intersect.

Originally outlined in his 1967 lecture "Des espacesautres," Foucault identified heterotopias as "real places—places that do exist and that are formed in the very founding of society." They embody paradox: being both outside and inside society.

Yet, Foucault's formulation is fragmentary and deliberately open. Building on this, scholars such as Johnson, Saldanha, and others argue that heterotopias serve as heuristic tools rather than fixed categories.

The concept of heterotopia, as introduced by Michel Foucault in his essay "Des EspacesAutres" (1984, 1967), constitutes a radical theoretical intervention in understanding the relationship between space, power, and cultural order. Heterotopias are real spaces that function as reversed or counter-sites relative to dominant social space. In contrast to utopias—which are imaginary or unrealizable—heterotopias exist and simultaneously perform social critique while reproducing systems of power.

Foucault identified six key principles of heterotopias that establish a kind of typology of spatial differentiation. First, every society creates its own heterotopias. Second, heterotopias change in function over time. Third, they coexist with other social spaces and either mirror or invert their relationships. Fourth, they exhibit heterochronicity—a relation to time that differs from the everyday. Fifth, they possess mechanisms of inclusion or exclusion, governed by specific rules. Sixth, heterotopias operate either as reflections of society or as different order's proposals.

These principles are not merely theoretical descriptors; they offer an analytical toolkit for understanding how power, knowledge, and subjectivity are spatially embedded. For instance, a museum or library accumulates time (heterochrony), while a prison or psychiatric institution activates mechanisms of exclusion and discipline (Hetherington, 1997).

The critical theoretical engagement with Foucault's six principles has revealed both political and aesthetic dimensions. Soja (1996) argues that these principles encapsulate the contradictions of urban experience under late capitalism. Heterotopia is not simply an architectural condition, but is produced through social practices, narratives, and materialities (Lefebvre, 1991). Critical geography and poststructuralist interpretations, such as those by Johnson, suggest that Foucault's heterotopia is a term used to describe different spatial and temporal disruptions. These disruptions imaginatively question and undermine certain ideas of what a utopia is. It ends by explaining how heterotopia challenges the space in which we live.

Lefebvre, when speaking of the "right to the city" and the city as oeuvre (work), essentially attempts to describe the terrain upon which the critical questioning of capitalist reproduction can unfold—the potential emergence of social practices not oriented toward production, sale, or consumption. He believed that the dominant arena where the capitalist system is both reproduced and potentially subverted is that of everyday life. Thus, for Lefebvre, space does not merely reflect existing (dominant) social relations; rather, it constitutes a constant battleground where, beyond bourgeois dominance, one encounters dissent, resistance, and alterity. He argued that although the production of capitalist space aspires to homogeneity and abstraction (totalizing space), in practice it leaves behind fragments.

In "La production de l'espace," Henri Lefebvre proposes a triadic model of spatial analysis consisting of spatial practices (the material and empirical dimensions of space shaped by social relations, needs, and power structures), representations of space (the conceived space of planners, scientists, and technocrats), and spaces of representation or representational spaces (space as directly lived and appropriated through everyday experience). While Lefebvre dedicates extensive attention to the theorization of space, his references to heterotopia are infrequent and dispersed, lacking a systematic definition. However, in texts such as "La révolutionurbaine," "The Right to the City," and "The Production of Space," he invokes heterotopia in contrast to isotopia, often positioning it as its conceptual and spatial opposite, alongside the notion of utopia. For Lefebvre, isotopia is the dominant spatial logic—normative and centralized—while heterotopia is defined as a counter-spatial logic, representing a spatial form that challenges or resists the dominant order represented by isotopia.

The reasons behind Lefebvre's use of the term heterotopia appear to diverge from those of Foucault (see: Foucault, space, and alterity). Lefebvre was a vocal critic of structuralist thought, including Foucault—for three main reasons. First, he argued that structuralism denied historical becoming, approaching social structures in a static, anti-dialectical manner. Second, it portrayed human agency as the predetermined outcome of structural forces, leaving little room for freedom. Third, it failed to bridge the gap between theoretical abstraction and lived social reality.

Nonetheless, despite their different starting points, both Lefebvre and Foucault shared a deep concern with space and alterity, and by the late 1960s, their theoretical trajectories appeared to be converging—what Edward Soja later described as "crossing paths."

In the same framework, Soja proposes a reading of Foucault's heterotopias through the lens of Lefebvre's spatial theory. He articulates a "trialectics" of spatial practices, representations of space, and spaces of representation, building upon Lefebvre's triadic model. Following Lefebvre, Soja critiques the tendency of spatial theory to oscillate between dualistic models—one privileging objectivity, materiality, and structure; the other, subjectivity and consciousness.

Lefebvre referred to this as the "double illusion" surrounding spatial discourse. In response, Soja introduces the concept of "Thirding-as-Othering," a way to think beyond traditional binaries and resist static, one-dimensional interpretations of space—whether object—or subject-centered. This thirding logic is not a synthesis or average of the two poles; nor is it an ultimate epistemic resolution. Rather, it is a dynamic and recursive approach to understanding the mutual constitution of subject and object, and more broadly, the social production of space.

In contrast, Marco Cenzatti attempts to advance this "Lefebvrian" analysis of heterotopia a step further. Similar to Soja, he associates heterotopia with "spaces of representation," while also proposing a new periodization. Drawing on Foucault's historical distinction in the first principle of heterotopology—where he refers to the heterotopias of crisis in "primitive" societies and the heterotopias of deviation in modern ones—Cenzatti introduces a third category: heterotopias of difference. According to him, under today's socio-cultural and economic conditions, heterotopias of deviation have ceased to be the dominant form, replaced instead by heterotopias of difference.

Cenzatti argues that in contemporary societies, the norms that define deviation (i.e., "the normal") have become far more flexible and dynamic. Deviation can no longer be understood as something static or fixed, but rather as something that is continually redefined. The social struggles that emerged from the late 1960s onward—led by various "minority" groups such as LGBTQ+ communities, Black activists, and feminists—were centered on claiming the "right to difference," ultimately shaping a new mode of social reproduction. Capitalism was not defeated but instead reorganized to integrate these new conditions. Post-Fordism, which replaced mass, homogenized production, became a flexible, adaptive economic model capable of incorporating and even profiting from this evolving social reality.

Under these new conditions, social identities that exist "outside" of the normative framework can now be more readily assimilated, while new excluded identities continue to emerge. In Cenzatti's logic, heterotopias can no longer be understood as fixed, spatially delimited institutions (e.g., prisons, clinics) that contain or exclude stable social subjects. Rather, similar to Soja's view in the era of multiple identities, heterotopias must be understood as a type of practice form of spatial appropriation by shifting groups at specific moments in time. When the social relations that constitute a space as heterotopic dissolve, that space ceases to function as a heterotopia.

A given physical location may thus acquire heterotopic qualities due to specific social relations that unfold within it during certain events (e.g., a marketplace, demonstration, uprising, celebration). Cenzatti emphasizes that he does not suggest older forms of heterotopia (crisis, deviation) have disappeared, but rather that heterotopias of difference have become the dominant mode. In summary, Cenzatti's heterotopias of difference represent ephemeral socio-spatial conditions occurring in interchangeable spaces involving shifting social subjects.

It appears that most analyses of heterotopia which adopt Lefebvre's approach to space (e.g., Soja, Cenzatti, Stickells, Lang) share a common characteristic: the desire to "release" heterotopia from the constraints of fixed places and to reconceive it as something diffuse and dynamic within social reality. At the same time, albeit implicitly, most of these interpretations tend to conceptualize heterotopias unilaterally as conditions of resistance. That is, they appear as peripheral or marginal spaces relative to the center (the isotopia), where "other" relations are formed in opposition to the dominant, but only in terms of their subversion.

Consequently, scholarly attention has focused on underground, marginal, and illicit aspects of social reality (Soja), in spaces and conditions that are not incorporated into mainstream forms of social reproduction but follow their own codes and logic.

Thus, while the connection to Lefebvre's "spaces of representation" is productively employed, the relative one-sidedness of many of these analyses tends to obscure some of the core attributes of heterotopia. Although they seek to politicize and more concretely define the role and function of heterotopias, they often leave unresolved certain ambiguities regarding their relationship to Foucault's original intentions and broader conceptual concerns—elements that remain critically important.

Moreover, the notion of spaces as sites of tension (e.g., among gender, race, class) has been addressed by feminist theory, which identifies heterotopias as potential arenas for subjective and political reconstitution. Thus, Foucault's spatial thinking offers a framework for understanding spatial multiplicity and cultural entanglement under conditions of global mobility, social inequality, and postmodern identities.

Foucault has shown, through his analyses, that the "disruptions" of order, and more specifically of the forms of knowledge, power, and truth, do not generally tend toward a "progressive" direction, neutral development, an integration. Rather, the opposite can be deduced as a conclusion from his work. Foucault devotes many studies to the attempt to deconstruct the "human" and socially superior face of the Enlightenment era. He seeks to show that the questioning of older forms of knowledge (e.g., metaphysics) is combined with new forms of domination, deeper, more detailed, and more stable, and is not the result of general progress of the "human spirit." Thus, in the introduction of 1966, while on the one hand he himself seems to agree with a more "elastic" interpretation of heterotopias (as regards the question of their spatial delimitation), on the other hand he does not treat them unambiguously as places (whether mental or real) that a priori gives rise to questioning, in favor of a perspective of emancipation. Rather, we could say that he describes them as places—rifts. Places in which, potentially, the other can be born. Heterotopias should not be interpreted as enclosed spaces of lawfulness, but as networks of places, practices, and situations that constitute multiform relations and interactions. They constitute fields that host and are simultaneously given meaning, by processes of different forms of "subjectifications," which are exposed to a large network of contradictions, connections, and inclusions. They are not only ideas about space, but also specific spatial (material) configurations that tend either to deepen power relations or to create fields of questioning and rupture with the order of social reality. Their internal conflicts, as well as the dynamic relations they develop with the rest of the environment, place them in the field of contingency and "potential" passage. They appear, in short, as "floating spaces of otherness." Although by general admission, it is a concept that until the publication of Foucault's lecture in 1984 was almost unknown and had not been a field of particular reflection and discussion, it seemed to hide an extraordinary dynamic. The interest in its content was identified with the search for theoretical tools that could capture and interpret the complexity of modern societies. Its vague "scale of application" made it particularly useful to geographers, urban planners, and architects. The important thing, however, is that beyond the scientific interest it presents, it appeared to be predominantly a political concept. Both because it brought to the surface important aspects of the functioning of the system, regarding its need for reproduction and "expansion" (e.g., prisons, clinics), but mainly because it highlighted new ways and new tools of social resistance. Resistances, which, as it turned out, are not limited only to denials of power, but are intertwined with the demand for a total overthrow of the order of social reality. Heterotopias show that the construction of the "other" is not an easy, teleological, and redemptive process, but a laborious one, where the new subjectivities that are produced and the new relationships that develop within them, determine their very content, their very ability to fulfill the needs that gave birth to them. Whether in their ephemeral or more stable version, they have the ability to condense social and political time, to highlight or even create new associations. They have the power, beyond participating dynamically in the present, to shape data, open passages, and experience the "other."

Antonio Gramsci, at the beginning of the twentieth century, referring to historical moments when the dominants are unable to project a universal social vision, spoke of the crisis of hegemony. For situations where the ordinary, the "normal" is questioned, where different/"other" social forms appear, where new fields of possibilities open up. Possibilities, not predetermined, but contingent, where the new may seem foreign and hostile to the old. Heterotopias, in their various forms, are places that operate in "non-hegemonic" conditions, spaces that do not participate in social reality in a smooth and passive way, but fields in which the conditions can be formed where "the old dies and the new struggles to be born." For precisely these reasons, Foucault's theory proves to be very contemporary as it expresses and interprets our transitional era, and due to this observation, the research was integrated into Foucauldian theory.

2.3 Today Dystopia: The Impact of Pandemic and Global Crisis in European Cities

Dystopian literature is a form of speculative fiction that began as a response to utopian literature. A dystopia is an imagined community or society that is dehumanizing and frightening. A dystopia is an antonym of a utopia, which is a perfect society (Claeys, 2020).

Among the social consequences of the coronavirus pandemic, it is worth highlighting a rather tangible increase in attention to the dystopian science fiction genre. Moreover, the assertion that the real world of the present and the future is a dystopia made manifest has gained increasing popularity in various recent theories and texts that forecast the world's development in the post-epidemic era (Barabanov, 2020). The surge of protest movements in the United States and elsewhere last summer has also spawned openly dystopian scenarios. In this regard, it's already an interesting prospect, both culturally and politically, to trace the main themes of the dystopia genre and its prospects for their realization in the modern world. Sales of George Orwell's "1984" and Margaret Atwood's "The Handmaid's Tale" have skyrocketed since 2016. Young adult dystopias—for example, Suzanne Collins' "The Hunger Games," Veronica Roth's "Divergent," Lois Lowry's classic, "The Giver"—were best-sellers even before, and with COVID-19, dystopias featuring diseases have taken on new life. Netflix reports a spike in popularity for "Outbreak," "12 Monkeys" (Barabanov, 2020).

Most dystopian works present a world in which oppressive societal control and the illusion of a perfect society are maintained through one or more of the following types of controls:

- Corporate control: One or more large corporations control society through products, advertising, and/or the media. Examples include Minority Report and Running Man.
- Bureaucratic control: Society is controlled by a mindless bureaucracy through a tangle of red tape, relentless regulations, and incompetent government officials. Examples in film include Brazil.
- Technological control: Society is controlled by technology—through computers, robots, and/or scientific means. Examples include The Matrix, The Terminator, and I, Robot.
- Philosophical/religious control: Society is controlled by philosophical or religious ideology often enforced through a dictatorship or theocratic government (Read, Write, Think, 2006).

So real life is so much like dystopian fiction. The impact of the pandemic seems to shape the real dystopias in European cities. In many European societies, the existential health threat of the virus has resulted in large-scale government interventions, often limiting basic social and political rights. The combination of these developments has created an economic shock, likely to surpass that of the financial crisis of the last decade (Contiades & Fotiadou, 2020).

The dominance of global neoliberalism has established and imposed the idea that the orthodox economy is that in which "governmental intervention is practically nonexistent, negative free markets" (Centeno & Cohen, 2012). While at the same time we have before our eyes in the society in which we live, freedoms that are extolled to the skies, with emphasis placed on economic freedom, though nobody cares (apart from fringe groups who are usually marginalized) about the inequalities that arise from that perception (Anselmi & Gouliamos, 2005; Sakellaropoulos, 2018).

In comparison with the territorial base of the nation states, the term "globalization" reminds one of swelling rivers that threaten to undermine border securities and destroy the construction. (Habermas, 2003). As a result, economic globalization is testing the "nation's" limits. It goes to the point of substituting it, since on many occasions global economic circles act like shadow governments.

The above thought is probably likely to reinforce the neo-Marxist theory that globalization is nothing more than another neo-liberal political program which is implemented in order to achieve global dominance (Anselmi & Gouliamos, 1998; Bitsani, 2014; Centeno & Cohen, 2012; Swyngedouw, 2011).

On the other hand, there is a big migration flow in Europe since the 1990s and still continues almost 30 years later all over the world (Bitsani, 2013a, b), contributing to the expansion of cultural and national host countries and creating phobic feelings. Migration, which leads to the inclusion of a cultural minority, is defined as one of the main factors which increase the social exclusion at the personal and group level.

Apart from the cosmopolitan neocolonial world there are also the "Beautiful Dirty things," just so that we remember the ever so relevant film by the English director Stephen Frears, which is about immigrants' everyday life. Sassen mentions the parallel "underground" world, the immigrants' world, the minority citizen's world, the world that the low-paid workers experience, it is these very people who contribute to global economy while at the same time they are socially excluded. Sassen (2008) characterizes the inhabitants of global cities as de-nationalized, based on the logic that the rich who are participants in the global capital stock, care for exactly just that, hence ignore the nation in which they work in, while the minorities are socially excluded and therefore are indifferent toward the nation in which they live, which is also indifferent toward them. The global city is a city of inequalities in which marginalization is as common as legibility, and poverty interchanges with ostentatious affluence (Anselmi & Gouliamos, 2005; Bitsani, 2014; Sassen, 2001). In this framework, internal colonialism with marginalized immigrant communities is also developing. These communities usually live in dystopian suburbs in conditions of misery resulting in them becoming more introverted and clustered persistently by their identity of origin. This fact creates fantastic utopian environments in their micro-societies, as an antidote to the dystopia of reality (Anselmi & Gouliamos, 2005).

In this horizon the pandemic has drawn a furrow between a "before" and an "after," a dramatic watershed, to say the least, between a world we thought we could control and a world with an uncertain profile. Since the beginning of 2020, in the space of a few weeks, most of our certainties have been shattered. The world crisis unleashed by Covid-19, the first "red zone" outside China, was established in Codogno and in nine other municipalities by order of 21 February 2020; the lockdown was then gradually extended, in Italy, to affect the entire national territory on 11 March; in the same weeks, lockdown measures were gradually adopted in other European countries and, after a while, also in non-European countries—it imposed the need to change the way of living in common, as it had previously only happened following wars, revolutions or the establishment of dictatorships and totalitarianisms.

So, we have been called to reflect on what has happened, on the causes of what we are experiencing, on the social, economic, and political consequences; we have been called to ask ourselves questions about the future that awaits us. A tiny virus has neutralized us and we struggle to be resilient, resist, and react.

After the shock, that of deserted, imaginary cities, the first lock, abandoned markets, the disappearing project, schools and universities closed, trains and planes almost all stopped or were half empty, in what direction can we hope for restart; And then we have to deal with the unprecedented experience of isolation, of separation from others in recent months, which has affected everyone, young and old: indoors, locked in the house, even separately outside, carefully avoiding each other on the street, and in department stores and even more so in small rooms.

The pandemic has created a widespread state of insecurity, for us, for others, for those who are close to us, for those who are far away from us. It is a generalized state of insecurity, from which one cannot really escape, alone, but together, all together.

Apart from disturbing economic stability, the crisis affected everyone within the affected areas. It posed a threat upon public people, elected politicians, policy makers, the private sector like enterprises, and also the residents. It was the source of economic decline but also the source of decline regarding identities. This occurred due to the construction of bad images associated with places that were struck by the crisis and the economic recession. On the other side, associations and individual citizens organized solidarity with the needy and provided food to the hungry, long before the public administration.

The challenge is serious, not only because most people live in the city, but also because the way cities work today raises serious questions in the face of a global crisis. The crisis may be the right opportunity for citizens, politicians, and policy makers to radically rethink the way we all live, consume, produce, and travel (Bitsani, 2013a, b). Cities are places where everything and everyone is connected, which makes cities lively and inspiring, but makes them vulnerable. People living in cities depend more than others on public and private services for transportation, food, and use of open spaces. The lock had shown how urban interconnection can be a vulnerability, in a situation where inadequate access to or interruption of services makes them vulnerable. At the same time, neighborhood initiatives that had emerged around the world have clearly shown that there is another way to live together in cities. What successful initiatives are worth pursuing in the future and how can they be maintained beyond the pandemic? What specific urban challenges need to be addressed to provide long-term and sustainable solutions for citizens?

2.4 Conclusion

This chapter examines the intertwined trajectories of utopian and dystopian thought within the history of urban theory, from Plato's Republic and More's Utopia to the spatial philosophies of Foucault and Lefebvre. It has shown that the city functions simultaneously as a material construct and a symbolic arena—where ideals of order, justice, and harmony coexist with exclusion, inequality, and control. Over the centuries, the utopian imagination has served as a critical instrument for rethinking collective life, while dystopian visions have exposed the fragility and contradictions of modern urbanity.

Drawing on Michel Foucault's notion of heterotopia, the chapter has argued that cities embody complex spaces of otherness that both reflect and subvert dominant social orders. Heterotopias reveal how power operates spatially through inclusion, surveillance, and exclusion, yet also how new possibilities of resistance and reinvention arise within those same structures. In dialogue with Henri Lefebvre's theory of the production of space and his concept of the right to the city, this reading situates urban environments as living processes rather than fixed forms—produced and reproduced through everyday practices, conflicts, and aspirations.

Contemporary interpretations by Soja, Cenzatti, and others extend these frameworks, portraying the heterotopia not as a static enclave but as a dynamic and ephemeral field of difference. Within this view, urban space becomes a heterotopic condition—fluid, plural, and contested—where diverse social actors continually negotiate meanings of belonging, identity, and power. Such a perspective illuminates how the city today remains a site of struggle between utopian visions of inclusion and dystopian realities of exclusion.

The global pandemic has made this dialectic particularly visible. European cities experienced both the collapse of mobility and the emergence of new solidarities; the same spaces of isolation became arenas of resilience and creativity. In this sense, the pandemic operated as a heterotopic rupture, revealing the paradoxes of contemporary urban life: the interdependence that sustains cities and the vulnerabilities that threaten them. It reaffirmed the relevance of Lefebvre's call for the right to the city—not only as access to urban resources, but as a collective right to reimagine and reshape space itself.

Ultimately, this chapter concludes that the future of the city depends on sustaining the dialogue between utopian imagination and dystopian critique. By embracing the heterotopic nature of urban life, its tensions, contradictions, and transformative potential, the city can be re-envisioned as a biocultural space of coexistence, creativity, and renewal. In this ongoing negotiation between ideal and real, the urban landscape continues to serve as both a mirror of contemporary crisis and a laboratory for imagining alternative, more equitable futures.

Conflicts of Interest The authors have no conflicts of interest to declare that are relevant to the content of this chapter.

References

Adorno, T. W., Perrin, A. J., & Jarkko, L. (2005). Opinion Research and Publicness: (Meinungsforschung und Öffentlichkeit)*. *Sociological Theory, 23*(1), 116–123. https://doi.org/10.1111/j.0735-2751.2005.00245.x (Original work published 2005)

Alsayyad, N. (2000). The cinematic city: Between modernist utopia and postmodernist dystopia. *Built Environment (1978-), 26*(4), 268–281. http://www.jstor.org/stable/23287791

Anselmi, W., & Gouliamos, K. (1998). *Elusive margins: Consuming media, ethnicity, and culture* (Essay series) (Vol. 29). Guernica Editions.

Anselmi, W., & Gouliamos, K. (2005). *Happy slaves: A duologue on multicultural deficit* (Duologue series 2). Athabasca University Press.

Anselmi, W., & Wilson, S. (2011). Technologies of memory, identity, and oblivion in Persepolis (2007) and Waltz with Bashir (2008). In M. Ginsberg & S. Jahani (Eds.), *Familiar and foreign: Identity in Iranian film and literature* (pp. 233–260). Brill.

Barabanov, O. (2020, September 14). The world as dystopia: A literary genre or reality? https://valdaiclub.com/a/highlights/the-world-as-an-anti-utopia-a-literary-genre/

Benevolo, L. (1980). *The history of the city*. MIT Press.

Berneri, M. L. (1950). *Journey through utopia* (1st ed.). Routledge. https://doi.org/10.4324/9780429343841

Bitsani, E. (2013a). Intercultural city identity and human intercultural cities (H.I.C.): A dynamic ontological model for the social co-existence and social cohesion of modern and post-modern cities [Book]. Nova Science Publishers.

Bitsani. E.(2013b). Theoretical Approaches to the Organizational Culture and the Organizational Climate: Exploratory Research Examples and Best Policies in Health Care Services. *Journal of Human Resource Management, 1*(4), 48–58. https://doi.org/10.11648/j.jhrm.20130104.11

Bitsani, E. (2014). The intercultural relations as a main factor of intercultural social capital for socio-economic development in multicultural societies. The research example of Trieste, Italy. *Archives of Business Research, 2*(4), 69. https://doi.org/10.14738/abr.24.347

Blumenfeld-Kosinski, R. (1998). Review of the book Sacred fictions: Holy women and hagiography in late antiquity, by L. L. Coon. *Medieval Feminist Forum, 26*(1). https://doi.org/10.17077/1054-1004.1702

Booker, M. K. (1994). *Dystopian literature: A theory and research guide*. Greenwood Press.

Burkard, R. E. (1999). Utopia. *European Journal of Operational Research, 119*(2), 224–234. https://doi.org/10.1016/S0377-2217(99)00127-7

Centeno, M. A., & Cohen, J. N. (2012). The arc of neoliberalism. *Annual Review of Sociology, 38*, 317–340. https://doi.org/10.1146/annurev-soc-081309-150235

Claeys, G. (2020). *Utopia: The history of an idea*. Thames & Hudson. ISBN:9780500295526.

Contiades, X., & Fotiadou, A. (2020). Comparative constitutional change: A new academic field (with Dr. A. Fotiadou). In X. Contiades & A. Fotiadou (Eds.), *Handbook of comparative constitutional change* (Routledge handbook) (pp. 1–21).

Foucault, M. (1976). *The history of sexuality: Volume I, An introduction* (R. Hurley, trans). Pantheon Books.

Foucault, M., & Miskowiec, J. (1986). Of other spaces. *Diacritics, 16*(1), 22–27. https://doi.org/10.2307/464648

Giroux, H. (2003). Utopian thinking under the sign of neoliberalism: Towards a critical pedagogy of educated hope. *Democracy & Nature, 9*(1), 91–105. https://doi.org/10.1080/1085566032000074968

Goodwin, B., & Taylor, K. (2009). *The politics of utopia: A study in theory and practice*. Peter Lang.

Gospodini, A. (2002). European cities in competition and the new 'uses' of urban design. *Journal of Urban Design, 7*(1), 59–73. https://doi.org/10.1080/13574800220129231

Habermas, J. (2003). Toward a cosmopolitan Europe. *Journal of Democracy, 14*(2), 86–100. https://doi.org/10.1353/jod.2003.0077

Harvey, D. (2000). *Spaces of hope*. University of California Press.

Hetherington, K. (1997). *The badlands of modernity. Heterotopia and social ordering*. Routledge. https://doi.org/10.4324/9780203428870. ISBN: 9780313291159.

Jacoby, R. (2000). *The end of utopia: Politics and culture in an age of apathy*. Basic Books.

Jameson, F. (2005). *Archaeologies of the future: The desire called utopia and other science fictions*. Verso.

Kostof, S. (1991). *The city shaped: Urban patterns and meanings through history*. Thames & Hudson.

Lefebvre, H. (1991). *The production of space* (D. Nicholson-Smith, Trans.). John Wiley & Sons Ltd. ISBN 9780631181774.

Levitas, R. (2013). *Utopia as method: The imaginary reconstitution of society*. Springer.

Massey, D. (2005). *For space*. SAGE Publications Ltd..

McConica, J. (2011). Thomas More as humanist. In G. M. Logan (Ed.), *The Cambridge companion to Thomas More* (Cambridge companions to religion) (pp. 22–45). Cambridge University Press. https://doi.org/10.1017/CCOL9780521888622.003

More, T. (2003). *Utopia* (P. Turner, Trans. & Ed.). Penguin Books.

Morris, A. E. J. (1994). *History of urban form before the industrial revolution* (3rd ed.). Routledge. https://doi.org/10.4324/9781315841199

Morton, T. (2013). *Hyperobjects: Philosophy and ecology after the end of the world*. University of Minnesota Press. http://www.jstor.org/stable/10.5749/j.ctt4cggm7

Mumford, L. (1962). *The story of utopias*. Viking Press. (Original work published 1922).

Muravchik, J. (2003). *Heaven on earth: The rise and fall of socialism*. Encounter Books.

Nikolaidou, S. (1993). *The social organization of urban space*. Papazisis.

Ostalska, K., & Fisiak, T. (2021). Utopia and dystopia in the 21st century: Feminism, intersectionality, and the rejection of binarism. In *The postworld in-between utopia and dystopia* (1st ed., p. 13). Routledge. https://doi.org/10.4324/9781003082958

Parker, M. (2002). Utopia and the organizational imagination: Eutopia. *The Sociological Review, 50*(1_suppl), 217–224. https://doi.org/10.1111/j.1467-954X.2002.tb03586.x. (Original work published 2002).
Ragon, M. (1975). *L'homme et les villes*. Casterman.
Sakellaropoulos, T. (2018). Social state and social policy in a historical perspective.
Sassen, S. (2001). *The Global City: New York, London, Tokyo* (REV-Revised). Princeton University Press. https://doi.org/10.2307/j.ctt2jc93q
Sassen, S. (2008). *Territory, authority, rights: From medieval to global assemblages* (2nd ed.). Princeton University Press. (Original work published 2006).
Servier, J. (1967). *Histoire de l'utopie*. Gallimard.
Shadurski, M. (2016). *Utopia as a world model: The boundaries and borderlands of a literary phenomenon [Utopiia kak model mira: Granitsy i pogranichiia literaturnogo iavleniia]*. IKRiBL. ISBN: 978-83-64884-57-3.
Shadurski, M. (2019). *The nationality of utopia: H. G. Wells, England, and the world state*. Routledge. https://doi.org/10.4324/9780429317743
Silva, C. N. (2003). Urban utopias in the twentieth century. *Journal of Urban History, 29*(3), 327–332. https://doi.org/10.1177/0096144203029003011. (Original work published 2003).
Soja, E. W. (1996). *Thirdspace: Journeys to Los Angeles and other real-and-imagined places*. Wiley-Blackwell. ISBN: 978-1-557-86674-5.
Stevenson, D. (2012). *The city*. Polity Press.
Swyngedouw, E. (2011). *Designing the post-political city and the insurgent polis* (Civic City Cahier 5). Bedford Press.
Wallerstein, I. (2006). *European universalism: The rhetoric of power*. Paradigm Publishers.

Chapter 3
The Context of Culture, Environment, and Sustainability: The European Cities as Cultural Landscapes

3.1 Sustainability Theories

Dating back to 1987, the concept of sustainable development, popularized by the Brundtland Report, emphasizes "meeting present needs without compromising the ability of future generations to meet their own needs" (World Commission on Environment and Development, our common future, U.N. 1987, p. 43). The report was founded on four core principles, namely intergenerational equity, the integrated approach, global cooperation, sustainability, and the focus on poverty. The first principle, intergenerational equity, asserts that the long-term impacts of our actions must be considered with respect to future generations. The second principle is that economic growth, environmental protection, and social equity are interconnected. As such, they must be addressed together. The third principle is that sustainability is a global challenge requiring cooperation across borders. The fourth and final principle, the focus on poverty, posits that poverty and environmental degradation are linked; the addressing of one often helps the other. The report provided the foundation for modern sustainability discourse, leading to major global conferences (such as the 1992 Earth Summit in Rio) and influencing the creation of the UN Sustainable Development Goals (SDGs). Critics contend that the Brundtland definition is vague and focused overly on economic growth (Redclift, 2005). But in the bottom line, as Drexhage and Murphy (2010) stated, sustainability is "a visionary development paradigm."

Resilience theory, as pioneered by C.S. Holling and his colleagues in 1973, centers on the capacity of social-ecological systems to withstand disruptions, adapt, and evolve when confronted with alteration. The text emphasizes the importance of maintaining diversity, adaptive capacity, and feedback mechanisms to enhance the long-term sustainability and stability of systems. The present study draws upon the research conducted by Russo & Scarnato (2018), who focused on urban resilience and the importance of cities being able to absorb and overcome disturbances. In his

E. Bitsani et al., *The Biocultural City as a Utopia or the Future Reality?*, Cities, Heritage and Transformation, https://doi.org/10.1007/978-3-032-17584-7_3

1994 study, Davison examined the interrelationship of protective factors in the development of resilience in individuals exposed to stressful living conditions. In the present study, Schroll (2009) examines the concept of resilience in the context of ecological and social research. In the present study, Daly (2020) undertakes a thorough investigation of the concept of resilience across a range of disciplines, placing particular emphasis on the themes of resource enhancement, regulatory adaptability, and recovery capability.

The ecological footprint theory, developed by Mathis Wackernagel and William Rees in 1996, is an evaluation of the human impact on the environment. It quantifies the biologically productive land and water required to sustain a population's consumption patterns and dispose of its waste. The objective is twofold: firstly, to promote sustainable resource use; and secondly, to reduce ecological overshoot. The program also identifies the ecological limits of human consumption and guides efforts to reduce resource consumption and ecological footprint. The text under consideration places significant emphasis on the necessity of living in accordance with the Earth's carrying capacity. Collins and Flynn (2015) introduced the Ecological Footprint concept and detailed its progress, emphasizing its simplicity and usefulness as a quantitative valuation method for regional development policy. Meanwhile, Mancini et al. (2018) elucidate the relationship between sustainable development and ecological principles and see the Ecological Footprint as an effective measure for evaluating ecosystem services. Dong-Jing's (2001) study examines the Ecological Footprint in Northwest China through an objective assessment of deficits and surpluses and explores the implications for sustainable development.

Natural Capitalism, as proposed by Hawken et al. (2003, cited in Rao, 2003), acknowledges the global economy's substantial reliance on natural resources and ecosystem services. The present paper offers a critical analysis of conventional "Industrial Capitalism," arguing that this capitalist scheme does not fully adhere to its own accounting principles, as it liquidates its capital and labels it as income. The study disregards the significance of the substantial resources it utilizes, encompassing natural resources, living systems, and the social and cultural systems that underpin human capital.

3.1.1 Recent Approaches

The planetary boundaries framework, proposed by Johan Rockström et al. (2009), identifies "key Earth system processes that maintain the stability of the planet, including climate change, biodiversity loss and freshwater availability." The concept delineates boundaries within which human activities must operate to avert irreversible environmental degradation and preserve a safe operating space for humanity. Biermann (2012) provides a comprehensive overview of the concept within the broader context of Earth system governance, underscoring the intricate political dynamics that surround its identification. Kim and Kotzé (2020) is an advocate for the implementation of a framework convention on planetary boundaries in international law. Cornell (2012) acknowledges the significance of boundaries in establishing global sustainability but emphasizes the requirement for additional clarification on their systemic characteristics.

The triple bottom line theory, also known as the "three pillars" or "three dimensions" of sustainability, and alternatively referred to as the 3Ps (People, Planet, Profit), is a concept which "broadens the conventional emphasis on economic performance to encompass social and environmental aspects" and was introduced back in 1994 by John Elkington. The objective of the initiative is to ensure that companies and institutions consider not only economic profitability, but also social responsibility and environmental stewardship in their decision-making processes. The theory under consideration places great emphasis on the integration of social, environmental, and economic considerations. In essence, it suggests that sustainable development should aim to achieve positive outcomes in all three dimensions concurrently. Recent studies in the field claim that the TBL framework is predicated on the notion of decision-making that is informed by considerations of social equity, environmental stewardship, and economic viability. Bitsani and Eletheriou (2015) emphasizes the significance of considering broader stakeholder concerns, environmental responsibility, and social equity along with financial performance. In Babitha (2013), the application of the TBL is examined in the Indian materials sector, investigating the correlation between environmental and social performance metrics and financial indicators. On the contrary, House (2011) has provided a critical evaluation of the TBL, highlighting its prioritization of economic accounting over social and environmental accounting. This approach, as House contends, fails to address the inherent tensions that arise between economic, environmental, and social priorities.

In conclusion, this theoretical framework suggests that it is vital for organizations to extend their understanding of success beyond financial metrics and to encompass social and environmental factors.

3.1.2 Economic and Technological Approaches

Scholars from a political economy approach posit that political economy is indispensable to comprehending the challenges and opportunities of contemporary renaissance. This contemporary renaissance is characterized by rapid technological change, globalization, and cultural transformation. The prevailing challenges of economic inequality and environmental sustainability are recognized, and social and cultural transformation is regarded as a potential opportunity (Nobre & Tavares, 2021).

Doughnut economics, as proposed by Kate Raworth (2017), presents a conceptual framework for an economic system that operates within a "doughnut"—a metaphorical representation of the social foundation for human well-being and the ecological ceiling for planetary boundaries. The objective is to establish an economy that is both regenerative and distributive, thereby ensuring the needs of all are met while remaining within the planet's environmental limits. The seven principles to which the organization adheres are as follows: In order to effect change, it is first necessary to alter the objective, taking into consideration the entirety of the table. The cultivation of human nature is also essential, as is a more profound understanding of the systems in place. The next step is to redraw the plans in order to redistribute, and then to create in order to regenerate. Finally, it is important to adopt an agnostic attitude toward growth. The concept of a "social foundation" is employed to denote the inner boundary of the doughnut symbolizing the guarantee of universal access to essential resources and a satisfactory quality of life. The outer boundary is defined as the environmental ceiling, representing the limits of the Earth's resources and ecosystems. The objective is to operate within the secure and equitable limits of the doughnut, thereby satisfying the requirements of humanity without surpassing the boundaries of the planet. In her 2018 critical review of Kate Raworth's "Doughnut Economics" book, Zhivitskaya (2018) expressed reservations regarding the book's transformative potential. By contrast, Ceil (2018) acknowledges the significance of doughnut economics in mitigating environmental degradation and underlines the importance of establishing a social foundation within the doughnut framework. Luukkanen et al. (2021) present the Sustainability Window approach as a tool to quantify and implement the doughnut economy, with Thailand serving as a case study. In her 2019 study, Ross examines Raworth's book and her proposal to revise economic theories to encompass both human well-being and planetary sustainability.

Other scholars emphasize the significance of technological advancement; Cabezas et al. (2003), underscore the necessity of maintaining an objective stance on sustainability, with a particular emphasis on the ecological, social, economic, and technological components of human-nature systems. As Yang et al. (2023) emphasize, the development of innovative technologies is of paramount importance in addressing the intricate interconnections between the environment, society, and the economy.

3.1.3 Cultural Ecology and SDGs Goals

The emergence of environmental concerns in the context of the global climate crisis has underscored the imperative for the cultivation of ecological awareness, as evidenced by Bernard et al. Cultural ecology, a methodological approach to the study of human-environment relationships, is a multidisciplinary field that considers culture as a key to understanding the evolutionary process. The term was coined by Steward in 1995, who defined it as referring to the natural and biological part of the human environment. The adjective "cultural" was then used to emphasize the interaction between society and nature in the context of the evolutionary process, with humans at its core (Steward, 1955; Piccardo & Canepa, 2020; Fang & Li, 2022).

The notion of an inherent interdependence between the fields of ecology and culture is pivotal for a comprehensive understanding of human development and its relationships. As Sutton and Anderson (2004) observe, "the field of cultural ecology is concerned with the identification of cultural adaptations." The settlement patterns which denote the location of people, their activities, residences, work premises, facilities, and sacred sites in a culturally significant way represent the primary focus of this analysis. Furthermore, it is the environmental context, specifically during periods of cultivation, that may foster the emergence of behavioral storage patterns, ultimately contributing to the cultural complexity of the system. It is evident that the utilization of cultural ecology as a methodological apparatus facilitates the reflection, address, and management of social, economic, and environmental issues, thereby supporting the achievement of sustainable development goals. The interaction between cultures and the environment exerts a direct influence on the manner in which societies confront and resolve the shared challenges of human existence (Piccardo & Canepa, 2020; Nsamenang, 2015).

3.1.3.1 SDGs into Cultural Perspective

The United Nations' 2030 Agenda for Sustainable Development is an international policy blueprint aimed at achieving sustainable development for all by 2030. It comprises a set of 17 Sustainable Development Goals (SDGs), which are a series of objectives designed to address the global challenges we face today. The program encompasses a broad spectrum of "social, economic and environmental issues, including poverty eradication, gender equality, climate action and sustainable cities." The Sustainable Development Goals (SDGs) provide a comprehensive framework within which countries and stakeholders can work toward sustainable development and address global challenges collectively.

The sustainability of modern cities is contingent upon the establishment of a cultural ecosystem consisting of people, culture, and nature in symbiosis (Chen et al., 2020; Fang & Li, 2022). In order to address the multiple challenges currently being experienced by cities, it is essential that a transdisciplinary strategy is employed, engaging stakeholders and decision-makers in the process of urban sustainable planning (Ahern, 2013). Smythe (2014) critiques the conventional three-way relationship model of sustainability, proposing a more comprehensive approach with an emphasis on meaningful work and lives.

Culture has been identified as a key factor in achieving economic, social, and environmental sustainability. This notion has been formally recognized by the 2030 Agenda for Sustainable Development, which acknowledges the pivotal role of culture in development for the first time. In the context of the 2030 Agenda, which encompasses 17 Sustainable Development Goals, Sustainable Development Goal 11, specifically addressing the subject of sustainable cities, underscores the pivotal function of culture in achieving sustainable urban development. This is particularly evident in the imperative to bolster endeavors aimed at safeguarding the world's cultural and natural heritage, as outlined by UNESCO.

The new economic engine for the cities should be built on the foundations of art, creativity, and culture, all of which should be recognized as clean energies, as Moreno (2020) claims. In this paradigm, the cultural heritage of the past is posited as a dynamic material draw, a resource that can be tapped to re-think and re-imagine the future. As posited by Moreno, Kern et al., the notion of cultural heritage is posited as a pragmatic conduit for the establishment of more coherent urban communities. This is achieved through the utilization of public spaces and shared memories as storytelling platforms. Furthermore, a relationship has been demonstrated between cultural heritage, urban sustainable development, and climate policy. The capacity of cultural heritage to demonstrate resilience and sustainability over time provides a clear framework for the creation of a socially, culturally, and environmentally sustainable city. As demonstrated in the works of Moreno (2020).

3.1.4 The Urban Condition: From Capitalist Urbanism to Informal Assemblages

Cities are the primary arenas where spatial ideologies materialize and become contested. David Harvey's (1973, 2000) work on capitalist urbanization remains influential, especially his concept of the city as a spatial fix and his call for the "right to the city." Cities, for Harvey, are instruments of capital but also spaces of potential utopian resistance. Postcolonial scholars like Ananya Roy (2009) and Abdou Maliq Simone (2016) deconstruct Western-centric urban models, showing that informality and improvization define the urban experience in much of the world. Dovey and Recio (2024) advance this argument by theorizing informal settlements and street vending as spatial assemblages. They argue for a new epistemology of urban informality—one that prioritizes flow, adaptation, and survival over fixed plans. Further, Bratton (2021) and Ritter et al. (2025) illuminate how cities are now shaped by computational governance, synthetic urban models, and predictive AI. These insights challenge static urban categories and open up questions about agencies, infrastructure, and digital inequality.

A comparative focus on European cities provides insight into how varied historical contexts, planning cultures, and socio-political frameworks affect the reconstruction of urban landscapes. Postcolonial urban theory adds further depth by exposing the lasting impacts of coloniality on urban space and governance.

- Brussels: Administrative Landscapes and Multicultural

Fragmentation Brussels represents the tensions of reconstructing urban space in a polycentric, multilingual context. The European Quarter has transformed from a historic neighborhood into a fortress of EU bureaucracy. Critics argue that this has created spatial segregation and diluted local identity (Beeckmans, 2013). Attempts to reintegrate civic life, such as the redevelopment of Place Jourdan, reveal the complexity of balancing international prestige with local needs.

- Amsterdam: Water, Heritage, and Sustainability

Amsterdam integrates heritage preservation with forward-looking urban ecology. The city's approach to reconstructing docklands (e.g., Eastern Docklands) has emphasized mixed-use development and architectural experimentation. However, gentrification and the commodification of heritage remain key issues (Doucet, 2014). Efforts to integrate sustainability through initiatives like Amsterdam Rainproof highlight the potential for climate-responsive urbanism.

- Oslo: Post-Industrial Transformation and Public Waterfronts

Oslo's Fjord City project demonstrates how former industrial spaces can be reconstructed into public realms. The Barcode Project, while visually striking, has sparked debate over architectural homogenization and social exclusion. Nonetheless, Oslo's emphasis on publicly accessible waterfronts suggests a Nordic model of inclusive design (Rogers, 2021).

- Athens: Crisis, Informality, and Resistance

Reconstruction in Athens must be understood in the context of austerity and migration. The Exarcheia district symbolizes resistance to top-down urbanism, with self-managed spaces emerging amidst state neglect. Postcolonial urbanism provides a useful lens here, as it reveals how crisis narratives often obscure local agency and alternative urban imaginaries (Kaika, 2017).

- Lisbon: Creative Economies and Urban Regeneration.

Lisbon has leveraged creative industries and tourism to reconstruct neighborhoods like Mouraria and Cais do Sodré. While these changes have revived urban life, they have also accelerated displacement and speculative investment (Tulumello, 2016). Postcolonial critique points to the erasure of Afro-Portuguese histories in these processes.

- Venice: Heritage, Tourism, and Environmental Fragility

Venice is a critical case where heritage and environmental vulnerability collide. The MOSE project and the struggle against over-tourism underscore the challenge of reconstructing space without undermining ecological and cultural sustainability. Venice exemplifies the need for sensitive, historically informed urbanism.

3.1.5 Global and Local: A Critical Interrelation

Global-local dynamics are increasingly entangled in the context of urban spatial theory. The older binary thinking that viewed the local as a discrete, bounded entity and the global as a homogenizing force has given way to a more relational, co-constitutive understanding. Massey (1994, 2005) proposes the concept of a "global sense of place," which reframes locality not as static or insular, but as a node in a wider network of flows. Rather than a retreat from global interconnectedness, local spaces emerge as terrains where global power relations materialized, contested, and potentially transformed.

This reframing is essential in understanding the urban experience in European cities. Brenner's notion of planetary urbanization highlights that even the seemingly peripheral or rural areas are drawn into global circuits of capital, logistics, and governance. Consequently, the local is not a passive recipient but an active participant in global restructuring. As cities experience the pressures of transnational investment, migration, and tourism, the assertion of local agency becomes not only desirable but politically urgent.

Barcelona's governance experiment under Barcelona en Comú is emblematic of this trend, where local political mobilization disrupted the global city script based on speculative real estate and touristification (Russo & Scarnato, 2018). Similarly, Berlin's anti-gentrification movements have successfully politicized housing as a right, rejecting the dominance of global financial interests in determining local living conditions (Bernt, 2019; Holm, 2014). These movements articulate the city not as a mere space of accumulation but as a contested site of citizenship and justice.

Furthermore, the valorization of local initiative is not without risks. As Harvey (2000) cautions, a naive celebration of localism can mask exclusionary practices or neoliberal co-optation. Escobar (2001) similarly argues that the politics of place must be situated within broader structures of power, lest they devolve into reactionary identity politics. Therefore, critical engagement with the locals must account for its relational dependencies and scalar entanglements.

The European context offers diverse manifestations of these tensions. In Amsterdam, the adoption of Raworth's (2020) doughnut economics model demonstrates how municipal strategies can simultaneously address global environmental thresholds and local social foundations. In Athens, post-crisis autonomous collectives offer bottom-up responses to austerity and institutional collapse (Dalakoglou, 2017). In Brussels and Lisbon, citizen engagement in planning challenges EU-led redevelopment models and mass tourism dynamics, respectively (Salet & Gualini, 2015; Tulumello, 2021).

Thus, the emphasis on the local, particularly in a globalized urban context, is not anachronistic but essential. It allows for alternative imaginaries and interventions to emerge, challenging the homogenizing tendencies of global neoliberal urbanism. Locality, when critically theorized, becomes crucible for democratic experimentation, ecological responsibility, and spatial justice.

Space, landscape, and city must be understood as dynamic, contested, and relational constructions deeply enmeshed in global systems of power and resistance. Classical frameworks from Lefebvre (1991), Harvey (1973, 2000), and Massey (2005) laid a foundation for critique; contemporary scholars have pushed the debate further to address questions of informality, ecological crisis, algorithmic mediation, and planetary urbanization. The result is a vibrant, plural, and interdisciplinary spatial theory that not only diagnoses injustice but imagines new forms of spatial justice and urban life.

3.2 The European City as Cultural Landscape: The Role of City's Cultural Heritage in the Context of Sustainable Development and Biocultural Approach

The concepts of "space," "place," and "landscape" are key axes of analysis in a variety of scientific fields, from geography and urban planning to sociology, anthropology, and cultural studies. Their theoretical distinction and interpretive correlation reveal not only the material dimension of the environment but also its cultural, historical, and experiential significance.

"Space" in scientific thought is often defined as an abstract framework within which physical, social, or economic relationships manifest. Henri Lefebvre (1991) distinguishes three forms of space: the perceived, the experienced, and the notional, noting that space is not neutral but socially produced. The concept of space functions as a background of possibilities—it is the field where relationships, structures, and forms can potentially evolve.

Over the years, from the geometric concept of measuring space, science has advanced to the concept of place. Based on the presence of man and the plurality of emotions, images, expectations, and dreams, abstract space has been transformed into a place in a region with an identity and unique characteristics that differ from region to region and constitute its physiognomy. Physiognomy, after all, is the ... "opinion that we form about the nature of an entity" (Bitsani, 2013). This is why each region bears a unique name, which is also a condenser of its physiognomy. And while space is material and easily understood, as it is tangible and measurable, place is at the same time intangible. Starting from the simple registration of the name, but also the intangible emotions that constitute a place, a country.

In contrast to the abstract nature of space, "place" emerges through human experience and emotional investment. Space becomes place when it is enriched with meaning, memory, and identity. Place constitutes an embodied and existential experience: it is inhabited, recalled, and signified.

"Landscape" extends beyond simple geographical depiction. As Denis Cosgrove (1984) argues, landscape is both a way of seeing and an object of control: it is a culturally charged representation of place. According to John Wylie (2007), the landscape is a "hybrid," combining natural features with social and aesthetic meanings. It appears as a synthesis of human intervention and natural dynamics, taking shape through art, tourism, politics, and memory.

The first official use of the term "Landscape" was made at the beginning of the twentieth century by the Geographer Otto Schulter, who defined two types of landscape, the natural (Urlandschaft), that is, the landscape that existed before humans caused changes, and the cultural (Kulturlandschaft), created by human civilization (Karpodini-Dimitriadis, 2009). However, as we have seen, the term landscape gradually acquired various interpretations in the international bibliography, going beyond the simple two-dimensional categorization of natural and cultural, giving basis to its unified concept. The most famous interpretation is that of the European Landscape Convention, also known as the Florence Convention1, presented in 2000 by the Council of Europe, based on which (art. 1) Landscape is "an area, as perceived by people, whose character is the result of the action and interaction of natural and/or human factors" (European Landscape Convention, ETS No.176, Florence, 2000, Article 1).

The anthropogenic landscape, by general admission, forms an identity with unique characteristics, and therefore inevitably acquires a cultural character, regardless of whether it has a particular historical or archaeological value. Thus, all landscapes have cultural references because they have been influenced by human action or perception in any way. Furthermore, the concept of landscape is not limited to the natural background or the spatial field of development of a culture, since it is not treated as an object but as a process in action, which shapes man and is shaped by him (Inglis, 1977; Ingold, 1993; Darvill, 1999). In other words, it is not identified with the imprint of culture, as if the movement (force) was completed by acquiring a solid, material form, but is the result of the very human daily activity in space, through which the changing forms of the landscape arise. A similar view is supported by Fowler, who says that the landscape is not simply the result of dynamics but is itself dynamic at every moment and the evolution of people and their activities constitutes both a motive and a product of this dynamic. The landscape, therefore, historical or not, is a carrier of culture, so it must be treated as such in every planning attempt. This means that in cases of shaping archaeological sites, whose cultural value is considered a given, the cultural parameter refers to the whole of the wider landscape beyond the historical one and its integration into the design is necessary (Gkoltsiou & Terkenli, 2012).

The transition from space to place and ultimately to landscape is a process of interpretation and cultural integration. Abstract and neutral space acquires social and cultural significance when it is transformed into a place, while its representation—aesthetic, artistic, or touristic—elevates it into a landscape. This three-dimensional approach allows us to grasp the complexity of the human-environment relationship.

Today's societies are characterized by complexity in contrast to the stability that characterized those of the pre-modern era. The conditions of globalization have as their characteristic elements the ever-developing technologies, new forms of communication, transactions, and movement. The profile of the activities of the daily life of the modern Western man is governed by intense mobility, speed, and complexity. These conditions have led to the creation of a civilization, the Western world in particular, which is characterized by a particular multiculturalism, cultural homogenization, compression of time, bombardment, and speed of exchange of messages and images. Within this framework of rearrangements, the subject of the postmodern period appears fragmented as elements such as class, gender, nationality, and place of origin have ceased to constitute a stable system of reference and identification of the person, which was the case pre-modernly. According to Hall (2003), the structural changes taking place in modern societies can lead to an "identity crisis" as they deprive the individual of a framework for stable reference and identification in his or her daily life. The transformations have led to the formation of a hybrid culture of global scope, the result of the mixing of various cultural elements. The result of this process is the overshadowing of the particularities of each local culture within the framework of the dominance of the global Western model. In this context, the identity of the place is being reshaped and reconstituted, and therefore the identity and consciousness of modern man, as well as his relationship with the place. In pre-modern societies, what characterized the relationship between subject and place was the rhizome, something that indicated a particularly strong bond. However, as modern data changes, the relationship of the individual with the place also changes. Today, the relationships of people with places are increasingly weakened, which can result in a loose bond with the place or even the loss of the sense of place.

The concepts of space, place, and landscape are not only epistemological constructs but also tools for analyzing the human-environment relationship. Their dynamic interaction helps in understanding cultural geography and ecological consciousness. Through the examples, it becomes clear that the interpretation of the environment is a matter of both experience and power—of both cultural and historical memory.

3.2.1 The Cultural Landscape

The term cultural landscape was first mentioned by the cultural geographer Carl Ortwin Sauer, who in 1925 wrote in "The Morphology of Landscape" the most famous reference to landscape: "Culture is the agent, the natural area is the medium, the culture landscape is the result" (Sauer, 1963). In this proposition we find, paradigmatically formulated, the basic position that leads to the overcoming of direction usually described as "environmental determinism," that is, in overcoming the view that the natural environmental conditions of a place can determine the social environment.

Sauer thus becomes a harbinger of directions such as cultural ecology, political ecology, or historical ecology. Sauer, influenced by anthropologists of the time, argues that the cultural landscape is formed in a natural place by a cultural group of people. Similar natural environments provide conditions for the formation of different cultural landscapes, due to different cultural processes in each area (Terkenli, 1996). Taking into account the dimensions of the historicity and geography of space, the cultural landscape is not defined only on the basis of specific physical dimensions but acquires meaning and symbolic value for humans. Symbolic or socio-spiritual elements of a landscape express a specific system of meanings, significances, and values, which are manifested through the articulation of space, such as linguistic expressions, religious symbols, social institutions, cultural peculiarities, economic activities, etc. (Terkenli, 1996). The connection of the landscape with the socio-economic and cultural characteristics of a region began to emerge strongly in the last decades of the twentieth century, to become widely accepted by all sciences that directly or indirectly deal with it and even to be strongly discussed in its nature as purely cultural (Tratsela, 2012). Several years ago, R. Brunet expressed the view that since the geographical landscape is the womb of human actions, the natural environment itself is culture (Brunet, 2001). Consequently, with the same logic with which we deal with a work of art or a monument, as evidence of our cultural identity, we must also apply it to the treatment of the landscape and the natural environment. Any human activity, any small or large human intervention, any important or insignificant work cannot exist and be understood outside the time and space in which they take place. Every human action manifests, evolves, and intervenes in a specific natural environment based on a continuous and uninterrupted relationship. The diachronic approach of cultural evidence and natural forms demonstrates the influence, but also the dependence of one on the other. The cultural landscape includes the symbolism and unique meaning of the place as well as the bonds that connect people with their environment (Taylor, 1999). Due to the interaction of man and environment and the creation of a sense of uniqueness of the space (sense of a place—genius loci), natural spaces become places with human meaning and identity. The term "cultural landscape," therefore, is not used to describe a specific type of landscape but to define a new way of perceiving space, which emphasizes the interaction of the human factor and nature over time. As Ken Taylor emphasizes, cultural landscapes as imprints of human history are the richest historical archives we have. They can tell us, if we learn to interpret and understand their stories, about the achievements and values of our ancestors, about our current values and ideals as well as those of future generations. They are a window to our collective past, to the exhibition of our culture. In other words, it is not only what we see but also a way of seeing. We see with our eyes, but we interpret with our minds (Taylor, 2003). The anthropogenic landscape, by general admission, forms an identity with unique characteristics, so it inevitably acquires a cultural character, regardless of whether it has a particular historical or archaeological value. Thus, all landscapes have cultural references because they have been influenced by human action or perception in any way. Furthermore, the concept of landscape is not limited to the natural background or the spatial field of development of a culture, since it is not treated as an object but as a process in action, which shapes man and is shaped by him (Inglis, 1977; Ingold, 1993; Darvill, 1999 in Tratsela, 2012). In other words, it is the result of human daily activity in space itself, through which the changing forms of the landscape arise. A similar view is supported by Fowler, saying that the landscape is not simply the result of dynamics but is the same dynamic at every moment and the evolution of people and their activities constitutes both a motive and a product of this dynamic. The landscape, therefore, historical or not, is a carrier of culture, so it must be treated as such in every attempt either at design or interpretation. This means that in cases of the development of archaeological sites, monuments, and cultural areas within the city, whose cultural value is considered a given, the cultural parameter refers to the entire wider landscape beyond the historical one and its integration into the design is essential.

However, the importance of recognizing the cultural landscape, that is, the cultural value that a place has for the societies associated with it, has appeared in recent decades to have a much more direct practical significance than its theoretical definition. The "cultural landscape" has been established since 1992 with a protection movement organized by the World Heritage Committee, which has allowed important movements, such as the Vienna Memorandum (the Vienna Memorandum signed only in 2003). This Memorandum, together with the Florence Charter adopted by the ICOMOS-IFLA International Committee in 1982, for the protection of historic gardens and landscapes, describes the official global acceptance of the correlation of cultural records with the landscape background of societies. Following the European Landscape Convention definition "every landscape has an important role to play in the public interest in cultural, ecological, environmental and social terms, and is a resource which is favourable to economic activity and the protection, management and planning of which can contribute to the creation of jobs; contributes to the formation of local cultures…" (Council of Europe, 2000).

This conception of landscape was later criticized as empiricist, parochial, and having an antiquarian "object fetchism" in landscapes and not concerned with the inner workings of culture (Mitchell, 2000). Stephenson (2007) proposes a framework for understanding the manner in which disparate cultural values—including spiritual, historical, and aesthetic values—are intricately interwoven within the tapestry of landscapes. The proposed model emphasizes a holistic approach to landscape management, incorporating diverse community perspectives in order to more accurately reflect the social and cultural significance of the locale (Stephenson, 2007). Moving from landscape to place, scholars define place as a social human construction of space that results from the attribution of meaning, aesthetic or symbolic value to it, and is usually associated with unique characteristics, particularities, and specific social identities (Thrift, 1999).

In this context, we need public spaces which are structured to overcome isolation and segregation and to regain the ideal of the ancient Greek city, where public space was a place of dialogue and balance between public and private life (Aristotle, ca. 350 BCE/tr. 2013).

The concept of space has emerged as a fundamental analytical category for the understanding of social relations, as human action is inscribed in space and spatially constituted (Stevenson, 2003) given that if cultural considerations of everyday life are combined with more structural understandings of the larger processes that continue to shape landscapes, the result is a richer understanding of contemporary landscape and culture. In this theoretical landscape is implicated in the structural oppression and marginalization of groups by class, gender, race, and ethno-cultural group (Stevenson, 2003).

It is crucial to avoid treating place as a pathetic part, because it is more than an object and we must see place as a specific derivative of the involvement between man and his environment. When cultural considerations of everyday life are combined with structuralist understandings of larger processes that shape landscapes, the result is a richer understanding of contemporary landscapes and cultures. Stevenson (2003) and Plieninger and Bieling (2012) explore how cultural landscapes—environments shaped by human activity—adapt and transform in response to social, ecological, and economic changes.

Currently, the climate crisis is having an impact on cultural landscapes, leading to an increased risk of alteration. In their study, Cook et al. (2021) present a systematic approach to assessing the impact of climate change on cultural landscapes and heritage sites. The overarching structure integrates a variety of factors, including environmental, social, and cultural elements, with the aim of identifying vulnerabilities and providing a framework for adaptive management. The necessity for proactive strategies to ensure the preservation of cultural heritage in the face of climate change is emphasized.

3.2.2 From Sauer to the Present: The Evolution of Cultural Landscape Theory

The concept of the cultural landscape stands as a foundational paradigm in human geography, offering a lens through which to interpret the dynamic interactions between human societies and the natural environment. Introduced and rigorously developed by Carl O. Sauer in the early twentieth century, cultural landscape theory has since evolved in response to critiques and the emergence of new theoretical frameworks. This part explores Sauer's original formulation of cultural landscape theory and traces its transformation through recent theoretical developments, highlighting critical, feminist, and post-humanist perspectives that have redefined our understanding of landscapes.

Carl Sauer, a prominent geographer associated with the Berkeley School, introduced the concept of cultural landscape in his seminal 1925 essay, "The Morphology of Landscape." Sauer argued that culture acts as an agent transforming the natural landscape into a cultural landscape. This perspective rejected environmental determinism, which posited that the physical environment solely shaped human behavior. Instead, Sauer emphasized human agency, historical processes, and the material imprint of culture on the land.

Sauer's methodology was empirical, historical, and regionally focused. Through fieldwork and archival research, he sought to reconstruct the evolution of landscapes over time. He advocated a careful, observational approach to landscape interpretation, underlining the role of long-term historical processes. His framework established the groundwork for cultural geography as a distinct discipline. However, despite these contributions, Sauer's work has been critiqued for insufficiently engaging with the broader political and theoretical contexts of cultural production, treating culture in a somewhat static and homogeneous manner.

In the latter half of the twentieth century, scholars began to criticize Sauer's approach for under-theorizing culture and neglecting the role of power, ideology, and representation. One of the most influential critics was Denis Cosgrove, who redefined landscape as a cultural image and a way of seeing, shaped by ideologies and social structures. In his 1984 work Social Formation and Symbolic Landscape, Cosgrove argued that landscapes are symbolic forms reflecting dominant worldviews, particularly those of Western capitalism and colonialism.

Don Mitchell further advanced the theory by conceptualizing landscape as a site of labor and social struggle. Influenced by Marxist geography, Mitchell emphasized that landscapes are not merely representations but material products of social processes, including class conflict and capitalist development. His later work (e.g., Mitchell, 2020) has also explored landscapes of homelessness and urban inequality, highlighting the ongoing relevance of cultural landscape theory in understanding contemporary social issues.

James Duncan and Nancy Duncan (2004) have contributed a cultural turn within landscape studies, investigating how landscapes are imbued with cultural meaning and aesthetic judgments that reflect and reproduce social hierarchies. Their work in suburban American landscapes demonstrates how race, class, and power manifest in spatial arrangements.

J.B. Jackson contributed to landscape theory by focusing on vernacular landscapes, the everyday spaces created and inhabited by ordinary people. This perspective democratized landscape studies, shifting attention away from elite or aestheticized spaces to those shaped by lived experiences. Jackson's work suggested that cultural meaning is embedded in routine and practical engagement with the environment, offering a counter-narrative to more formal or institutional representations of space.

Feminist geographers like Gillian Rose and Doreen Massey introduced gender and positionality into landscape analysis. Rose critiqued the masculinist biases of traditional geographic knowledge, highlighting how visual representations often exclude or marginalize female perspectives. Massey emphasized the relational and dynamic nature of space, arguing for a multiplicity of spatial narratives. Her concept of space as a "simultaneity of stories-so-far" reflects the ongoing, contested, and plural nature of place-making.

Recent feminist interventions have further nuanced the role of affect, emotion, and embodiment in landscape experiences. Authors such as Lorraine Dowler (2018) have argued that landscapes are also emotional terrains, particularly in the context of conflict, memory, and trauma.

More recently, post-humanism and non-representational theories have questioned the anthropocentric bias of earlier landscape theories. Scholars like Sarah Whatmore and Nigel Thrift advocate for a view of landscapes as assemblages where human and non-human actors co-constitute space. Donna Haraway's (2016) concept of "staying with the trouble" further contributes to this discourse by emphasizing entanglement, multispecies relations, and situated knowledge in an era of ecological crisis.

The evolution of landscape theory has profound implications for both academic inquiry and practical applications. In urban planning, for example, the incorporation of cultural, social, and ecological dimensions leads to more inclusive and sustainable design practices. Landscape urbanism, influenced by contemporary theories, seeks to understand cities as ecological systems where human and non-human processes interweave.

In heritage and conservation studies, landscape is increasingly viewed not just as scenery but as a living, evolving entity that embodies memory, identity, and conflict.

Cultural landscapes designated by UNESCO often reflect complex histories and contested meanings, prompting debates over whose narratives are preserved and whose are excluded.

Additionally, landscape theory has found relevance in studies of migration, globalization, and climate change. As global flows disrupt traditional spatial arrangements, landscapes become key sites for negotiating belonging, adaptation, and resistance. Understanding landscape as a dynamic and socially constructed space helps scholars and policymakers address the cultural dimensions of environmental challenges.

From Sauer's early twentieth-century conceptualization to contemporary critical and post-humanist reinterpretations, cultural landscape theory has undergone significant transformation. While Sauer laid the groundwork by emphasizing human agency in shaping the environment, recent scholars have enriched theory by incorporating insights from critical theory, feminism, and post-humanism. Today, cultural landscapes are understood not as static backdrops, but as dynamic, contested, and multispecies terrains shaped by complex socio-material processes. These evolving perspectives underscore the enduring relevance of landscape as a concept for analyzing the intersections of culture, power, and environment.

3.2.3 The Interaction between Urban Cultural Landscape and Urban Cultural Heritage: The Cities as Cultural Landscape

The term urban space, like open-air space, refers to collective, open areas that facilitate communication and social activity within the urban fabric, as opposed to natural or rural settings (Bitsani & Kalomenidis, 2007). Urban landscapes emerged toward the end of the Middle Ages, specifically between the eleventh and fourteenth centuries (Bitsani & Kalomenidis, 2007). Urban cultural heritage is increasingly conceptualized as a form of cultural landscape living, evolving palimpsest of natural and human interventions. Rather than isolating historical monuments or architectural icons, this approach emphasizes the spatial, social, and symbolic dimensions of urban environments, acknowledging their embedded cultural values, practices, and meanings. This expanded notion of heritage recognizes cities not just as containers of heritage but as cultural landscapes in themselves, shaped by and shaping identity, memory, and power (Taylor & Lennon, 2012; Bandarin & van Oers, 2012).

Modern landscapes evolve rapidly, reflecting continuous changes in the relationship between humans and nature (Terkenli, 2001). These changes often have negative impacts, as contemporary phenomena such as land abandonment, de-agriculturalization, deforestation, urbanization, and industrialization mainly harm biodiversity and human well-being. Consequently, a societal demand arises to either limit landscape transformation or redirect these changes toward sustainable solutions (Bürgi et al., 2017; Moreira et al., 2006). The evolution of cultural landscapes constitutes not only a cultural phenomenon but also a socio-economic process, influenced by interactions at local, regional, national, and even international levels (Gong et al., 2022). Factors such as climate change and socio-cultural variables also shape cultural environments over time (Mercuri, 2014).

Despite the dynamic and constantly shifting nature of cultural landscapes, antiquities and cultural monuments remain fixed, timeless landmarks due to their historical significance. Scientific interest in the transformation of cultural landscapes has grown across Europe. The HERCULES project (Sustainable Futures for Europe's HERitage in CULtural LandscapES) aims to secure the long-term viability of European cultural landscapes.

The concept of the Historic Urban Landscape (HUL) extends beyond a city's historical center to include its broader built environment—both historical and modern—its infrastructure, spatial organization, geographical context (topography, geomorphology, hydrology, natural features), and embedded social, cultural, and economic traditions (Bitsani & Eletheriou, 2015). Within this urban landscape, ancient monuments and historical buildings are key features that convey the identity and history of a place and its people.

Urban landscapes are among the most powerful conveyors of a settlement's character, encompassing both tangible elements (e.g., land use patterns, spatial arrangements, drainage, vegetation, circulation systems, and architectural forms) and intangible values (e.g., festivals, rituals, traditional practices, and places of collective memory) (O'Donnell, 2008).

Urban regeneration fosters socio-economic development through interventions aimed at addressing environmental and social challenges. Cultural heritage often plays a central role in such initiatives. The Vienna Memorandum (2005b) introduced the concept of "historic urban landscape" as part of efforts to recognize the social and political history embedded in urban environments (UNESCO, 2005b; Taylor, 2015). Similarly, the 2011 UNESCO Recommendation on the Historic Urban Landscape emphasizes that urban heritage—including both tangible and intangible aspects—is a fundamental resource for sustainable urban development (UNESCO, 2011b). Essentially, it introduces a transformative framework that integrates urban heritage within broader dynamics of city life, such as social development, economic pressures, and environmental change. The HUL approach explicitly treats urban fabric as a cultural landscape, extending conservation priorities to include intangible heritage, community values, and the everyday use of space.

Cultural landscapes in urban environments are significant due to the dense human presence. In Europe, most people live in cities with rich historical backgrounds and diverse socio-economic activities (Bitsani, 2013). This global trend of urban population growth results in more cultural landscapes becoming part of urban settings (Taylor, 2015). From a scientific perspective, local communities are integral to cultural landscapes; historical, political, and cultural factors influencing communities are part of humanity's interaction with place (Taylor, 2015). Since the 1980s, heritage management has increasingly focused on linking monuments with local communities, marking a "social turn" in the field (Lekakis, 2025). Today, there is broad consensus that heritage must be managed in connection with local populations (Poulios, 2015).

Notably, each cultural landscape is unique due to its direct relationship with local society. Therefore, any development project in such a landscape must consider the needs of the local community. While cultural landscapes are constantly evolving, communities must be shielded from disruptive changes (Poulios, 2015). Recent years have seen growing community-led initiatives to preserve especially intangible heritage. However, these efforts sometimes lack inclusive representation (Poulios, 2015).

This shift toward community-centered heritage is also reflected in key international conventions. The 1975 Amsterdam Declaration underscored society's role in monument preservation, and the 1985 Granada Convention emphasized public awareness (Council of Europe, 1985). The 1992 Valletta Convention added that such awareness should begin in childhood (Council of Europe, 1992).

The 2005 Faro Convention repositioned people—individuals and communities—at the center of heritage management. Articles 2 and 9 underscore community ownership of heritage and its sustainable use (Council of Europe, 2021). The convention advocates for equal participation of civil society and private entities in heritage stewardship, asserting that heritage gains meaning through human presence.

The way local communities experience and shape landscapes is central to understanding cultural spaces such as arenas of social practice, ecological structures, and historical narratives. Community participation is now viewed as essential for safeguarding both tangible and intangible heritage. Sustainable development depends on community well-being and culturally informed practices, as confirmed by various legal frameworks. According to the four-pillar sustainability model, key components are social equity, environmental responsibility, economic growth, and cultural development (Poulios, 2012).

Since the 1990s, landscape research has become increasingly linked to sustainability studies. Heritage landscapes serve as critical sources of identity and insight into past land use systems, informing contemporary ecological planning. They are also vital for fostering resilient societies through cultural identification, education, and recreation (Crumley et al., 2017).

UNESCO's 2011 Recommendation reiterates that urban heritage contributes to sustainability and supports social cohesion and economic growth in a rapidly changing world (UNESCO, 2011).

A living cultural heritage site maintains its original function and responds spatially to changing societal conditions at local, national, and global scales.

Recognizing cultural heritage as a public good—enshrined in Greece's Constitution—implies that its protection must involve the community. This principle aligns with the value-centered and "living heritage" management models (Poulios, 2012). In the former, society actively participates under expert guidance. In the latter, heritage evolves with society, emphasizing community relevance and engagement (Poulios, 2012).

Cultural landscapes also support sustainable development by promoting ecological practices such as traditional land use, which helps conserve biodiversity

(Mitchell et al., 2009). Both tangible and intangible heritage can serve as levers for social, educational, and economic growth.

3.2.4 *Urban Cultural Heritage*

Urban cultural landscapes are understood as multi-layered constructs comprising architectural, spatial, cultural, and ecological elements. They embody historical continuities and ruptures, where heritage is not a static object but a dynamic process (Rey-Pérez & Pereira Roders, 2020). As such, the city becomes a textual landscape, constantly rewritten by social interactions, planning regimes, and cultural performances. Contemporary urban heritage discourse has moved decisively away from the object-centered model. Scholars argue that viewing heritage only through monuments and aesthetics neglects lived experiences, social practices, and local narratives (Labadi & Logan, 2015). Cities contain not only preserved architectural forms but also dynamic layers of intangible heritage—languages, rituals, street art, and spatial practices—that form part of their cultural landscape.

This shift is evident in urban policy where heritage-led regeneration aims to preserve community character while promoting sustainable development. However, such initiatives often risk cultural commodification or gentrification, unless grounded in community participation and inclusive narratives (Gravari-Barbas & Guinand, 2017). The challenge is to balance place continuity with urban change, a core tenet of the cultural landscape approach.

In addition, urban cultural landscapes function as repositories of collective memory and identity. These spaces are often contested, especially in postcolonial or multicultural cities, where heritage becomes a site of negotiation (Bitsani, 2016; Waterton & Smith, 2010). Memory is not evenly distributed in urban space; dominant histories are often spatialized in city centers, while subaltern or minority narratives remain peripheral or invisible (Bitsani, 2016; Winter, 2014).

Reclaiming marginalized heritages and recognizing plural histories within urban landscapes is crucial to inclusive urban heritage planning. For example, cultural mapping and participatory heritage assessments are tools that surface non-official heritage values, allowing communities to re-inscribe meaning onto space (Bitsani, 2013; Logan et al., 2016).

3.2.5 *Urban Archaeological Sites as a Crucial Part of a City's Cultural Landscape*

As Harrison (2013) points out, archaeological remains constitute an integral component of a city's "heritage ecology," a system wherein objects, places, and practices interact to shape the manner in which the past is experienced in the present.

Urban excavation sites, whether visible or subterranean, frequently become focal points for identity formation, tourism, and local storytelling. The integration of archaeological heritage in contemporary urban design, as evidenced by projects such as the Al Zubarah Fort in Qatar and the layered presentations at Rome's Largo di Torre Argentina, demonstrates the manner in which these sites mediate the relationship between the ancient and the modern (Winter, 2014). These efforts frequently serve to enhance local pride, while concurrently shaping state narratives and, on occasion, excluding alternative historical accounts.

(a) Politics of Space: Preservation vs. Development

The role of urban archaeological sites within cultural landscapes is characterized by intricate tensions between the imperatives of heritage preservation and the demands of urban development. As cities densify and globalize, the preservation of archaeological sites often comes into conflict with infrastructure expansion, the provision of housing, and the interests of real estate developers. Labadi (2013) emphasizes that the integration of heritage conservation into planning, or its treatment as an impediment, is contingent on governance structures, economic pressures, and public perceptions. Post-2010 case studies, including the excavation under Istanbul's Yenikapı Marmaray station, illustrate both the potential and controversy of urban archaeology. The unearthing of Byzantine-era remains during a significant transport initiative has been known to prolong development for extended periods, thereby inciting both public and academic discourse concerning the values held by cities and the manner in which they strike a balance between economic advancement and historical integrity (Özdemir Işık & Demir, 2017).

(b) Economic and Social Dimensions of Archaeological Landscapes

Urban archaeological sites have the potential to function as drivers of sustainable development, tourism, and education when incorporated within inclusive heritage strategies. Poulios (2015) posits that community-based heritage conservation engenders socio-economic benefits by empowering local actors and acknowledging diverse forms of value, including spiritual and indigenous associations with urban space. However, when heritage is commodified without sensitivity to context, it may contribute to cultural gentrification or "heritage-washing" (Skrede & Berg, 2018), where urban archaeological aesthetics are used to justify exclusionary redevelopment. A genuinely sustainable cultural landscape approach necessitates engagement with living communities, rather than merely static monuments.

(c) Community Participation and Multivocality in Urban Archaeology

Contemporary heritage discourse is characterized by an increasing demand for inclusive, participatory approaches to the management of urban archaeological sites. As Waterton and Smith (2010) contend, heritage encompasses not only the preservation of tangible and intangible assets, but also the active performance and interpretation of these by diverse publics. Initiatives such as the "Archaeology in the City" project in Sheffield, UK, illustrate the potential of urban archaeology to function as a conduit for social justice, educational endeavors, and the decolonization of urban discourses. Innovative tools, including digital storytelling, augmented reality, and citizen-led excavations, have emerged as ways to democratize heritage and reconnect urban populations—especially youth and underrepresented groups—to the archaeological layers of their cities (Perry, 2019).

From this perspective, urban archaeological sites are considered integral to the cultural landscapes of cities, providing a crucial link between historical depth and contemporary urban life. These elements function as anchors of memory, identity, and civic imagination, while simultaneously posing critical inquiries concerning equity, sustainability, and the future of urban environments. The integration of these sites into modern cities necessitates a multi-scalar, interdisciplinary approach, combining planning, community engagement, critical heritage theory, and inclusive governance. As cities face accelerating change, the safeguarding of the cultural landscapes embedded in urban archaeology is no longer a luxury, but a necessity for meaningful, connected, and resilient urban development.

3.3 Living Cultural Heritage and Management Models

In the case of living cultural heritage, the definition has shifted from the mere use of monuments and heritage sites to the continuity of traditions and practices, particularly following UNESCO's 2005 Convention on the Protection and Promotion of the Diversity of Cultural Expressions. Different states have since interpreted living heritage as encompassing ongoing traditions, skills, and even techniques (Court & Wijesuriya, 2015; Yang et al., 2023).

Living cultural heritage is thus characterized by continuity: of the original function of a monument or site, of its intended purpose, and of the core community evolving through time. As these communities and their environments change, so too do the expressions—both tangible and intangible—associated with the heritage, distinguishing it from archaeological remains.

The community at the core of a living cultural space is responsible for preserving the heritage using traditional or accepted practices. In this context, change is considered part of heritage continuity, not a disruption to be avoided. Therefore, communities should actively participate in decision-making concerning urban landscapes and the natural environment, especially when designated as World Heritage Sites.

Intangible cultural heritage acts as a bridge between traditional and contemporary cultural values. It includes oral traditions, artisanal skills, artistic, social, and ritual customs, and knowledge systems passed down through generations (UNESCO Courier, 2019). It also encompasses practices and professions that help communities maintain and transmit their heritage (European Heritage Days, 2021).

Until recent decades, cultural heritage management focused primarily on conserving historical buildings, archaeological sites, and artworks, emphasizing the tangible past. However, as cities now host over half the world's population and millions of tourists annually, they face increasing pressure to ensure sustainable futures amid climate change, over-tourism, and events like the COVID-19 pandemic (Leiden-Delft-Erasmus Universities, 2021).

Urban heritage cities must reconcile historic preservation with contemporary societal needs. While city design and character attract tourism and investment, residents today live in environments shaped by decisions of the past yet must address modern challenges. These necessitate internal adaptations that change urban appearance and function. In World Heritage Cities, such changes require careful consideration of residents' needs, investors' demands, and tourists' expectations.

To meet these demands, three key heritage management models have emerged:

- The "object-centered model," originating in the nineteenth century with archaeology's rise, emphasizes material preservation. It values historical, aesthetic, and scientific significance, aiming to prevent deterioration via technical restoration plans (Poulios, 2015).
- The "value-centered model," developed from the 1980s, emerged as diverse groups beyond archaeologists sought roles in heritage stewardship. It focuses on values assigned by stakeholder communities and priorities safeguarding both tangible and intangible elements. In this model, heritage derives significance from community values and is managed through participatory planning (Poulios, 2014).
- The "living heritage model," internationally recognized in the 1990s and prevalent in non-Western contexts, emphasizes ongoing creation and adaptation. Here, communities maintain heritage continuity, seeking social and cultural benefits while using traditional or modern tools (Wijesuriya et al., 2013). Although this model sees heritage as part of a community's living identity, it has been criticized for prioritizing groups based on the degree of their connection to the heritage, which may challenge democratic inclusiveness.

The governance of urban heritage is characterized by a series of tensions, primarily between the pursuit of development and the imperative of conservation, the exercise of state control and the agency of local communities, and the demands of global tourism and the preservation of local authenticity (Bitsani & Thymi, 2022) The cultural landscape perspective demands interdisciplinary collaboration and multi-stakeholder engagement (Poulios, 2014). However, urban planning processes frequently prioritize economic growth over cultural continuity, leading to heritage loss or superficial preservation.

In urban areas experiencing rapid urbanization, heritage sites may be subject to protection measures. However, the surrounding contexts are often transformed to such an extent that the functional and symbolic connections that define them as cultural landscapes become obscured (Pendlebury, 2012). A genuinely landscape-oriented approach would entail the integration of heritage values into zoning, land-use planning, and urban design, thereby ensuring the coexistence of historical continuity with contemporary life.

3.4 Critical Tensions and European-Specific Challenges

Urban landscapes are subject to a number of pressures, including gentrification, climate change, and postcolonial contestation. As Zukin (2020) revisits, the process of commodification transforms cultural heritage into consumable spaces, often marginalizing long-term residents. This dynamic is evident in European cities such as Lisbon, where the process of heritage regeneration has resulted in displacement (Historic England, 2024). Concurrently, postcolonial cities have been engaged in critical reappraisals of colonial legacies. The reevaluation of imperial monuments in Brussels and Paris reflects an urgent call to reinterpret heritage as inclusive memory work (Abdurahiman, 2025).

From an ecological perspective, heritage is becoming increasingly aligned with the concept of urban resilience. Pauleit et al. (2011) argue for the integration of ecological infrastructure with cultural layering to ensure sustainability, especially in waterfront or climate-vulnerable historic areas. The integrity of urban cultural landscapes is being jeopardized by modern urbanization and global real estate markets.

(i) ***Tourism-Driven Gentrification* vs. *Resident Rights***

A comparison of vibrant heritage districts with their gentrified trajectories reveals the sharp trade-offs faced by European cities. While tourism has the potential to revitalize, it is important to acknowledge the adverse effects that it can also have, including social segregation and a loss of local authenticity. This critique has been drawn out by HUL studies in European contexts. Recent literature acknowledges that heritage-led regeneration often results in touristification and displacement. Discussions pertaining to spatial politics have demonstrated that the process of converting historic centers into tourist hubs frequently fails to consider the needs of the local population, thereby resulting in the effective exclusion of the original inhabitants. This tendency is especially evident in globally popular heritage destinations.

(ii) **Gentrification and Commodification of Heritage**

Heritage districts have been observed to become sites of gentrification, whereby cultural landscapes are commodified for the purposes of tourism or elite consumption, resulting in the displacement of the area's original inhabitants (Zukin, 2020). This process has the effect of undermining the social dimension of heritage and eroding intangible cultural continuity.

(iii) **Decolonizing and Reclaiming Post-Imperial Spaces**

In contemporary urban settings characterized by post-colonial transitions, cities formerly associated with former empires, for example, Brussels and Paris, are undertaking revaluations of colonial memorials and landscapes. Recent European scholarship has examined the question of whether HUL strategies provide frameworks for the reinterpretation, recontextualization, or repurposing of these heritage forms in the context of postcolonial memory work. A critical postcolonial perspective posits the question of how cities can decolonize their landscapes. In the context of contemporary architectural discourse, there is an ongoing reassessment of colonial-era architecture, with a focus on the preservation, removal, adaptation, or reinterpretation of built symbols of imperialism. This reevaluation is not limited to a specific geographical area but is a global phenomenon that is being discussed in various cities, including Brussels. These debates are indicative of a more sophisticated ethical dimension in the discourse surrounding urban heritage.

(iv) **Urban Ecology and Heritage Resilience**

European discourse frequently underscores sustainability in heritage practice. In their seminal work, landscape ecologists Pauleit et al. (2011) place significant emphasis on integrative frameworks that address ecological, socio-economic, and planning vectors in European city-regions. HUL tools now integrate flood defense, green infrastructure, and heritage layering to foster resilient urban systems.

(v) **Climate Change and Environmental Pressure**

Climate change has been shown to present a threat to the existence of urban heritage, particularly in coastal cities with rich historical legacies, such as Venice and Alexandria. The preservation of cultural landscapes in such contexts necessitates integration.

3.4.1 Toward Inclusive and Sustainable Urban Heritage Landscapes

The future of urban cultural heritage as a cultural landscape lies in fostering resilience, inclusivity, and adaptability. The UN Sustainable Development Goals, particularly SDG 11.4, recognize the need to protect and safeguard the world's cultural and natural heritage, underscoring its relevance to urban sustainability.

Integrating heritage into urban development requires place-sensitive planning, where cultural landscape analysis guides infrastructure investment, green space design, and housing policy (Bitsani, 2016). It also calls for heritage literacy among planners, developers, and citizens alike—an awareness of how heritage shapes identity, community well-being, and spatial justice.

Understanding urban cultural heritage as a cultural landscape provides a robust, multidimensional lens through which to interpret and manage the complexities of contemporary urban environments. It situates heritage not as an obstacle to development, but as an enabler of sustainable, meaningful, and inclusive urban futures. As cities grow and transform, this perspective can help balance the tensions between preservation and progress, identity and innovation, memory and modernity.

3.5 Biopolitics and the Cultural Heritage Landscape of the City: A Foucauldian Analysis

European cities are not merely dense accumulations of architecture and infrastructure—they are historic palimpsests, layered with centuries of spatial, cultural, and political transformation. It is a well-documented fact that scholars in the field of urban morphology, such as Conzen and Whitehand, have long emphasized the incremental evolution of street networks, land parcels, and building forms. This is due to the fact that such phenomena are able to serve as a record of the changing needs and values of successive eras. In the contemporary era, these insights are being refined by the tools of urban morphometrics, a field that utilizes geospatial analytics to map and measure the historical layering of urban form. In metropolitan areas such as London, evidence of historical urban development is evident, with vestiges of Roman roads, medieval street grids, and Victorian infrastructures contributing to contemporary urban infrastructure. This embedded legacy serves as a nexus, connecting the past with the present spatial logic of the city.

Foucault's concept of biopolitics centers on how modern states manage life through disciplines and regulatory controls (Foucault, 2008). In the urban context, governmentality manifests in the design of public space, environmental management, and health-related ordinances aimed at guiding the behavior of populations. Cities are not neutral backgrounds for human activity; they are deeply political constructs that organize life. Drawing on Foucault's theories of biopower—power to regulate life at both individual and population levels—we claim that the city's cultural landscape is a biopolitical field where power, culture, and space intersect (Foucault, 2007, 2008).

Foucault (2008) elucidates a pivotal shift from sovereign power ("to let die or make live") to biopower, characterized by regulatory controls intended to maintain and optimize life. Biopolitics operates at the scale of populations via statistical knowledge, health infrastructures, urban planning, and surveillance systems. Foucault's concept of governmentality refers to a governance logic extending beyond state coercion, involving self-regulation enabled by expert knowledge, maps, and spatial planning (Foucault, 1977). Spatial arrangements—grid systems, zoning, mapping—become techniques of governmentality aimed at shaping behavior, order, and risk (Miller & Rose, 2008).

Foucault's notion of heterotopia characterizes urban sites—parks, museums, malls—as real "other spaces" that create and reinforce social norms even as they appear as neutral public amenities (Foucault, 1997). The layout of cities—boulevards, parks, sewer systems—functions to renew, sanitize, surveil, and control populations, turning the urban fabric into a material apparatus for biopower operation (Miller & Rose, 2008). Beautification projects and curated public art mask disciplinary rationales while reinforcing hegemonic order: aesthetic discourse becomes a vehicle for governing conduct under the guise of cultural enrichment. The city can be read as a cultural text (Zukin, 2020). Green spaces, parks, and monuments are not merely aesthetic; they are forms of spatial inscription where cultural norms and state ideologies are performed and enforced.

Research on urban garden design in eighteenth-century colonial contexts reveals how public greenery acted as disciplinary mechanisms: gardens were sites for instilling order and "civilized" behavior, reflecting the colonial governmentality of the period (Certomà, 2013). Contemporary "smart cities" extend biopolitical control via sensors, data analytics, and health surveillance. Foucault's early reflections on urban planning prefigure the logic of today's data-driven governance. Foucault acknowledges that where power exists, resistance also arises. Urban heterotopias—art collectives, squats, community gardens—offer counter-spaces that challenge hegemonic norms, creating alternative subjectivities and reclaiming the city as a site of life and freedom.

Building on Foucault, theorists such as Braun (2014) and Luke (1999) introduced the concept of populations. Urban greening, recycling programs, and sustainability initiatives are seen as methods for constructing responsible environmental citizens. Achille Mbembe (2019) expands the analysis by introducing necro politics: the power to determine who may live and who must die. In the city, necropolitics reveals itself through zones of abandonment, under-resourced areas, and spatial neglect that create unequal conditions for life.

The cultural heritage landscape of the city is deeply implicated in biopolitical agendas designed to regulate populations and bodies through spatial-cultural interventions. A Foucauldian lens reveals how aesthetics, planning, and discipline operate together in "making live"—and control. Yet the city also gestures toward sites of resistance where new forms of life and culture might flourish. Recognizing the palimpsest of power and counter-power in cities is essential for envisioning emancipatory urban futures.

However, comprehending urban forms as intricate layers extends beyond mere physical analysis. Following the adoption of UNESCO's 2011 Recommendation on the Historic Urban Landscape (HUL), a substantial transformation has been observed in both academic discourse and urban policy. The concept of heritage is no longer regarded as being confined to preservation in isolation. Conversely, the HUL approach advocates for a dynamic, integrated model that conceptualizes heritage as an integral component of the urban ecosystem. Sonkoly (2020) characterizes this as a transition toward multi-scalar governance, whereby heritage, urban planning, and ecology function collectively across various levels of decision-making.

This conceptual evolution has gained practical traction in recent years. The notion of cities as cultural palimpsests, in which tangible structures, symbolic meanings, and lived experiences intersect, has been developed further by scholars such as Taylor (2018) and Zhang (2024). In this perspective, heritage is not confined to monuments or architectural styles; rather, it is intricately interwoven with the quotidian rhythms of life, the collective memories of communities, and the evolving character of the physical environment. These layers, both visible and invisible, collectively constitute the experiential fabric of the urban landscape.

The HUL framework has also been translated into actionable strategies. In their 2023 study, Landry et al. developed a set of thirteen practical guidelines with the aim of operationalizing HUL. The guidelines focus on adaptability, sustainability, and the principles of the circular economy. This approach encourages cities to view heritage not as a constraint, but as a resource for innovation and resilience. The mounting emphasis on adaptive reuse, climate resilience, and stakeholder governance, as emphasized by Min, signifies a more profound integration of heritage into the prevailing challenges of contemporary urbanization.

This perspective is pivotal in its repositioning of the city as a cultural landscape, defined as a space shaped by the interaction of human and natural systems, imbued with layers of meaning and memory. Urban landscape theory has evolved through thinkers such as Taylor (2018), who argue for a holistic understanding that bridges the physical and the intangible, the structural and the symbolic. This approach eschews the "museumification" of cities, instead acknowledging their fluid and lived nature, thus embracing both change and continuity.

As Cosgrove (1998) reminds us, landscapes are not neutral entities—they are the product of cultural construction and political ideology. The configuration of urban form is indicative of historical power dynamics, social values, and collective aspirations. In this sense, the interaction between urban cultural landscapes and urban heritage is characterized by both mutual reinforcement and, at times, tension. As urban areas encounter the challenges posed by globalization, migration, and climate change, the imperative extends beyond mere preservation to encompass adaptation, integration, and the facilitation of meaningful, inclusive future development.

The result is a more sophisticated and nuanced understanding of the urban condition, which perceives cities not only as places to live and work, but as evolving narratives where past and present converge, and where heritage actively influences future trajectories.

The role of urban heritage in the construction and maintenance of cultural identity is critical. As Ashworth et al. (2007) contend, heritage is a selective process through which societies narrate their past to serve present purposes. These narratives are firmly embedded within urban landscapes, manifesting in sites, rituals, and spatial symbolism that collectively serve to reinforce community identity and continuity.

For instance, historic districts such as the Medina of Fez or the Old Town of Edinburgh exemplify a cultural layering that combines urban morphology, material heritage, and lived traditions, thus rendering them quintessential urban cultural landscapes.

Urban landscapes also function as repositories of memory, and as such, are subject to contestation. In postcolonial urban contexts, the preservation of heritage often mirrors prevailing narratives, potentially marginalizing alternative or indigenous spatial practices (Holtorf, 2013). A pertinent example of this phenomenon is the reconstruction of Warsaw following the Second World War, which can be regarded as a political act of reclaiming Polish identity through architectural memory. This process serves to illustrate the manner in which urban heritage is intertwined with cultural politics (Pendlebury, 2009).

The recent academic landscape highlights a shift in focus from technical preservation to adaptive, inclusive, and critically reflective heritage practices. Cities are now seen as complex layers of cultural value, requiring governance that considers socio-economic justice, environmental change, and historical complexity. The evolving toolkit, which includes HUL, geospatial analysis, decolonial critiques, and sustainability frameworks, enables a richer, more engaged, and more ethically attuned approach to urban heritage.

The evolution of urban heritage management reflects a broader shift toward reflexivity, justice, and inclusivity. Cities are now regarded as living infrastructures, with heritage integrated into contemporary urban life rather than preserved in isolation. A 2024 report by Historic England, based on rigorous metrics, confirms a positive correlation between investment in local heritage and growth in the creative industries, resulting in increased economic productivity and civic pride (theguardian.com). Built heritage therefore acts as a form of cultural capital, providing economic and social benefits.

In this context, it becomes evident that the cultural heritage landscape of the city is deeply implicated in biopolitical agendas designed to regulate populations and bodies through spatial-cultural interventions. A Foucauldian lens reveals how aesthetics, planning, and discipline operate together in "making live"—and control. Yet the city also gestures toward sites of resistance where new forms of life and culture might flourish. Recognizing the palimpsest of power and counter-power in cities is essential for envisioning emancipatory urban futures.

3.6 Conclusion

This chapter has examined the European city as a living biocultural landscape shaped by the interaction of culture, environment, and sustainability. It has been argued that urban space should be understood not merely as a physical environment but as a dynamic field where memory, identity, and ecology converge. Sustainable urbanism, therefore, requires an integrated approach—one that unites cultural values, social equity, and ecological balance within a shared spatial framework.

Drawing on Lefebvre's theory of the production of space, the chapter has demonstrated that the city is both the product and the expression of social relations, continually reshaped through politics, economics, and everyday life. In parallel, Foucault's notion of biopolitics reveals how urban environments operate as instruments of governance, yet also as arenas of resistance and reinvention. Together, these perspectives highlight the dual nature of the city as a space of control and creativity, conformity, and transformation.

The pandemic underscored these tensions, exposing the fragility of global urban systems while also inspiring new practices of solidarity, localism, and care. European cities have shown that sustainability extends beyond environmental management to encompass cultural resilience and collective responsibility.

Ultimately, this chapter concludes that the future of the city depends on embracing its biocultural character—where heritage, ecology, and human agency are intertwined. Through this lens, the European city can be re-imagined as a space of coexistence and renewal, capable of harmonizing preservation with change and fostering sustainable, inclusive urban futures.

Conflict of Interest The authors declare that the research was conducted in the absence of any commercial or financial relationships that could be construed as a potential conflict of interest.

References

Abdurahiman, S. (2025). Urban heritage sociocultural impact assessment (UHSCIA): Scale development and psychometric validation. *Built Heritage, 9*, 35. https://doi.org/10.1186/s43238-025-00206-y

Ahern, J. (2013). Urban landscape sustainability and resilience: The promise and challenges of integrating ecology with urban planning and design. *Landscape Ecology, 28*, 1203–1212. https://doi.org/10.1007/s10980-012-9799-z

Ashworth, G. J., Graham, B., & Tunbridge, J. E. (2007). *Pluralising pasts: Heritage, identity and place in multicultural societies*. Pluto Press.

Babitha, V. (2013). Triple bottom line of material sector. In *Proceedings of the ICETEM 2013* (pp. 74–77). International Journal of Advanced Trends in Computer Science and Engineering. ISSN 2278-3091

Bandarin, F., & van Oers, R. (2012). *The historic urban landscape: Managing heritage in an urban century*. Wiley-Blackwell.

Beeckmans, L. (2013). Editing the African city: Reading colonial planning in Africa from a comparative perspective. *Planning Perspectives, 28*(4), 615–627. https://doi.org/10.1080/02665433.2013.828447

Bernt, M. (2019). How post-socialist is gentrification? Observations in East Berlin and St. Petersburg. *Critical Sociology, 45*(2), 177–192. https://doi.org/10.1080/15387216.2016.1259079

Biermann, F. (2012). Planetary boundaries and earth system governance: Exploring the links. *Ecological Economics, 81*, 4–9. https://doi.org/10.1016/j.ecolecon.2012.02.016

Bitsani, E., & Kalomenidis, P. (2007). Integrated interventions for the management of cultural landscapes in contemporary cities. *Archives of Economic History, 19*(1), 351–371.

Bitsani, E. (2013). *"Intercultural city identity" and "human intercultural cities" (H.I.C.): A dynamic ontological model for the social co-existence and social cohesion of modern and post-modern cities*. NOVA Science.

Bitsani, E. (2016). Cultural heritage and the historic memory as a main factor for sustainable development of modern city in the framework of crisis. The case study of the museum cluster 'Museumsinsel' in Berlin. In *1st InHeriT Conference on Cultural Heritage as Economic Value in Athens, May 11–13, 2016, within the framework of European research program InHeriT Promoting Cultural Heritage as a Generator of Sustainable Development ERASMUS+, LLP.*

Bitsani, E.,& Eletheriou , A. (2015). The management of natural World Heritage sites as an essential component of cultural tourism and sustainable development: The Tokaj–Hollókö case study in north-eastern Hungary: From a national/local past towards an international/global future. In N. Columbus (Ed.), *Tourism development and management* (Chapter 9, pp. 125–167). Nova Publishers.

Bitsani, E., & Thymi, I., (2022). Biocultural Approach and Digital Strategy of Cultural Organizations: Towards a Renewed Proximity through the Pandemic in Public Administration Review vol. 2 no.

Bratton, B. H. (2021). AI urbanism: A design framework for governance, program, and platform cognition. *AI & Society, 36*(4), 1307–1312. https://doi.org/10.1007/s00146-020-01121-9

Braun, B. P. (2014). A new urban Dispositif? Governing life in an age of climate change. *Environment and Planning D: Society and Space, 32*(1), 49–64. https://doi.org/10.1068/d4313

Brunet, R. (2001). Models in geography? A sense to research (Modèles en géographie ?). In *12ème Colloque Européen de Géographie Théorique et Quantitative, St-Valéry-en-Caux, France, 7–11 septembre 2001* (p. 204). https://doi.org/10.4000/cybergeo.4288

Bürgi, M., Verburg, P. H., Kuemmerle, T., & Müller, D. (2017). Analyzing dynamics and values of cultural landscapes. *Landscape Ecology, 32*(10), 2077–2081. https://doi.org/10.1007/s10980-017-0573-0

Cabezas, H., Pawlowski, C. W., Mayer, A. L., & Hoagland, N. T. (2003). Sustainability: Ecological, social, economic, technological, and systems perspectives. *Clean Technologies and Environmental Policy, 5*(3), 167–180. https://doi.org/10.1007/s10098-003-0214

Ceil, C. (2018, July 28). *A report on Doughnut economics in the 21st century*. Available at SSRN https://ssrn.com/abstract=3520434 or https://doi.org/10.2139/ssrn.3520434

Certomà, C. (2013). Expanding the 'dark side of planning': Governmentality and biopolitics in urban garden planning. *Planning Theory, 14*(1), 23–43. https://doi.org/10.1177/1473095213506202

Chen, T. L., Kim, H., Pan, S. Y., Tseng, P. C., Lin, Y. P., & Chiang, P. C. (2020). Implementation of green chemistry principles in circular economy system towards sustainable development goals: Challenges and perspectives. *Science of the Total Environment*. https://doi.org/10.1016/j.scitotenv.2020.136998

Collins, A., & Flynn, A. (2015). *The ecological footprint: New developments in policy and practice*. Edward Elgar Publishing. https://doi.org/10.4337/9780857936967

Cook, B. I., Mankin, J. S., Williams, A. P., Marvel, K. D., Smerdon, J. E., & Liu, H. (2021). Uncertainties, limits, and benefits of climate change mitigation for soil moisture drought in southwestern North America. *Earth's Future, 9*(9). https://doi.org/10.1029/2021EF002014

Cornell, S. (2012). On the system properties of the planetary boundaries. *Ecology and Society, 17*(1). http://www.jstor.org/stable/26269022

Cosgrove, D. (1984). *Social formation and symbolic landscape*. Croom Helm.

Cosgrove, D. (1998). *Social formation and symbolic landscape* (2nd ed.). University of Wisconsin Press.

Council of Europe. (1985). *European Convention on the Protection of the Architectural Heritage of Europe* (Granada, 3 October 1985; Council of Europe Treaty Series No. 121). https://www.coe.int/en/web/conventions/full-list/-/conventions/treaty/121

Council of Europe. (1992). *European Convention on the Protection of the Archaeological Heritage* (Revised) (Valletta, 16 January 1992; Council of Europe Treaty Series No. 143). https://www.coe.int/en/web/conventions/full-list/-/conventions/treaty/143

Council of Europe. (2000). *European Landscape Convention*. https://www.coe.int/en/web/landscape/about-the-convention

Council of Europe. (2021). *European heritage days 2021 celebrate "heritage: All inclusive!"* https://www.europeanheritagedays.com/News/European-Heritage-Days-2021-celebrate-Heritage-All-Inclusive

Court, S., & Wijesuriya, G. (2015). *People-centred approaches to the conservation of cultural heritage: Living heritage*. ICCROM (International Centre for the Study of the Preservation and Restoration of Cultural Property). https://www.iccrom.org

Crumley, C. L., Kolen, J., de Kleijn, M., & van Manen, N. (2017). *Sustainability Science, 12*(6), 881–893.

Dalakoglou, D. (2017). The city at a time of crisis: Transformations of public spaces in Athens. *City, 21*(3–4), 438–450.

Daly, L. M. (2020). Resilience: An integrated review. *Nursing Science Quarterly, 33*(4), 330–338. https://doi.org/10.1177/0894318420943141

Darvill, T. (1999). The historic environment, historic landscapes, and space-time-action models in landscape archaeology.

Dong-jing, C. (2001). Ecological footprint in Northwest China. *Journal of Glaciolgy and Geocryology*. https://www.semanticscholar.org/paper/Ecological-Footprint-in-Northwest-China-Dong-jing/8b346db8bdd4319a617acf2cc395068cb6567594

Doucet, B. (2014). A process of gentrification: Waterfront redevelopment in Amsterdam and Hamburg. *European Urban and Regional Studies, 21*(1), 6–22. https://doi.org/10.1177/0969776412439195

Dovey, K., & Recio, R. B. (2024). Inventraset assemblages: The spatial logic of informal street vending, transport and settlement. *Urban Studies, 61*(12), 2265–2289. https://doi.org/10.1177/00420980231223060. (Original work published 2024).

Dowler, L. (2018). Embodying power and space: Woman, war and protest. *ACME: An International Journal for Critical Geographies, 17*(4), 1134–1152.

Drexhage, J. R., & Murphy, D. G. (2010). *Sustainable development: From Brundtland to Rio 2012. Background paper prepared for consideration by the High-Level Panel on Global Sustainability at its first meeting 19 September 2010*. ID: 157381054.

Duncan, J. S., & Duncan, N. G. (2004). *Landscapes of privilege: The politics of the aesthetic in an American suburb*. Routledge.

Escobar, A. (2001). Culture sits in places: Reflections on globalism and subaltern strategies of localization. *Political Geography, 20*(2), 139–174.

European Heritage Days. (2021). *Living heritage*. https://www.europeanheritagedays.com/

Fang, Q., & Li, Z. (2022). Cultural ecology cognition and heritage value of Huizhou traditional villages. *Heliyon, 8*(12). https://doi.org/10.1016/j.heliyon.2022.e12627

Foucault, M. (1977). *Discipline and punish: The birth of the prison*. (A. Sheridan, Transl.). Random House Vintage Books.

Foucault, M. (1997). Of other spaces: Utopias and heterotopias (J. Miskowiec, Trans.). In N. Leach (Ed.), *Rethinking architecture: A reader in cultural theory* (pp. 330–336). Routledge. (Original work published 1984).

Foucault, M. (2007). *Security, territory, population: Lectures at the Collège de France, 1977–1978* (M. Senellart, Ed., G. Burchell, Trans.). Palgrave Macmillan.

Foucault, M. (2008). *The birth of biopolitics: Lectures at the Collège de France, 1978–1979* (M. Senellart, Ed., G. Burchell, Trans.). Palgrave Macmillan.

Gkoltsiou, A., & Terkenli, T. S. (2012). An interdisciplinary analysis of tourist landscape structure. *Tourismos: An International Multidisciplinary Journal of Tourism, 7*(2), 145–164.

Gong, Y., Zhang, Y., Zhou, Y., Zhou, J., & Wang, X. (2022). *Sustainability, 14*(4), 4.

Gravari-Barbas, M., & Guinand, S. (2017). *Tourism and gentrification in contemporary metropolises: International perspectives* (1st ed.). Routledge. https://doi.org/10.4324/9781315629759

Hall, S. (2003). Cultural identity and diaspora. In P. Williams & L. Chrisman (Eds.), *Colonial discourse and post-colonial theory: A reader* (pp. 222–237).

Haraway, D. (2016). *Staying with the trouble: Making kin in the Chthulucene*. Duke University Press.

Harrison, R. (2013). *Heritage: Critical approaches*. Routledge.

Harvey, D. (1973). *Social justice and the city*. Edward Arnold.

Harvey, D. (2000). *Spaces of hope*. Edinburgh University Press.

Hawken, P., Lovins, A., & Lovins, L. H. (2003). *Natural capitalism: Creating the next industrial revolution*. Little, Brown and Company.

Historic England. (2024, November 14). *Historic England reveals its Heritage at Risk Register 2024*. https://historicengland.org.uk/heritage-at-risk

Holm, A. (2014). Berlin's gentrification mainstream. In K. Lees, H. Bang, & A. Wyly (Eds.), *Gentrification and resistance in Berlin* (pp. 27–45). Routledge.

Holtorf, C. (2013). On pasts and lost futures: Heritage and the temporality of history. *International Journal of Heritage Studies, 19*(6), 609–618. https://doi.org/10.1080/13527258.2012.700799

House, R., Taylor, C., Livingston, J., Watt, A., & Minster, M. (2011). A crooked stool: The rhetoric of the triple bottom line. In *2011 IEEE International Professional Communication Conference*, pp. 1–4.

Inglis, F. (1977). Nation and community: A landscape and its morality. *The Sociological Review, 25*(3), 489–514. https://doi.org/10.1111/j.1467-954X.1977.tb00301.x. (Original work published 1977).

Ingold, T. (1993). The temporality of the landscape. *World Archaeology, 25*(2), 152–174. http://www.jstor.org/stable/124811

Kaika, M. (2017). Between the bulldozer and the beast: Rethinking the intellectual work of urban geography in the age of smart urbanism. *Environment and Planning D: Society and Space, 35*(3), 496–511. https://doi.org/10.1177/0263775816658292

Karpodini-Dimitriadis, E. (2009). *The spirit of agricultural landscapes, civilization, memory and messages*. Cult-Rural, culture 2000, framework Programme in support of culture.

Kim, R. E., & Kotzé, L. J. (2020). Planetary boundaries at the intersection of Earth system law, science and governance: A state-of-the-art review. *Review of European, Comparative & International Environmental Law, 29*(1), 3–15. https://doi.org/10.1111/reel.12383

Labadi, S. (2013). *UNESCO, cultural heritage, and outstanding universal value: Value-based analyses of the World Heritage and Intangible Cultural Heritage Conventions*. AltaMira Press.

Labadi, S., & Logan, W. (Eds.). (2015). *Urban heritage, development and sustainability: International frameworks, national and local governance*. Routledge.

Landry, C., et al. (2023). Operational guidelines for integrating heritage in landscape planning. *Journal of Urban Planning and Development*.

Lefebvre, H. (1991). *The production of space* (D. Nicholson-Smith, Trans.). Oxford: Blackwell.

Leiden-Delft-Erasmus Universities. (2021, April). Edition April 2021. https://www.leiden-delft-erasmus.nl/

Lekakis, S. (2025). Cultural policy and public engagement with modern architectural heritage in Greece: An empirical analysis. *Cities, 162*, 105925. https://doi.org/10.1016/j.cities.2025.105925

Logan, W., Craith, M. N., & Kockel, U. (Eds.). (2016). *A companion to heritage studies*. Wiley-Blackwell.

Luke, T. (1999). Eco-managerialism: Environmental studies as a power/knowledge formation. In F. Fischer & M. Hajer (Eds.), *Living with nature* (pp. 103–120). Oxford University Press.

Luukkanen, J., Vehmas, J., & Kaivo-oja, J. (2021). Quantification of doughnut economy with the sustainability window method: Analysis of development in Thailand. *Sustainability, 13*(2), 847. https://doi.org/10.3390/su13020847

Mancini, M. S., Galli, A., Coscieme, L., Niccolucci, V., Lin, D., Pulselli, F. M., Bastianoni, S., & Marchettini, N. (2018). Exploring ecosystem services assessment through ecological footprint accounting. *Ecosystem Services, 30*, 228–235. https://doi.org/10.1016/j.ecoser.2018.01.010

Massey, D. (1994). *Space, place, and gender*. University of Minnesota Press.

Massey, D. (2005). *For space*. Sage.

Mbembe, A. (2019). *Necropolitics*. Duke University Press.

Mercuri, A. M. (2014). *Quaternary International, 353*, 1802.

Miller, P., & Rose, N. (2008). *Governing the present: Administering economic, social and personal life*. Polity Press.

Mitchell, D. (2000). *Cultural geography: A critical introduction*. Blackwell.

Mitchell, D. (2020). *Mean streets: Homelessness, public space, and the limits of capital*. University of Georgia Press.

Mitchell, N., Rössler, M., & Tricaud, P. (2009). *World heritage cultural landscapes: A handbook for conservation and management*. UNESCO World Heritage Centre.

Moreira, F., Queiroz, A. I., & Aronson, J. (2006). Restoration principles applied to cultural landscapes. *Journal for Nature Conservation, 14*(3–4), 217–224. https://doi.org/10.1016/j.jnc.2006.05.007

Moreno, L. D. (2020). Sustainable city storytelling: Cultural heritage as a resource for a greener and fairer urban development. *Journal of Cultural Heritage Management and Sustainable Development, 10*(4), 399–412. https://doi.org/10.1108/JCHMSD-05-2019-0043

Nobre, G. C., & Tavares, E. (2021). The quest for a circular economy final definition: A scientific perspective. *Journal of Cleaner Production, 314*, 127973. https://doi.org/10.1016/j.jclepro.2021.127973

Nsamenang, A. (2015). Ecocultural theories of development. In J. D. Wright (Ed.), *International encyclopedia of the social & behavioral sciences* (2nd ed., pp. 838–844). Elsevier. https://doi.org/10.1016/B978-0-08-097086-8.23205-4

O'Donnell, P. M. (2008). Urban cultural landscapes & the spirit of place [Presentation]. ICOMOS Quebec 2008. Preservation of the Spirit of Place. Available from https://publ.icomos.org/publicomos/jlbSai?html=Pag&page=Pml/Not&base=technica&ref=C27C5AA57146EC14CA9B176C9899E6CB

Özdemir Işık, B., & Demir, S. (2017). Integrated multi-criteria decision-making methods for the sustainability of historical–cultural structures on the Trabzon Coastline. *Sustainability, 9*(11), 2114. https://doi.org/10.3390/su9112114

Pauleit, S., Liu, L., Ahern, J., & Kazmierczak, A. (2011). Multifunctional green infrastructure planning to promote ecological services in the city. In I. Douglas, R. Goode, M. Houck, & R. Wang (Eds.), *Handbook of urban ecology* (pp. 272–285). Oxford University Press. https://doi.org/10.1093/acprof:oso/9780199563562.003.0033

Pendlebury, J. (2009). *Conservation in the age of consensus*. Routledge.

Pendlebury, J. (2012). Conservation values, the authorised heritage discourse and the conservation-planning assemblage. *International Journal of Heritage Studies, 19*(7), 709–727. https://doi.org/10.1080/13527258.2012.700282

Perry, S. (2019). The enchantment of the archaeological record. *European Journal of Archaeology, 22*(3), 354–371. https://doi.org/10.1017/eaa.2019.20

Piccardo, C., & Canepa, M. (2020). Cultural ecology: Paradigm for a sustainable man-nature relationship. In W. Leal Filho, A. Azul, L. Brandli, P. Özuyar, & T. Wall (Eds.), *Partnerships for the goals. Encyclopedia of the UN sustainable development goals*. Springer. https://doi.org/10.1007/978-3-319-71067-9_26-1

Plieninger, T., & Bieling, C. (2012). *Resilience and the cultural landscape: Understanding and managing change in human-shaped environments*. Cambridge University Press. https://doi.org/10.1017/CBO9781139107778

Poulios, I. (2012). Discussing strategy in heritage conservation: Living heritage approach as an example of strategic innovation. *Heritage Management, 1*(2), 325–340.

Poulios, I. (2014). Discussing strategy in heritage conservation: Living heritage approach as an example. *The Historic Environment: Policy & Practice, 5*(1), 15–26.

Poulios, I. (2015). *The past in the present: A living heritage approach - Meteora, Greece*. Ubiquity Press.

Rao, P. K. (2003). *Sustainable development: Economics and policy*. Blackwell Publishing.

Raworth, K. (2017). *Doughnut economics: Seven ways to think like a 21st- century economist*. Random House Business. ISBN: 9781847941374.

Raworth, K. (2020, July 16). *So you want to downscale the doughnut? Here's how*. Kate Raworth. https://www.kateraworth.com/2020/07/16/so-you-want-to-create-a-city-doughnut

Redclift, M. (2005). Sustainable development (1987–2005): An oxymoron comes of age. *Sustainable Development, 13*(4), 212–227. https://doi.org/10.1002/sd.281

Rey-Pérez, J., & Pereira Roders, A. (2020). Historic urban landscape: A systematic review, eight years after the adoption of the HUL approach. *Journal of Cultural Heritage Management and Sustainable Development, 10*(3), 233–258. https://doi.org/10.1108/JCHMSD-05-2018-0036

Ritter, H., Herzog, O., Rothermel, K., et al. (2025). City models: Past, present and future prospects. *Frontiers of Urban and Rural Planning, 3*, 7. https://doi.org/10.1007/s44243-025-00057-2

Rockström, J., Steffen, W., Noone, K., Persson, Å., Chapin, F. S., Lambin, E., et al. (2009). A safe operating space for humanity. *Nature, 461*(7263), 472–475. https://doi.org/10.1038/461472a

Rogers, D. (2021). Waterfront regeneration in Oslo: Publicness, architecture and social equity. *Urban Studies, 58*(4), 729–748. https://doi.org/10.1177/0042098020943490

Roy, A. (2009). The 21st-century Metropolis: New geographies of theory. *Regional Studies, 43*(6), 819–830. https://doi.org/10.1080/00343400701809665

Russo, A. P., & Scarnato, A. (2018). "Barcelona in common": A new urban regime for the 21st-century tourist city? *Journal of Urban Affairs, 40*(4), 455–474.

Salet, W., & Gualini, E. (2015). Framing strategic urban projects: Learning from current experiences in European cities. *European Planning Studies, 23*(6), 1113–1131.

Sauer, C. O. (1963). *Land and life: A selection from the writings of Carl Ortwin Sauer* (J. Leighly, Ed.). University of California Press.

Schroll, H., Kjærgård, B., & Thorn, P. (2009). Resilience is more than an elastic jump. *Journal for Transdisciplinary Environmental Studies, 8*(1) http://www.journal-tes.dk/vol_8_no_1/artiekl%201%20Henning%20Schroll.pdf

Simone, A. (2016). City of potentialities: An introduction. *Theory, Culture & Society, 33*(7–8), 5–29. https://doi.org/10.1177/0263276416666915

Skrede, J., & Berg, S. K. (2018). Cultural heritage and sustainable development: The case of urban densification. *The Historic Environment: Policy & Practice, 10*(1), 1–20. https://doi.org/10.1080/17567505.2019.1558027

Smythe, K. R. (2014). An historian's critique of sustainability. *Culture Unbound: Journal of Current Cultural Research, 6*, 913–929. https://doi.org/10.1007/s10980-012-9799-z

Sonkoly, G. (2020). *Historic urban landscape: A conceptual analysis*. ResearchGate.

Stephenson, J. (2007). The cultural values model: An integrated approach to values in landscapes. *Landscape and Urban Planning, 84*(2), 127–139. https://doi.org/10.1016/j.landurbplan.2007.07.003

Stevenson, D. (2003). *Cities and urban cultures*. Open University Press. Retrieved from the Greek edition: Stevenson, D. (2007). Cities and urban civilizations (A. Vogiatzis, Transl.). Kritiki publications. (Original work published in 2003).

Steward, J. (1955). *Theory of culture change: The methodology of multilinear evolution*. University of Illinois Press.

Sutton, M. Q., & Anderson, E. N. (2004). *An introduction to cultural ecology* (1st ed.). Routledge. https://doi.org/10.4324/9781003135456

Taylor, J. B. (1999). A Historical Analysis of Monetary Policy Rules. In Taylor, J. B., (Eds.), *Monetary Policy Rules* (pp. 319–348). University of Chicago Press, Chicago. https://doi.org/10.7208/chicago/9780226791265.001.0001

Taylor, A. (2003). Birdlife. *Landscapes: The Journal of the International Centre for Landscape and Language, 2*(1) https://ro.ecu.edu.au/landscapes/vol2/iss1/9

Taylor, K. (2015). The historic urban landscape: Managing heritage in an urban century.

Taylor, K. (2018). Connecting cultural landscape and historic urban landscape: Politics of similarity. *Built Heritage, 2*(1), 53–67.

Taylor, K., & Lennon, J. (2012). Cultural landscapes: A bridge between culture and nature? *International Journal of Heritage Studies, 17*(6), 537–554. https://doi.org/10.1080/13527258.2011.618246

Terkenli, T. S. (1996). *The cultural landscape: Geographical approaches (in Greek)*. Papazissis and University of the Aegean.

Terkenli, T. S. (2001). Towards a theory of the landscape: The Aegean landscape as a cultural image. *Landscape and Urban Planning, 57*, 197–208. https://doi.org/10.1016/S0169-2046(01)00203-8

Thrift, N. (1999). Steps to an ecology of place. In D. Massey, J. Allen, & D. Sarre (Eds.), *Human geography today* (pp. 295–321). Blackwell.

Tratsela, M. (2012). *Wind turbines and landscape transformation: A necessity turned to an artifact*. https://www.academia.edu/32178405/Wind_turbines_and_Landscape_Transformation_A_necessity_turned_to_an_artifact_pdf

Tulumello, S. (2016). Neoliberalism and urban planning in a 'peripheral' city: Reflections on Lisbon's housing policies. *Urban Research & Practice, 9*(3), 253–272. https://doi.org/10.1080/17535069.2016.1197303

Tulumello, S. (2021). Urban austerity, tourism, and social movements in Lisbon: From resistance to redefinition? *Urban Geography, 42*(2), 163–185.

UNESCO. (2005a). *Operational Guidelines for the Implementation of the World Heritage Convention*. Paris.

UNESCO. (2005b). *Vienna Memorandum*. https://whc.unesco.org/archive/2005/whc05-15ga-inf7e.pdf

UNESCO (2011a). *Education for Sustainable Development: An Expert Review of Processes and Learning*. http://unesdoc.unesco.org/images/0019/001914/191442e.pdf

UNESCO. (2011b). *Recommendation on the Historic Urban Landscape*. Paris: UNESCO. https://whc.unesco.org/en/hul/

UNESCO. (2019a). The UNESCO Courier, July–September 2019 (Vol. 2019, Issue 3). https://en.unesco.org/courier/2019-3

UNESCO Courier. (2019b). *What is Intangible Cultural Heritage?* https://en.unesco.org/courier/

Wackernagel, M., & Rees, W. E. (1996). *Our ecological footprint: Reducing human impact on the Earth*. New Society Publishers. https://www.researchgate.net/publication/234092351_An_overview_on_ecological_footprint_and_sustainable_development_A_chat_with_Mathis_Wackernagel

Waterton, E., & Smith, L. (2010). The recognition and misrecognition of community heritage. *International Journal of Heritage Studies, 16*(1–2), 4–15. https://doi.org/10.1080/13527250903441671

Wijesuriya, G., Thompson, J., & Young, C. (2013). *Managing cultural world heritage (world heritage resource manual)*. UNESCO. https://whc.unesco.org/en/managing-cultural-world-heritage/

Winter, T. (2014). Heritage conservation futures in an Asian century. *Journal of Cultural Heritage, 15*(3), 317–321. https://doi.org/10.1016/j.culher.2013.07.003

World Commission on Environment and Development. (1987). *Our common future*. Oxford University Press. https://sustainabledevelopment.un.org/content/documents/5987our-common-future.pdf

Wylie, J. (2007). Landscape. In S. Holloway & G. Valentine (Eds.), *Key ideas in geography*. Routledge.

Yang, M., Chen, L., Wang, J., et al. (2023). Circular economy strategies for combating climate change and other environmental issues. *Environmental Chemistry Letters, 21*, 55–80. https://doi.org/10.1007/s10311-022-01499-6

Zhang, L. (2024). Layered evolution in Sheshan urban countryside Historic District. *Heritage Science nature.com*

Zhivitskaya, M. (2018). *Book review: doughnut economics: seven ways to think like a 21st-century economist by Kate Raworth*. https://www.semanticscholar.org/paper/Book-review%3A-doughnut-economics%3A-seven-ways-to-like-Zhivitskaya/c99cd4f0dbeff1a74d5af8bd6a7d3415f5cd41a4

Zukin, S. (2020). *The innovation complex: Cities, tech, and the new economy* (Online edition). Oxford Academic. https://doi.org/10.1093/oso/9780190083830.001.0001

Chapter 4
Public Policies for Sustainable Development and City Typology: Cities' Models

In the era of late modernity, the understanding and management of various collective issues were addressed through public power action. This can occur at different levels, including central government, local government, or even through international organizations that facilitate coordination. The actions taken by these entities involve implementing public policies, which encompass diagnosing the problematic situation, planning interventions, executing programs, and evaluating their outcomes.

Environmental, economic, and social sustainability form a critical framework of objectives that led the United Nations to establish a common set of priorities, known as the 17 Sustainable Development Goals. These goals aim to ensure the balanced development of modern societies. The framework was designed to serve as a foundation for public policies across all modern states, guiding them toward the desired direction for sustainable development.

In this chapter, we will first explore the modern approaches in public policy that have been developed to address complex problems. Next, we will analyze the fundamental components that constitute the framework of the 17 Sustainable Development Goals established by the United Nations. Finally, we will outline the basic models of cities, highlighting the unique challenges and potential solutions each model presents. This will emphasize the distinct public policy approaches needed to promote sustainability in urban areas.

E. Bitsani et al., *The Biocultural City as a Utopia or the Future Reality?*, Cities, Heritage and Transformation, https://doi.org/10.1007/978-3-032-17584-7_4

4.1 Sustainability as a Polycrisis and a Wicked Problems Field

4.1.1 Addressing the Complexities of Sustainability

The sustainability of modern societies, especially in large urban centers that function as key hubs in today's social and economic networks, relies on various factors.

The seventeen United Nations Sustainable Development Goals (SDGs) represent the diverse aspirations necessary for balanced development of modern societies, benefiting all members. These pursuits span various policy areas, including the fight against poverty, promoting social equilibrium, and addressing health, employment, and education for all. They also encompass balanced economic development, technological innovation for the common benefit, responsible consumption, environmental protection, and climate change. Simply listing these aspirations highlights the complexity and interdependence of the challenges modern societies face.

The complexity of these factors affecting social, environmental, and economic sustainability requires the development of new conceptual and methodological approaches. How do different sectors of society perceive and respond to the concept of sustainability and its various dimensions? What specific technical and policy strategies can be implemented to address these complex interconnections? Additionally, how can we establish concepts and methodological frameworks that accurately understand and tackle the main challenges within sustainability?

Public policy refers to large-scale interventions that regulate, provide, reproduce, or transform a specific area of socio-economic activity. Such interventions typically allocate significant social resources, making it essential for them to be effective, efficient, and of high quality while ultimately serving the collective good. To achieve these goals, a consistent and coherent methodology is necessary, one that is regularly tested against real-world scenarios and either validated through success or adjusted in response to failure. This ongoing process embodies the interaction between theory and practice.

To illustrate this, we can refer to the classic model of organizational learning developed by David Kolb (1984). According to his theory, collective or organizational learning occurs through a cyclical process that includes the following steps: (a) recording experiences (practice), (b) reflecting on experiences (analysis), (c) forming rules (theory), (d) implementing new and improved actions (practice).

An appropriate and robust theoretical framework leads to effective practice. Conversely, systematic evaluation and analysis of that practice can strengthen and refine the theory, creating a positive cycle of successful policies and enhanced political-administrative capacity. On the other hand, inadequate theories result in ineffective practices, while poor analysis of those practices leads to insufficient theoretical frameworks. This creates a negative cycle that fosters failures in political and administrative capacities.

New tools are, therefore, essential for comprehending these intricate issues at hand and for implementing policies that effectively address the challenges faced by contemporary societies, particularly those impacting their most vulnerable segments. Policy science today emphasizes two main areas—conceptual and methodological frameworks—in order to comprehend and address the complexities of contemporary issues: polycrisis and wicked problems. In the following section, we will examine these concepts to emphasize their significance in addressing social, environmental, and economic challenges while promoting sustainability, particularly in urban areas.

4.2 The Concept of Polycrisis

A first suitable theoretical framework for addressing the complex sustainability challenges is polycrisis theory.

The idea that crises in the modern world should be viewed as interconnected and mutually reinforcing was first introduced by the French philosopher and sociologist Edgar Morin. In his 1999 book "Homeland Earth: A Manifesto for the New Millennium," co-authored with Anne Brigitte Kern, he describes this situation as a "polycrisis."

The authors argue that hierarchizing "critical" problems is insufficient for addressing the major challenges of contemporary societies. Such a fragmented approach fails to consider the interconnectedness of multiple threats. Problems cannot be isolated; they must be examined from various perspectives across different regions. There is no single vital problem but rather a complex inter-solidarity of issues like health, overpopulation, and economic instability. These intertwined crises form what can be described as "polycrisis sets" (pp. 73–74).

The authors present the argument that within such an environment policy problems' prioritization can become too limited and reductionist. Contemporary societal challenges are interconnected and require a multifaceted approach. Rather than a single vital issue, there is a complex interplay of multiple crises—health, overpopulation, environment, nationalism, and economic instability—forming a "polycrisis set" of overlapping issues.

Adam Tooze, a historian at Columbia University, more recently reintroduced the term "polycrisis" in an article for the Financial Times on October 28, 2022. He stated that a problem becomes a crisis when it challenges our ability to cope and threatens our identity. In a polycrisis, the various shocks may be disparate, but they interact in a way that makes the overall situation even more overwhelming than the sum of its parts.

More recently the concept has gained significant traction in global policy and academic discourse with institutions such as the United Nations and the World Economic Forum (WEF) adopting the term. Time magazine even dubbed polycrisis the "buzzword of the day" at the WEF's January 2023 meeting in Davos (Serhan, 2023).

The growing use of the term signals an important shift in how scholars and decision-makers understand and articulate interconnected global threats. Beyond its definition as coinciding crises, polycrisis denotes a condition where interconnected crises intensify each other, producing greater systemic risks than each crisis individually (Lawrence et al., 2022). In the face of such challenges, societies may encounter a broad spectrum of outcomes, ranging from social unrest and civil violence to complete collapse. However, it is noteworthy that certain societies exhibit a higher degree of resilience, thereby ensuring the maintenance of fundamental social functions (Hoyer et al., 2023).

One important characteristic of polycrises is that the combined negative effects of multiple crises are different from—and usually more severe than—the individual negative effects of each crisis when the systems they affect are not so closely interconnected (Tooze, 2022; Lawrence et al., 2022).

The Cascade Institute characterizes global polycrisis as involving multiple entangled crises that threaten human survival, describing it as a condition in which fast-moving trigger events interact with slow-moving systemic stresses, leading to widespread disequilibrium (Jia et al., 2024). This interconnectedness often manifests through causal pathways such as common stress, domino effects, and inter-systemic feedback. These mechanisms contribute to synchronized disruptions across the global system, highlighting the necessity of a holistic understanding of complex global interrelations.

Polycrisis analysis is interrogating the very concepts of "risk" and "crisis," as well as understanding the dynamics of how different crises interact across time and space. Jia et al. (2024) emphasize that seemingly independent crises might function as interconnected nodes, producing cascading effects when combined. This insight prompts scholars to reexamine existing frameworks and differentiate polycrisis from traditional crisis models. The term serves as a valuable lens to explore multidimensional challenges, enabling more nuanced and integrative approaches to global governance and policy formation (Serhan, 2023). Lawrence argues that the world is entering a worsening "polycrisis"—a situation in which multiple, interconnected crises amplify, accelerate, and synchronize across global systems (Lawrence et al., 2024).

The concept of polycrisis presents two key challenges: the need for a systematic approach to enhance understanding and the lack of sufficient documentation regarding the interconnections between individual crises. Lawrence et al. (op.cit) observe that leaders often intuitively recognize the importance of addressing these crises collectively, and the term "polycrisis" encapsulates this idea. While the definition of a crisis can be politically sensitive, the potential misuse of the term can be reduced by relying on objective standards based on data and real-world events, rather than mere intuition.

Crises can originate in various policy sectors or at different levels of governance, manifesting in distinct public arenas. However, it is the interaction between these elements that gives rise to a polycrisis. This synthesis may involve relationships between different policy domains, governance areas, or even a combination of issues arising from multiple policy areas and levels of government simultaneously. This underscores the systemic nature of polycrisis.

Lawrence et al. argue that the interrelation in modern societies and economies that contributes to contemporary polycrises, including environmental, economic, cultural, and immigration issues, is intensified by "neoliberal hyperconnectivity" that creates domino effects across governance and policies. Neoliberal economic practices prioritize efficiency and market access over social and environmental protections, leading to global homogenization and, consequently, to global crises. These interconnected systems amplify risks such as financial instability, pandemics, economic inequality, and extremist ideologies.

The rivalry among strong states in the global security system complicates efforts to restructure the global energy system, which is vital for reducing environmental harm. As the agricultural, energy, and financial sectors have become more complex and specialized, their adaptability has decreased, leading to increased rigidity. Economic crises stemming from debt, inflation, and shortages of essential resources can fuel public discontent and empower populist politicians, undermining the rule of law. Such systems are prone to failure, and a breakdown in one can trigger failures in others, as many global systems currently demonstrate (Lawrence et al., 2024).

Klein et al. (2023) add an important dimension to the discussion, arguing that the polycrisis we currently face is primarily due to a lack of sufficient governance capacity. Key obstacles to preventing a polycrisis include the inability to accurately identify existing and potential problems, as well as a lack of coordination and effective measures to address the interconnected nature of these issues. Unfortunately, there has been minimal political or managerial effort dedicated to tackling this challenge, leading to confusion regarding the understanding and definition of emerging and interconnected problems. Consequently, alternative approaches that could potentially be effective are not being considered.

4.3 Sustainability in a Polycrisis Framework

The necessity for novel developmental qualities and societal commitment to sustainable development in the face of polycrisis is emphasized by Siirilä and Salonen (2024). The authors advocate for inclusive approaches that engage all sectors of society, highlighting the importance of collective action and shared responsibility in navigating the complexities of global crises.

Utilizing an autoethnographic approach, Takkinen and Heikkurinen (2024) examine the role of sustainability experts in a polycrisis context. The study explores how experts' experiences and reflections on their practices can inform more nuanced and context-aware approaches to sustainability. It challenges traditional notions of expertise and advocates for a more critical engagement with technological solutions.

Seshadri et al. (2025) discuss the systemic risks associated with the transition to low-carbon energy systems within the context of a global polycrisis. The authors emphasize the necessity of aligning net-zero transitions with strategies that reduce interconnected crises, underscoring the imperative for integrated policy approaches that encompass environmental, economic, and social dimensions.

Despite its potential, the academic field of polycrisis research remains in its early stages. Critics argue that the term is often employed as a broad rhetorical device, lacking clear analytical rigor (Zaki et al., 2024). For the concept to mature, scholars suggest focusing on three avenues: refining theoretical foundation, conducting empirical analyses the fully leverage the concept, and fostering engagement with practitioners to translate insights into actionable strategies (Zaki et al., 2024).

4.4 Sustainability Issues as Wicked Problems

The polycrisis approach, as previously mentioned, emphasizes that the modern world is confronted with several interconnected crises. These include climate change and its environmental impacts, the social consequences that arise from these issues, the predominantly negative effects of economic globalization, and the threats posed by radical technological changes.

According to the World Social Report 2025 (UN, 2025), societies worldwide are facing significant challenges, with high inequalities and many struggling in precarious jobs. Rising insecurity is fostering distrust and threatening global solidarity. A recent Gallup World Poll from 2023–2024 reveals that 60% of people worldwide are struggling with life satisfaction, while 12% are suffering. This is exacerbated by ongoing crises, climate change threats, and escalating conflicts. Therefore, urgent action is needed to break the cycle of environmental deterioration, economic insecurity, social inequality, and declining institutional and political trust.

The main challenge faced by the international community and individual states is that the complexity of these issues categorizes them as "wicked," making them particularly difficult to resolve. When confronted with this specific category of complex public problems, traditional methods of designing and implementing public policies appear insufficient.

"Wicked" problems, that can also be referred to as "Insolvable" issues in another context, are challenges that cannot be effectively resolved due to the lack of appropriate methods applied by the administrative sub-systems involved. Addressing these problems not only requires specific techniques for solving them but also involves structuring them correctly. This includes how the problems are understood, described, and ultimately defined. Therefore, a variety of methods, approaches, and a new philosophy for handling these issues are necessary.

Historically, the "rational" or "one-best-way" methodology, primarily based on neoclassical economists' perceptions, was regarded as the most effective means of addressing all types of issues, including public ones. This methodology consists of the following stages:

- Definition of the Problem ("Problem structuring").
- Clearly define and prioritize collective preferences.
- Identify and record all possible alternatives.
- Clearly define and prioritize the criteria for assessing the suitability of solutions.

- Select the best option from the alternatives.
- Implement the chosen solution, evaluate its effectiveness, make corrections as needed, and finalize the solution.

The "wickedness" approach presents a distinct philosophy compared to the "one-best-way" model. It emphasizes understanding the nature of problems and how to define them. This model was introduced by Horst Rittel and Melvin Webber (1973) who argue that while science can effectively address "controlled" problems, referred to as "lab problems," it struggles with policy issues that cannot be clearly and consensually defined. In a pluralistic society, reaching, for example, an objective definition of what constitutes social justice and an acceptable distribution of resources is considered impossible. Consequently, there can be no universally accepted notion of the public good. Public policies addressing social issues—among others—cannot simply be deemed entirely right or wrong; they can only be judged as good or bad, satisfactory or unsatisfactory, depending on specific social groups or lifestyle categories.

Social problems encompass a wide range of issues, including environmental and epidemiological concerns, which gain social significance as they affect communities and economies. These problems become "intractable" because their resolution is influenced by varying definitions of social justice, particularly in relation to the fair and collectively acceptable distribution of social resources.

Rittel and Weber highlight that the potential consensus surrounding distribution problems is undermined by social pluralism and the differing values that arise from the increasing diversity of society's members. The traditional "mechanical" or "Newtonian" scientific rationality does not apply to the "wicked" problems found in open social systems. Consequently, policy approaches based on linear rationalism and "Newtonian mechanistic physics" are unsuitable for modern interactive open systems. In these systems, a multitude of competing interests and values makes it "impossible to define uniform goals" and creates significant difficulties regarding rationality. According to the advocates of this concept, insolvability arises at the intersection of goal setting, problem definition, and social impartiality and justice.

So we see how Rittel and Webber establish a direct connection between policy wickedness and social problems, as well as various socially contentious issues. They emphasize that rather than seeking universal or technically correct solutions, we should focus on managing the conflicting definitions of what constitutes a socially just and desirable solution to a specific problem. The key difference between "wicked" problems and "controlled" ones lies in the fact that finding solutions to wicked problems requires a different type of reasoning that cannot be adequately addressed by the classical Newtonian paradigm of science and engineering.

Subsequent works of significance have expanded the wicked problem framework in various fields, including public policy, environmental governance, and organizational theory. For example, as Conklin (2006) noted, wicked problems were integrated into the domain of collaborative design, emphasizing the need for stakeholder engagement and iterative processes (p. 101). Moreover, the study by Head and Alford (2013) explored how public managers confront intractable problems through the mechanisms of adaptive leadership and collaborative governance. The wicked problem framework is notable for its emphasis on complexity, context, and the importance of interdisciplinary and participatory approaches.

Newman-Storen's (2014) study explores the significance of leadership in confronting intractable sustainability challenges. The author proposes a theoretical integration of creativity and leadership theory to develop innovative solutions for complex sustainability challenges. In his article, he emphasizes the importance of adaptive leadership in navigating the uncertainties inherent in wicked problems.

In their 2020 study, Campbell and Zellner sought to explore the manner in which the dynamics of intractable problems have the capacity to subvert conventional problem-solving methodologies within the context of urban governance. The authors engage with complex thinking and its potential to address urban challenges, delineating three distinct approaches. Firstly, complexity is described as a symptom of urban systems. Secondly, the dynamics of complex urban systems are analyzed. Thirdly, appropriate planning strategies that account for complexity are proposed as a means of intervention. The present study explores the application of this framework to the domain of urban sustainability issues, with a particular focus on the role of innovative governance approaches in addressing entrenched problems. Yearworth (2015) conceptualizes sustainability as a "super-wicked" problem, emphasizing the challenges posed by time constraints, the absence of a central authority, and the tendency to disregard future considerations. The paper discusses the implications of these characteristics for engineering methodologies and suggests that addressing sustainability requires a shift toward more adaptive and systems-oriented approaches. Earle and Leyva-de la Hiz (2020) investigate the difficulties inherent in the pedagogy of sustainability education, particularly in the context of confronting intractable challenges. The integration of design thinking and emerging technologies as pedagogical tools to facilitate student engagement with complex sustainability issues is a key theme. The article under discussion highlights the necessity for educational approaches that foster critical thinking and problem-solving skills in the context of wicked problems. DeFries and Nagendra (2017) present a discussion on the intricacies of ecosystem management, conceptualizing it as a "wicked problem" due to the multifaceted interconnection between ecological, social, and economic factors. The authors advocate for the implementation of adaptive management approaches that incorporate diverse knowledge sources.

Over time, through various approaches like those mentioned above, the concept of wickedness broadened beyond Rittel and Webber's original definition to encompass all types of complex problems, regardless of the nature of their complexities. Critics argue that the wicked theory lacks focus and clear definition, providing limited practical guidance for decision-making and potentially causing policy paralysis by overstating complexity. Additionally, the lack of a structured methodology has been shown to hinder the achievement of practical outcomes in applied contexts (Head & Alford, 2013).

Lippi and Tsekos (2024a) suggest that we can only resolve this issue by returning to the original interpretation of the concept. It's important to recognize that all public policy problems are, at their core, social issues. These problems arise precisely because they impact specific communities. Applying the principles outlined by Rittel and Webber regarding the complexity of such problems, particularly the idea of "wickedness," which stems from conflicting values and interests, suggests that these issues must be seen as potentially intractable, even though they often have technical dimensions.

The technical dimensions, however, should in principle be considered as "controlled" sub-problems. Therefore, classical rational approaches—and especially those that accept finite rationality and the search for relative satisfaction under given conditions (Simon, 1976; Lindblom, 1959)—are sufficient to solve them, even when the technicalities are particularly complex. Complexity increases the difficulty of dealing with and requires additional efforts and innovative solutions, but it remains within the scope of formal scientific rationality. The technical dimensions of policy problems can only become "intractable" if they are positively or negatively affected by the interests and values of the social actors concerned. If the disagreement of experts and policymakers on the facts is generated by—quantitatively and qualitatively—insufficient evidence, this can be resolved with more and better evidence (i.e., with more and better documentation) and with further, deeper, innovative, and/or more flexible application of the rational scientific approach. On the contrary, if the disagreement on the facts is the result of divergences of value judgments and/or alignment with different collective (and/or sometimes individual) interests, a different approach will be needed, linked to the political nature of wickedness (Lippi & Tsekos, 2024b).

Wickedness usually arises not just from complex technical factors; rather, it stems from a breakdown in the shared understanding of public problems due to differing values, political beliefs, and conflicting interests. This is where the uniqueness of the concept lies: it relates to complexities that stem not from technical issues but from a lack of consensus. The challenges in finding solutions arise not from technical obstacles, but from difficulties in building agreement around the proposed solutions.

The method of dealing with wickedness, therefore, consists in the search for all possible channels of convergence and the corresponding configuration of political functions in order to favor a structured and documented dialogue both in the field of government (i.e., within the political-administrative system) and in that of governance (i.e., in the field of interactions between the political-administrative system and civil society). It is also necessary to develop a political culture that facilitates convergences, even the minimum possible ones, or the avoidance of artificial differences, at a minimum. The above certainly concerns contrasts that can be manipulated in win-win situations. Problems that arise in the context of structural contradictions (zero-sum games) should—and until the overthrow of the conditions that produce them—be considered not as "intractable" but as practically "insoluble."

4.5 UN Policies Concerning Sustainability

Back in 1972 in Stockholm was held the first major gathering for environment, in which the United Nations issued the Stockholm Declaration that stressed the need to protect the environment and to integrate its protection to development planning (United Nations, 1972). The declaration was widely accepted and is considered to be the foundation of United Nations Environment Program and as Birnie et al. (2009) argue it was a basic step in the evolution of international environmental law and policy.

In 1974, the correlation between stratospheric ozone depletions and ODS was recognized (Velders et al., 2007) and a series of regulations were issued. Farman et al. (1985) were the first to report on the dramatic seasonal depletion of the ozone layer over Antarctica, a phenomenon that has since become known to the general public as the "Ozone Hole." The authors found out that chlorofluorocarbon, or CFC, chemicals are the main cause of ozone loss when they react with other gases like NOx. This led to worldwide action to protect ozone layer, which was successfully achieved in 1987 when countries signed the Montreal Protocol to stop using chemicals which were destroying ozone layer (UNEP, 1987). It aimed at phasing out the production and use of ozone-depleting substances, such as chlorofluorocarbons—CFCs (UNEP, 2020). In his 2019 paper, Huggins presents the argument that the Montreal Compliance System incorporates administrative procedures that promote rule of law values, including transparency, accountability, and fairness. He contends that this enhances the legitimacy and effectiveness of international environmental law enforcement. That led to its success in reducing ozone depletion and protecting the Earth's ozone layer, which is evident, as latest scientific assessments (WMO, 2022; Riordon, 2024, NASA, n.d.) showed that if compliance to the agreement continues, the ozone levels would return to pre-80 s levels by 2066.

The Brundtland report "Our Common Future," which was commissioned by the UN, was chaired by Gro Harlem Brundtland, former Norwegian Prime Minister. Its objective was to address the increasing concerns surrounding environmental degradation and development. The term "sustainable development" was introduced, defined as "development that meets the needs of the present without compromising the ability of future generations to meet their own needs" (WCED, 1987). The report was instrumental in affecting a shift in global discourse toward sustainability, with the result that policy frameworks and international agreements were influenced. As stated by Redclift (2005), the Brundtland Report is comprehensively analyzed in academic literature for its role in shaping sustainable development paradigms. Redclift also claimed that it is evident that issues of global environmental justice are as pressing as they were when the concept of "sustainable development" was in its nascent stage. However, the novel material realities of science and the environment in the twenty-first century necessitate a re-engagement with their social consequences, a factor which is predominantly disregarded by the (market) liberal consensus (Redclift, 2005).

Various other global agreements have been established over the years to promote sustainability and combat environmental challenges. Ratified in 1989, the Basel Convention is a globally recognized agreement designed to reduce the production of hazardous waste and ensure its environmentally responsible disposal. The convention also regulates the transportation of hazardous waste materials between countries, focusing on restricting such movements from developed to developing nations. However, as Ahmed (2020) pointed out, the objective of regulating shipments of hazardous waste from industrialized to developing countries is not being achieved. One reason for this failure is the problematic geographical imagination at work in the treaty (Lepawsky, 2015).

The Convention on Biological Diversity (CBD), adopted in 1992, is recognized as a comprehensive global agreement that aims to preserve the variety of living organisms on the planet while promoting sustainable development (UN, 1992a, 1992b, 1992c; Ekardt et al., 2023). However, the extent to which the CBD upholds biodiversity amid population growth and economic pressures is uncertain (Höft, 2008). As Ekardt et al. (2023) claimed, since this obligation to halt biodiversity loss has been violated, states might be subject to legal action before international or domestic courts for their actions or inactions that contributed to global biodiversity loss.

Agenda 21 is a comprehensive action plan for sustainable development that was adopted at the United Nations Conference on Environment and Development (UNCED) in Rio de Janeiro in 1992. It outlines strategies for integrating environmental and developmental concerns into decision-making processes at all levels. Agenda 21 was a 300-page blueprint attempting to integrate environmental protection with economic development for the first time at this scale (UN, 1992a, 1992b, 1992c). The "common but differentiated responsibilities" principle (UN, 1992a, 1992b, 1992c: P7) and the precautionary principle were politically crucial, as the historical responsibility for developed countries was acknowledged and at the same time the necessity to take action to prevent environmental degradation was strengthened despite the lack of scientific certainty (UN, 1992a, 1992b, 1992c: P15). While not legally binding, Agenda 21 has been widely implemented through national and local policies, though its effectiveness has been debated (Drexhage & Murphy, 2010). Critiques and evaluations of Agenda 21 focus on its voluntary nature and the challenges in translating its principles into concrete actions (Drexhage & Murphy, 2010).

In the same Summit, the United Nations Framework Convention on Climate Change (UNFCCC) was adopted in 1992 and serves as the foundation for addressing climate change issues (UN, 1992a, 1992b, 1992c). The primary objective of the UNFCCC is to stabilize greenhouse gas concentrations in the earth's atmosphere at levels sufficient to prevent harmful human interference with the climate system (Karybekova, 2018). It facilitates negotiation and cooperation between nations and provides an overall framework for global efforts to address climate change. In 1997, the UNFCCC adopted the Kyoto Protocol, which sets binding targets for industrialized countries to reduce greenhouse gas emissions (UN, 1998). Although it has largely been superseded by the Paris Agreement, it has played a critical role in the development of international climate change policies.

The United Nations Millenia Declaration (UN, 2000) concentrated among others on establishing peace and security, eradicating extreme poverty and hunger, protecting the vulnerables, ensuring environmental sustainability, and developing global partnerships for development. The MDGs garnered extensive acceptance and established a framework for international development cooperation. The program's success in garnering global attention and focus on quantifiable development targets is noteworthy. However, it is important to acknowledge that progress varied significantly across regions and goals, as Sachs (2012) highlighted in his analysis of development achievements and critique of the goals' conceptual frameworks and measurement approaches.

The preceding decades witnessed a proliferation of climate diplomacy, exemplified by the 1992 Rio Earth Summit and the 1997 Kyoto Protocol that led in 2015 to the establishment of the Paris Agreement, that signified a pivotal transition by encompassing both developed and developing countries within a cohesive framework (Bodansky, 2016). The agreement was ratified under the auspices of the United Nations Framework Convention on Climate Change (UNFCCC), with central aim to limit global temperature rise to well below 2 °C above pre-industrial levels, with efforts to pursue a 1.5 °C threshold (UN, 2015a). Key provisions include nationally determined contributions (NDCs), regular progress reviews through a global stocktake, and financial support from developed developing nations (Rajamani & Bodansky, 2019), in an unprecedented bottom-up structure, allowing states to set their own targets. According to Laurent Fabius (2016), "a fair agreement: it reflects responsibilities as well as economic realities between countries, while recognizing that they can evolve over time; and it makes sure that no country is left behind." The repercussions of the Paris Agreement have varied. While the conference succeeded in establishing a global consensus and inaugurating a series of novel climate policy initiatives, the long-term efficacy of these measures remains uncertain. Critics have highlighted the document's overreliance on voluntary commitments, which are devoid of robust compliance mechanisms, thereby potentially undermining its stated objectives (Keohane & Oppenheimer, 2016). Furthermore, analyses indicate a persistent discrepancy between current NDCs and the pathways necessary to achieve the 1.5 °C target, raising concerns about policy ambition and implementation (Rogelj et al., 2016). Notwithstanding, the Agreement has catalyzed substantial subnational and non-state action, underscoring the role of networks that transcend nation-states in global climate governance. Despite its limitations, a considerable number of researchers contend that the Paris Agreement has redefined climate diplomacy by incorporating dynamic accountability and iterative ambition as foundational principles (Hale, 2016).

The Sustainable Development Goals (SDGs) were established by the United Nations in 2015 (UN, 2015b). They emerged from the Brundtland Commission's 1987 report, which introduced sustainable development as a framework balancing economic growth, social inclusion, and environmental protection, as we have seen above. The 17 (SDGs) are accompanied by specific targets and indicators, with a total of 169 established for consideration and aiming to address global challenges such as poverty, inequality, and climate change, advocating for an integrated approach to development (UN, 2015b). Success will be evaluated on the basis of the contributions provided by governmental entities, civil society organizations, and private businesses in achieving the established goals (Fig. 4.1).

However, there have been critiques regarding the alignment of the SDGs with neoliberal ideologies. Adelman (2018) contends that the SDGs perpetuate an anthropocentric and growth-driven development model, which may compromise ecological sustainability and social justice. Despite the global acceptance of the importance of SDGs, according to researchers an embedded economic hierarchy was found as well as a tension on the kind of transformation and "how it can be pursued" (Lencucha et al., 2023). Macnaughton and Frey (2018) argue that SDGs treat labor as a tool for economic growth and not as a human right.

In their study, Bogers et al. (2022) claim that the impact of the SDGs on policy integration appears to be constrained, as international organizations tend to utilize the SDGs primarily to rephrase existing activities, policies, and programs. In summary, although the SDGs are frequently cited by numerous international organizations, they are not achieving one of their primary objectives: enhancing policy integration and "breaking down the silos" of global sustainable development (Bogers et al., 2022).

From another point of view, Grek et al. (2023) claim that the SDGs employ indicators to transform multifaceted social and environmental issues into quantifiable data. This approach may potentially result in the dismissal of cultural and moral complexities, thereby offering a limited and numerical representation of complex social and environmental issues.

Fig. 4.1 Sustainable goals UN (https://sdgs.un.org/goals)

The **Kunming-Montreal Global Biodiversity Framework** (GBF) was the result of a four-year consultation and negotiation process. It was formally adopted during the fifteenth meeting of the Conference of the Parties (COP 15). This Framework constitutes a significant milestone in the global effort to achieve the Sustainable Development Goals, building upon the Convention's prior Strategic Plans in pursuit of this objective (CBD, 2022). It delineates a robust, forward-looking strategy aimed at realizing the global vision of a world living in harmony with nature by the year 2050. The Framework is predicated on four overarching goals for the year 2050 and 23 specific targets for the year 2030 (Stephens, 2023). The adoption of the Kunming-Montreal Global Biodiversity Framework by all Parties entailed a commitment to establish national targets for its implementation. Additionally, all other actors have been invited to formulate and disclose their respective commitments (CBD, 2022). But in order to achieve the objectives outlined in the Framework, it is imperative that adequate funding, capacity-building, and the transfer of technology be made available (Stephens, 2023). Among the challenges that GBF faces are that its goals are not SMART and that it doesn't have a holistic approach integrating local and indigenous knowledge to expert insights (Li et al., 2023). In the same spirit, Hughes and Grumbine (2023) claimed that the indicators used are not appropriate to measure implementation.

4.6 EU Policies for Sustainable Development

As previously discussed, there is a growing sense of insecurity in the world at present. The consequences of climate change, technological advancements, and demographic shifts are rapidly transforming contemporary societies and ways of life.

The European Union's sustainability policies are examined to clarify the measures employed by the EU to achieve its sustainability objectives. European policies for sustainable development encompass a broad range of initiatives and strategies aimed at promoting environmental, social, and economic sustainability across the European Union (EU) member states.

4.6.1 European Green Deal

The European Green Deal (EGD), initiated in 2019, functions as the EU's overarching strategy for achieving net-zero greenhouse gas emissions by 2050. It sets out an interim target to reduce emissions by at least 55% by 2030 compared to 1990 levels (European Commission, 2019). These targets have been legally enshrined through the European Climate Law, which establishes binding climate targets (European Commission, 2021a).

The European Green Deal (EGD) has catalyzed significant investments in renewable energy, energy efficiency, and green technologies, positioning the European

Union (EU) as a global leader in climate action. Moreover, it has prompted member states to align their national policies with EU-wide sustainability objectives.

The different elements of the Green Deal are shown in the table below (Fig. 4.2):

The European Green Deal outlines key measures in a roadmap, such as:

4.6.1.1 RePowerEU

In response to the crisis of dependence on Russian fossil fuels, the REPowerEU package and the Renewable Energy Directive (RED III) foresee a 42.5% share of renewables by 2030 (European Commission, 2022b).

Analytically, the transition to sustainable energy is advancing at different speeds across EU Member States and has been accelerated by recent crises, including the Russian invasion of Ukraine, the ongoing effects of the pandemic, supply chain disruptions, and inflationary pressures. To achieve this, all Member States must adopt sustainable energy and reduce their dependence on homegrown and imported fossil fuels in order to make an organized transition toward energy security, decarbonization, and economic growth (Perez et al., 2019).

The Intergovernmental Panel on Climate Change (IPCC, 2014) has stated that sustainable energy has the potential to reduce greenhouse gas emissions. Consequently, the energy transition is a priority for the European Union, with the twin aims of reducing emissions and decreasing dependency on imported fossil fuels. The transition to renewable energy and sustainable development requires all economic and human stakeholders to adapt and cooperate in order to find effective solutions. Maintaining a balance between production and consumption, the economy and the natural environment, and economic growth and quality of life is of the

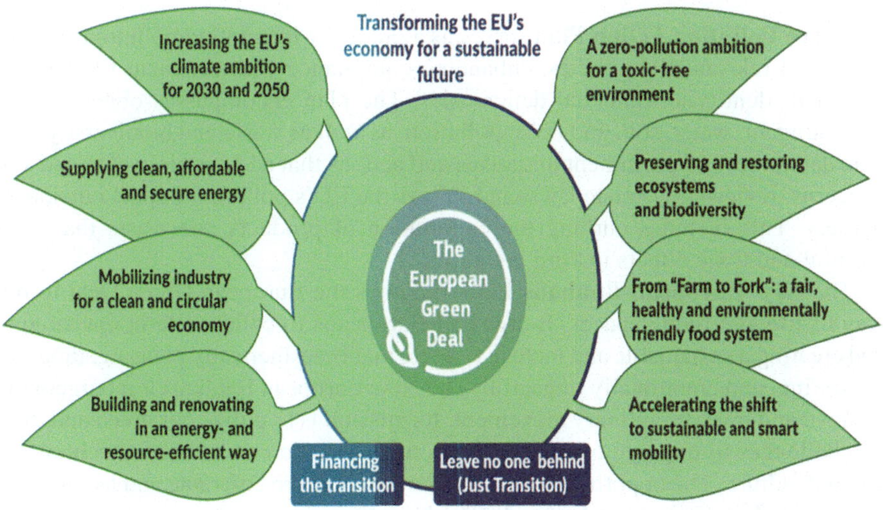

Fig. 4.2 European Green Deal: Reaching for the stars—GIS Reports (gisreportsonline.com)

utmost importance at national and global levels (Vîrjan et al., 2023). Hennicke et al. (2019) argue that the "European idea" can be imbued with a future-oriented essence through an energy transition. However, it is imperative that policies are established, and further research is conducted into regional disparities, storage solutions, and effective market design in order to maximize the socio-economic and environmental benefits of low-carbon electricity (Enrich et al., 2024). Think tanks, like Agora, suggest that to ensure the optimum functioning of the market in future, measures must be taken. Among others, the long-term price signals should be strengthened in order to provide the necessary incentives for clean investments to be made; market flexibility should be enhanced through better grid integration and demand response; regulatory frameworks should be improved in order to protect consumers whilst ensuring fair competition; and enhanced coordination among member states and more clearly defined roles for renewable energy sources in market design (Agora Energiewende, 2023).

4.6.1.2 The EU Biodiversity Strategy for 2030

The EU Biodiversity Strategy for 2030 aims to cease and reverse biodiversity decline and encompasses initiatives to safeguard and revitalize ecosystems, promote sustainable agriculture and forestry, combat invasive alien species, and integrate biodiversity-related matters into other policymaking domains. From grand rainforests to miniature gardens, from expansive blue whales to tiny fungi, biodiversity refers to Earth's extraordinary array of life.

4.6.1.3 Strengthening the Environment

The **Zero Pollution Action Plan** serves as a guide for integrating pollution prevention in all relevant EU policies, enhancing compliance with pertinent EU legislation, and identifying potential deficiencies. The plan encompasses objectives for mitigating air, water, soil, and noise pollution, as well as waste and biodiversity. The plan details several prominent initiatives and actions that address air, water, and soil concerns, reviews EU waste laws, and lowers the EU's pollution impact outside its borders. This includes curbing the exportation of products and waste that have harmful and toxic effects in third countries.

The **New European Bauhaus** Initiative links the European Green Deal to our way of life and living spaces. It urges all Europeans to collaborate in envisioning and creating a sustainable and inclusive future that is aesthetically pleasing, thought-provoking, and emotionally impactful. The New European Bauhaus is a burgeoning creative and transdisciplinary movement. Its mission is to utilize our green and digital challenges to improve our lives, connecting the spheres of science, technology, art, and culture. The opportunity to collaborate in order to co-create and tackle intricate societal issues is extended to all. The New European Bauhaus is encouraging a movement to facilitate and guide the transformation of our societies based on three interdependent values: sustainability, inclusivity, and quality of experience. (European Commission, 2021b).

The **Farm to Fork Strategy** is central to the European Green Deal, with the objective of creating fairer, healthier, and more environmentally friendly food systems. The European Commission aims to guarantee the safety of food and the welfare of animals and plants within the EU through Farm to Fork measures, ensuring appropriate monitoring and effective internal market implementation. By 2050, the Farm to Fork Strategy aims to ensure that agriculture, fisheries, and aquaculture play their part in achieving a climate-neutral Union. It addresses the impact of food systems on climate change, environmental degradation, and biodiversity loss. Furthermore, promoting sustainable food choices among consumers can enhance health outcomes and reduce societal expenses (EC, 2022). The F2F strategy also aligns with the EU's biodiversity strategy (Wesseler, 2022).

4.6.1.4 Net-Zero Industry

The Green Deal Industrial Plan (GDIP), launched by the European Commission in February 2023, aims to bolster EU competitiveness in clean technologies—such as batteries, solar panels, electrolyzers, wind turbines, and carbon capture—through four strategic pillars: regulatory simplification, public funding, skills development, and reinforcement of value chains and trade (EC, 2023; Voskoboinikova, 2023). While it responds to the U.S.'s Inflation Reduction Act, critics note that it repurposes existing funds rather than creating new money, risks distorting competition in favor of Germany and France, lacks clarity on targeted technologies, and compromises environmental and social oversight by relaxing reporting requirements (Voskoboinikova, 2023). Analysts further argue that its funding—far below the €520 billion per year needed for net-zero tech deployment—remains insufficient and overly reliant on private capital mobilization, while subsidies may fragment the internal market and fail to ensure a just social transition (EEA, 2024).

Adopted on 13 June 2024 by the European Parliament and the Council, Net-Zero Industry ACT (NZIA) establishes a framework to boost the EU's manufacturing capacity for net-zero technologies. The regulation sets targets, streamlines permitting processes, and facilitates access to finance and innovation for key technologies. It amends related administrative services into the Single Digital Gateway, improving support for businesses and stakeholders involved in the net-zero transition (European Parliament & Council of the European Union, 2024).

The European Commission's Carbon Border Adjustment Mechanism (CBAM) is a pioneering EU policy designed to prevent "carbon leakage" by ensuring that imported carbon-intensive goods face the same carbon price as domestically produced equivalents. Currently in a transitional reporting phase (2023–2025), CBAM requires importers of cement, iron & steel, aluminum, fertilizers, electricity, and hydrogen to monitor and report embedded greenhouse gas emissions—without financial obligations. Beginning in 2026, importers will purchase CBAM certificates tied to the EU Emissions Trading System to offset emissions embedded in their imports. The mechanism aims to equalize carbon costs between EU and non-EU producers, encourage global industrial decarbonization, and comply with WTO regulations. Simplification efforts—like a 50-ton de minimis exemption and streamlined processes—are underway to lessen administrative burdens on businesses (OECD, 2025).

Critics contend that the EGD's overreliance on market-based mechanisms and technological solutions may fail to adequately address underlying social inequalities and could disproportionately impact vulnerable populations. Furthermore, the impact of entrenched interests, such as the fossil fuel industry, has been posited as a potential factor that could potentially mitigate the rigor of the EGD's policies.

4.6.2 Circular Economy Action Plan

The concept of a circular economy is expanding, with the aim of promoting a more sustainable economy through recycling and reusing products, thus creating a circular approach. It represents an advanced economic model that seeks increased economic benefits, reduced resource consumption, and less environmental pollution. Moreover, it is a crucial method of enhancing urban environmental air quality. To attain enduring and consistent enhancement of the urban atmosphere, we ought to transform the city into a circular economy and society (Zhu et al., 2022). The circular economy, a substitute to conventional linear economic models that rest on a take-make-dispose approach, is gaining momentous popularity (Brusselaers et al., 2022). In accordance with the principle of subsidiarity, public policies aimed at promoting the circular economy should be developed at the local level of government. This allows regional authorities to utilize a significant extent of opportunities to devise, implement, and oversee such policies (Arauzo-Carod et al., 2022). Circular economy policies aim to coordinate economic operations based on the principle of "reduce, reuse, and recycle" in urban areas. The objective is to design and construct a city that is economically efficient, socially harmonious, and environmentally sustainable (Zhu et al., 2022). Sustainability and the circular economy are interdependent components that are critical to the current public policy framework. There is widespread consensus regarding the circular economy and sustainable principles which serve as guidelines for consumers, companies, and institutions. However, the multidimensional nature of these issues necessitates comprehensive policies, strategies, territorial considerations, and implementation approaches to achieve the intended outcomes (Arauzo-Carod et al., 2022). Various policy measures are suggested to increase the appeal of circular solutions, such as deregulation, political targets, information dissemination, procurement protocols, taxation adjustments, quotas, and additional voluntary programs (Johansson, 2021). The Circular Economy Action Plan (CEAP) was endorsed by the European Commission in March 2020 and is a vital component of the European Green Deal, which acts as a new pathway toward sustainable growth in Europe. The most recent action plan introduces schemes covering all product life cycle stages. Its key areas of interest comprise product design, implementation of circular economy protocols, promotion of sustainable consumption, and attainment of waste reduction alongside sustained resource retention in the EU economy. In order to effectively address various challenges, both legislative and non-legislative measures will be implemented, where EU-level interventions can contribute substantially to the cause (European Commission).

4.6.3 Sustainable Finance

The EU has introduced regulations and initiatives to promote sustainability in the financial sector. These include, in particular, the **European Taxonomy Directive,** which sets standards for determining the environmental sustainability of activities, and the Sustainable Financial Markets Disclosure Directive, which requires disclosure of the sustainability impact of investments by financial market participants. Sustainable finance involves integrating environmental, social, and governance (ESG) factors into financial sector investment decisions, resulting in increased investment in sustainable economic activities and projects over the long term. Environmental aspects encompass climate change mitigation and adaptation, as well as broader issues like preserving biodiversity, preventing pollution, and promoting a circular economy. Social considerations cover a range of issues, including inequality, inclusivity, labor relations, investment in people, their skills and communities, and human rights. Effective governance, both public and private, is crucial for the incorporation of social and environmental factors into decision-making. This involves the implementation of sound governance structures, maintaining positive employee relations and remunerating executives judiciously. Sustainable finance necessitates enhanced awareness and lucidity pertaining to the possible hazards that can destabilize the financial system's sustainability. This is only achievable through apt governance, implemented by corporate and financial actors to manage such risks (Claringbould et al., 2019).

In the context of EU policy, the term sustainable finance refers to finance that bolsters economic growth while simultaneously mitigating environmental harm, thereby contributing to the attainment of the climate and environmental goals of the European Green Deal, while ensuring social and governance considerations are taken into account. Sustainable finance requires transparency about ESG-related risks that may impact the financial system. Such risks should be mitigated through proper governance of financial and corporate actors. Notably, sustainable finance can harness critical resources toward climate action and the global transition to a low-carbon economy (Ameli et al., 2025).

4.6.4 Just Transition Mechanism

The Just Transition Mechanism aims to help the regions and industries most affected by the transition to a low-carbon economy by providing financial and technical support to and technical support to ventures that foster economic diversification, reskilling, and employment in sectors subject to significant alterations. The focal point of economic policy and administration will take into account both sustainability and the welfare of citizens. These policies reflect Europe's commitment to achieving the UN Sustainable Development Goals by prioritizing social justice and prosperity as the foundations for building a resilient society with the highest standards of well-being worldwide. To this end, the initiatives presented here support the action plan for implementing the European Pillar of Social Rights, including Equal Opportunities and Job Mobility; Economic Conversion and Job Creation; Promotion of Equality, Inclusion, and Social Protection; Combatting Poverty and Exclusion, etc.

4.6.5 Social Sustainability

Article 3(3) of the Treaty on European Union (TEU) states that sustainable development is a fundamental goal of the European Union (EU). The European Pillar of Social Rights and its accompanying initiatives introduced policies for social sustainability to the European Union (EU). The absence of social sustainability from the European vocabulary was exacerbated by the effects of the 2008 financial crisis. The subsequent criticisms sparked renewed interest in establishing a socially sustainable Union, which culminated in the Pillar and its accompanying initiatives (Alexandris Polomarkakis, 2019). The 2030 Agenda for Sustainable Development (United Nations, 2015a, 2015b) has the potential to deliver a holistic and integrated response to the cross-cutting challenges that Europe will face in the future. Emerging challenges necessitate a much more integrated approach and cross-sectoral policy initiatives. Access to basic social and health services, employment and business opportunities, and infrastructure (transportation, broadband, and so on) is unequally distributed among the population. Remote, rural, and disadvantaged areas are falling behind, with a consistent proportion of their children, working age, and elderly people unable to access essential services that ensure a decent quality of life and acceptable health conditions. The 2030 Agenda for Sustainable Development provides an opportunity to reinforce the EU's overall commitment to social, economic, and environmental cohesion.

The EU prioritizes social sustainability through schemes like the European Pillar of Social Rights, which establishes principles and entitlements to foster fair and inclusive societies. Its provisions guarantee just working conditions, access to superior education and medical services, and social protection for all inhabitants. An invaluable cornerstone for achieving personal fulfillment, employability, and active and responsible citizenship, education is fundamental. The European Pillar of Social Rights proclaims the right to quality and equitable education, training, and lifelong learning as its first principle. The Union is currently redefining its development strategy toward sustainability, with the green and digital transitions as its transformative factors. Education lies at the core of the European way of life, supporting the social market economy and democracy with freedom, diversity, human rights, and social justice. The European Commission proposes the strengthening of current policies and the development of the European Education Area through six dimensions: (a) quality, (b) inclusion and gender equality, (c) the green and digital transitions, (d) teachers and trainers, (e) higher education, and (f) the geopolitical dimension. Education is a fundamental basis for personal fulfillment, employability and active, responsible citizenship (European Commission).

Even though they have promised, more action from the Member States is required, along with more specific, legally binding proposals, as without them this social revival may once more yield to economic hegemony.

4.6.6 Sustainable Mobility

The strategy for sustainable and intelligent mobility, formulated by the European Commission, aims to achieve a significant undertaking with 82 initiatives. This master plan planned for the next few years serves as the groundwork for making the EU's transport system ecologically friendly, digital, and impervious to future crises. A crucial aspect of sustainable development is making all modes of transportation more environmentally friendly, with considerate emphasis on eco-friendly alternatives and incentives to facilitate the transition. Specific targets, such as at least 30 million zero-emission cars on European roads by 2030 and almost all cars, trucks, buses, and heavy-duty vehicles to be zero-emission by 2050, as well as 100 European cities to be carbon neutral by 2030, are part of the initiative toward a smart and sustainable future for Europe's transport system (European Commission). Developing and implementing decarbonized urban mobility solutions is crucial amidst the growing number of European cities confronting climate emergencies and extreme weather events. These solutions must effectively tackle the critical challenges faced by cities and their residents. The EU needs to promote sustainable mobility through policies that combine urban and mobility planning and actively involve citizens at all stages of the transformation process, from design to implementation (Tsavachidis & Petit, 2022). Similarly, Banister (2008) and Litman (2012) argue that transitioning to sustainable mobility requires a paradigm shift in planning.

Accordingly, other measures such as Transport and Promoting Sustainable Alternative Fuels will contribute to the Green Deal in the coming years (2021–2027).

4.6.7 Digital Europe

The digital transition is critical to the EU's economic development and strategic independence. The EU is working on a number of policies to help Europe transition to a digital future. A crucial component of Europe's digital transformation vision is the enhancement of education and skills. In shaping Europe's digital future for the next five years, the focus will be on three specific objectives. These objectives aim to guarantee that digital solutions facilitate the continuation of Europe's progress toward a digital transformation that honors our citizens by upholding our values. The EU's digital strategy aims to make this transformation work for people and businesses while also contributing to the EU's goal of achieving a climate-neutral Europe by 2050. Additionally, it will establish Europe as the flagship in the global discourse. Technology serves individuals, a just and competitive economy, and an open, democratic, and sustainable society (European Commission). Digitalization has the potential to provide solutions to many of Europe's and Europeans' problems, as well as opportunities such as job creation, advancement of education, increased competitiveness and innovation, combating climate change, and enabling a green transition.

According to the first report on the State of the Digital Decade (27/9/2023), we need to do more to accelerate the digital transition. The report evaluates the EU's progress toward the 2030 goals, with a focus on digital skills, infrastructure, and the digitalization of businesses and public services. Given the complexity of the issue of the impact of technological change on the socioeconomic sphere and its adaptation to current challenges, there is a need for collaboration among Member States, policy measures such as the Gigabit Infrastructure Act, and investments (Małkowska et al., 2021).

Despite strong institutional commitment, however, implementation is lagging behind, with many targets estimated to require acceleration or at risk of delay.

4.6.8 EU and UN's 2030 Agenda

The European Commission remains dedicated to the United Nations' 2030 Agenda. Sustainable development aims to meet present demands while preserving future generations' ability to satisfy their own needs. This concept is extensively grounded in European policies, and the SDGs are continuously integrated throughout the entire scope, supporting the Member States' initiatives. The 17 Sustainable Development Goals (SDGs) are included in the United Nations' 2030 Agenda.

The European Council's Strategic Agenda for 2019–2024 includes the following Political Guidelines:

A European Green Deal; An economy that works for people; A Europe fit for the digital age; Protecting our European way of life; A stronger Europe in the world; A new push for European democracy (Fig. 4.3).

Critical to Europe's post-pandemic recovery is the Next Generation program. The EU's long-term budget, together with the Next Generation EU (NGEU) instrument, constitutes the biggest-ever recovery package in Europe's history. These funds will be allocated toward addressing the continent's most pressing issues and assisting vulnerable populations.

More than 50% of the long-term budget and **Next Generation EU** support modernization, for example, through:

Fig. 4.3 European Commission Priorities (European Commission)

- Research and innovation, through Horizon Europe
- A fair climate and digital transitions, through the Just Transition Fund and the Digital Europe program
- Preparedness, recovery and resilience, through the Recovery and Resilience Facility, rescEU, and a new health program, EU4Health.
- Modernizing traditional policies such as cohesion and the Common Agricultural Policy to maximize their contribution to the Union's priorities.
- Tackling climate change, with 30% of EU funding, the highest ever share of the European budget.
- Protecting biodiversity and gender equality

4.7 City Typology: Cities' Models

4.7.1 City as "Polis"

"From the Middle Ages to the contemporary of fourth industrial revolution era, social, cultural, and economic development in Europe has always been dependent upon the city. The common history that European cities share has given them a similar facet" (Bitsani, 2013a, p. 1).

From antiquity, through the Middle Ages, and into the present day, cities have been confronted with internal upheavals and external threats such as economic crises, poverty, social instability, crime, and intense natural phenomena like floods and earthquakes (Bosch, 2019). Nevertheless, many cities have managed to preserve, despite these pressures, certain essential characteristics—such as cultural heritage, and/or to reconstruct themselves and thrive once again.

Urban space reflects the combination of all elements necessary to functionally meet the needs of the societies that develop within its boundaries. It also reminds us that urban planning and development are outcomes of strategic thinking and craftsmanship. Modern cities face numerous and complex pressures arising from natural factors, human activity, and socio-economic conditions. Urban planning practices and strategies aimed at resilience and sustainability are key tools in addressing these risks and, consequently, in protecting and maintaining cities and their buildings.

In particular, buildings of high aesthetic and historical value—such as preserved structures, monuments, and cultural sites—as well as residential ensembles like historic city centers and traditional settlements, constitute cultural assets of utmost importance for protection (Rodwell, 2007). Simultaneously, they can serve as sources of knowledge and methodological frameworks for achieving urban resilience and sustainability (Meier et al., 2007).

Cities have evolved into centers of human activity. From the Neolithic village and the early cities of the ancient world to the megacities of the twenty-first century, the transformation of cities has been brought about by a series of economic, social, and intellectual developments such as political revolutions, industrial revolution, religious changes, urbanization, free market, technological revolution, etc. All these developments contributed to the change of society in time and place (Ritzer & Stepnisky, 2020). The word society comes from the verb "socio" which in ancient Greek means to share—to do something together with others, which is the reason why cities have more "anthropogenic" and less "natural" elements than the countryside (Bithas, 2001). Aristotle gives us the definition of "polis" as a form of higher social coexistence.

The most famous—and most debatable—model of city's categorization is the one of Sjoberg: (1) folk-preliterate; (2) feudal; (3) preindustrial; (4) urban-industrial. This framework positions cities along a continuum that represents societal advancement. In the context of feudal periods, European cities were characterized by a comparatively limited level of development in comparison to the more advanced urban centers of the time, such as towns. The dwellings were unadorned, and the thoroughfares lacked paving. The consolidation of political power and the expansion of states during the medieval revival resulted in the growth of pre-industrial cities. The focal point of these cities was an imposing complex of religious and governmental structures, including cathedrals. As Europe transitioned into the urban-industrial age, high-rise buildings became a prominent feature of city landscapes. In metropolitan areas, financial and commercial institutions, which are often characterized by their skyscraper-like architecture, have come to dominate the urban landscape (Sjoberg, 1965). The eighteenth century marked the beginning of the Industrial Revolution, a period that would continue until approximately the mid-nineteenth century. This era is widely regarded as the pinnacle of the industrial age, and it was during this time that the hallmarks of the modern state began to take shape, leading to the emergence of the capitalist model of social development. Recent developments in industry have been marked by the emergence of new sectors experiencing rapid growth, shifts in international trade, advancements in communication infrastructure, the creation of employment opportunities, and the shaping of contemporary society. Nevertheless, these developments have not yet resulted in the establishment of a sustainable environment. As demonstrated, these forces also affect changes to the socio-economic structure.

Urbanization has been shown to have a multifaceted impact on residential patterns, with cities becoming not only employment centers but also economic and cultural hubs (Stevenson, 2003). As early as the beginning of the twentieth century, Simmel (1907) was already discussing the senses of people in metropolises, where the accidental touch can turn into a stimulus for misunderstanding. In the modern era, the eminent German sociologist employs a sensory approach to analyze the social environment, highlighting the psychological implications for individuals in contemporary urban settings. The Chicago School, a prominent institution in the field, has demonstrated a particular interest in the nature of urban life and its associated challenges. The city of Chicago has been utilized as a laboratory to study urbanization and its associated problems (Valasaki, 2020).

4.7.2 Compact City Model

The Compact City model is a key concept in contemporary urban planning, advocating sustainable development through strategic densification of urban areas. This model emerged as a response to the adverse consequences of urban sprawl, including excessive land consumption, car dependency, rising infrastructure costs, and environmental degradation. In contradistinction to the conventional approach of urban expansion in a horizontal manner, the Compact City model advocates for vertical growth, efficient land use, and the integration of diverse urban functions within a delineated spatial framework (Newman & Kenworthy, 1999).

The model's fundamental principle is to encourage high-density, mixed-use development, where residential, commercial, and recreational spaces coexist (Cervero, 1996). This approach has two key benefits: it minimizes the need for long commutes, and it encourages the use of sustainable transportation modes. It is argued that this model supports the creation of walkable and cyclable urban environments, complemented by robust public transportation systems. Collectively, these measures have the potential to reduce reliance on private vehicles, decrease greenhouse gas emissions, and improve urban air quality (Gehl, 2010; Litman, 2015).

The Compact City model exhibits several salient features that promote sustainable urban living. High population density, for instance, helps make the most of available land and infrastructure, thereby allowing public services to be delivered more efficiently (Barton et al., 2003). The integration of residential, commercial, and recreational components within a single neighborhood fosters vibrant and pedestrian-friendly communities. Furthermore, compact cities are designed to support walking and cycling, making active and environmentally friendly transport options more practical and appealing.

The model under discussion presents a number of advantages. From an environmental perspective, it enhances sustainability by curbing urban sprawl and reducing emissions. From a social perspective, it has the potential to promote equity by ensuring more equitable distribution of access to services and opportunities (Barton et al., 2003). From an economic standpoint, the model offers efficiency through the reduced cost of infrastructure and the stimulation of local economic activity (Cervero, 1996). Finally, the model has a positive effect on public health by encouraging more active, less sedentary lifestyles (Gehl, 2010).

Notwithstanding its sustainable aspirations, the Compact City model has several critical limitations. A key concern is the relationship between population density and infrastructure investment. When investment is inadequate, population density can lead to overcrowding, reduced quality of services, and a decline in urban liveability (Jenks & Burgess, 2000). Furthermore, the densification and mixed-use development that characterizes Compact City projects has been shown to increase property values, potentially resulting in gentrification and the displacement of lower-income demographics, thereby compromising social equity (Neuman, 2005). In terms of mobility, the model posits behavioral shifts toward sustainable transport modes; however, such transitions are often impeded by entrenched car cultures, inadequate transit infrastructure, and spatial constraints (Banister, 2008). The successful implementation of compact city principles is dependent on the presence of cohesive governance and long-term planning. Fragmented institutional frameworks and inadequate policy coordination frequently result in inconsistent applications, thereby impeding the model's effectiveness in achieving its sustainability objectives (Burton et al., 2000). This enhances Harvey's (2008) argument that compact city model is often linked to neoliberal urban policies that prioritize market-driven solutions over community needs and social welfare. In the same spirit, Smith et al. (2020) argue that access to public transport is not equitably distributed across all groups.

4.7.3 Green City Model

The Green Cities Model is a progressive urban development model that emphasizes environmental sustainability, economic vitality, and social well-being. The model is grounded in the principles of ecological urbanism and sustainable planning, with the objective of integrating green infrastructure, renewable energy sources, efficient public transportation, and eco-friendly architecture into the urban fabric.

A fundamental objective of the model is to reduce the environmental impact of cities by minimizing carbon emissions, managing natural resources responsibly, and enhancing biodiversity within urban areas. This objective is pursued through strategic land use planning, the incorporation of green spaces such as parks and green roofs, and the promotion of sustainable building practices. Moreover, the model prioritizes inclusive governance and community engagement to ensure that urban sustainability efforts are equitable and responsive to the needs of diverse populations. The promotion of compact, mixed-use development and active modes of transportation such as walking and cycling is integral to Green Cities' objective of enhancing the quality of urban life while addressing the challenges posed by climate change and rapid urbanization. The Green Cities Model is, therefore, a blueprint for reimagining urban environments as resilient, liveable, and ecologically responsible spaces.

The Green Cities Model faces several limitations in achieving sustainable development goals (SDGs). Key challenges include high initial costs for implementing green infrastructure and sustainable technologies, which can be prohibitive for developing cities with limited financial resources (Roseland, 2012). The economic viability of these investments is also uncertain, which can lead to resistance from stakeholders. Another key challenge pertains to the difficulty of ensuring equitable outcomes; affluent populations often have greater access to green benefits, while marginalized groups may face barriers to essential services such as green spaces and sustainable transportation (Anguelovski et al., 2018). Moreover, the model's inability to effectively address urban growth and sprawl has been identified as a key challenge, with the potential to hinder efforts to preserve ecosystems and manage resources in a sustainable manner (Seto et al., 2012). Finally, the implementation of the model when it is aligned with neoliberal development may produce inequalities under the disguise of sustainability through displacement of vulnerable communities (de Godoy et al., 2024). Consequently, achieving the SDGs necessitates a more inclusive, context-specific approach to urban sustainability.

4.7.4 Fifteen-Minute City Model

The reorganization of urban environments through the implementation of proximity-based planning models, most notably exemplified by the "15-minute city" concept, constitutes a substantial paradigm shift within the domains of urban policy, urban planning, and sustainable development. This emerging model promotes a spatial restructuring of urban life that privileges accessibility, inclusivity, and human-scale development, aligning with broader objectives of climate resilience and quality of life enhancement.

The concept was first introduced by Professor Carlos Moreno in 2016, and it synthesizes and extends principles of polycentric urbanism and chrono-urbanism (Moreno, 2016). Moreno's framework is anchored on four foundational principles: proximity, density, diversity, and digitalization. The concept of proximity necessitates spatial reconfiguration, with a view to ensuring that everyday needs can be met within localized catchment areas, thereby reducing reliance on motorized transport and promoting neighborhoods-based economies. Density, defined as the number of people living in a given area, is an important factor in the efficient delivery of public services and the sustainability of private business. Diversity emphasizes mixed-use development to foster socioeconomic inclusivity and functional integration within neighborhoods. Finally, digitalization, through the implementation of smart technologies and the integration of participatory platforms, has emerged as a pivotal element in enhancing the real-time engagement, governance, and adaptability of urban systems (Allam et al., 2022a).

The 15-minute city concept is predicated on the principle of decentralizing essential services—including, but not limited to, employment, healthcare, education, commerce, recreation, and culture—into urban cells that are accessible within a 15-minute walk or bicycle ride from one's place of residence. This proximity-driven framework addresses the increasing disjunction between urban functions that has characterized much of twentieth-century urban planning. In this period, monofunctional zoning and extensive car-centric infrastructure have produced spatial and temporal fragmentation (Papadopoulos et al., 2023; Zhang et al., 2022).

As a direct response to the socio-environmental challenges of modern urbanization, the 15-minute city model has gained traction among international policy networks. Significantly, the C40 Cities Climate Leadership Group, an alliance of nearly 100 mayors committed to addressing climate change, has adopted the 15-minute city as a strategic pillar in its agenda for just and sustainable cities. Since 2020, there have been over 18 pilot implementations in cities across the globe, emphasizing the growing policy commitment to the model (C40 Knowledge, 2020).

This human-centered model of urbanism entails a paradigm shift in urban planning priorities, moving from infrastructural expansion to the lived urban experience. By centering human well-being at the core of urban design, it aspires to promote social interaction, self-actualization, and equitable access to urban resources. The concept of temporal accessibility, which refers to the reduction of time required to access key urban functions, emerges as a metric of both equity and sustainability. This reframing of time as a central currency of urban life has been demonstrated in numerous studies (Allam et al., 2022b; Poorthuis & Zook, 2023; Khavarian-Garmsir et al., 2023).

The 15-minute city concept has the potential to reinforce socio-spatial inequalities by privileging areas that are already well-resourced. This may lead to urban fragmentation and exclusion, as well as a lack of economic agglomeration benefits. Consequently, it may inadvertently constrain economic agglomeration benefits by limiting inter-neighborhood connectivity. Furthermore, the model has the potential to overlook the needs of marginalized groups who require access to wider metropolitan opportunities. Furthermore, the implementation of this model may conflict with existing urban fabrics, necessitating costly retrofitting and governance realignment.

4.7.5 Smart Cities

According to the European Commission, a smart city is a place where traditional networks and services are made more efficient by using digital solutions for the benefit of residents and businesses. The OECD defines smart cities as "cities that harness digitalization and engage stakeholders to improve people's well-being and build inclusive, sustainable and resilient societies." This definition underlines that digitization and digital innovation are not ends in themselves, but aim to improve people's lives, achieve greater inclusion, sustainability, and resilience (OECD, 2020). Mohanty (2016) wants to give a simpler definition state that "...a smart city is a place where traditional networks and services are made more flexible, efficient and sustainable by using information, digital and telecommunication technologies to improve operations of the city for the benefit of its inhabitants...." By defining a city as smart, we mean the ways in which managers use technological solutions to improve their urban environment, but also how they monitor, analyze, plan, and govern the city through the data they receive, creating opportunities, focusing on durability and its sustainability. The technology company Cisco has estimated that a typical Smart City of one million inhabitants generates 180 million gigabytes of data per day ...." Arguably, through this process the city itself, commercial entities, governments, and individuals benefit from a deeper understanding of the city's functions (Paiho et al., 2022). A smart city's primary mission is to deliver value through what it chooses to do with technology, not just how much technology it can have. Characteristics that define a city as smart could include:

(a) Infrastructures that are based on technology
(b) What environmental initiatives
(c) The operating system of public transport
(d) Urban planning
(e) The use of resources by the people who live in it

Given the above characteristics, we understand that in order to give the characterization of a city "as smart" there should be a multi-composite process where it aims to manage the challenges of modern urbanization. Therefore, in order to create a smart environment, we should aim at the management of pollution and the reduction of carbon emissions, at the efficient disposal of waste, at the energy transition, at the safety of public spaces, and in general at the improvement of the quality of life of citizens. A typical example of improving the quality of life of a city's residents and reducing carbon emissions is the reduction of traffic congestion where through intelligent transportation systems there can be smooth traffic flow and road safety (Saleem et al., 2022). In other words, through the use of new technologies we create specific fields of action, i.e., "smart government," "smart economy," "smart environment," "smart life," "smart mobility," and "smart people" that work in combination (Paiho et al., 2022; Cavada et al., 2014).

Smart government: Focus on institutional and physical infrastructure, participatory decision-making, public and social services, and transparency, in order to get the maximum possible benefit. We should build a governance where it is able to respond to the rapid development of cities (Tyas et al., 2018).

Smart economy: By referring to the smart economy we mean an innovative economy based on a cooperative model of production, distribution, and consumption. The principles of a smart economy include innovation, creativity, research, entrepreneurship development, a strong and competitive economic sector, and above all thinking locally and acting regionally and globally (Tyas et al., 2018). All sectors operate in the light of cooperation and sharing.

Smart environment: Protection of natural heritage, natural resources, and biodiversity. Preservation of city ecosystem, enhancement of public green spaces, cleanliness, and recreational opportunities for all.

Smart living: Strong and shared values, safety for all, improved quality of life for residents, open and reliable public spaces, high-quality public services. Promotion of cultural and natural heritage art, enhancing the aesthetics of the residents' everyday life.

Smart mobility: Supporting not only the mobility of vehicles but also of people, promoting alternatives through movement and interconnecting spaces. Ensuring mobility for people with different abilities.

Smart people: Characteristics they should have are ability and a high index of human development, flexibility, creativity, and resistance to change. The smart city will attract people who are skilled and worthy in their field, will have integrated education and literacy through higher education in all aspects of life, will have brought together open-minded, multicultural people who will actively participate in sustainable urban development the space. According to the OECD (2020), through the opportunities offered to cities by the digital transition, artificial intelligence, cloud computing, and Big Data, the quality of life of their inhabitants can be improved by facilitating their access to goods and services but not strengthen participatory policymaking. The value of the smart city model lies in its ability to create a strong relationship between the public and all private sector actors as well as citizens. New technologies and governance models of digital platforms are increasingly permeating urban sectors and spheres of urban life. This phenomenon is known as platformization, which in turn has given rise to the phenomena of the platform society, where platforms have permeated the core of urban societies (Allam et al., 2022). The collection of data and any kind of information collected is used for the effective management of assets, resources, and services. Of course, on the contrary of all this, we should mention that all these complex networked systems also run the risk of security from malicious attackers who at any time can take advantage of all this wealth of data that exists and gain access to data banks, affecting the confidentiality of information, integrity of systems, and the inability to control user identity. All these attacks can affect various aspects of smart cities causing irreparable problems and disrupting the smooth flow of services (Sookhak et al., 2019). Modern means of dealing with these risks are constantly evolving as technology evolves, such as IoT-based systems, always with data security in mind. In conclusion, smart cities can be greener, more efficient, safer, and friendlier by including smart transport infrastructure, smart energy, smart care, and smart technology. The potential is there but much more is needed in order to have the environment where digital technologies will translate into better public services for residents with better use of natural resources and reduced environmental impact (Mohanty et al., 2016).

4.7.6 Ecological Cities

The eco-city seems today to be the only cure for the disease of excessive urbanization, where it will enable the redesign of urban space making sure it is healthy, safe, cared for, alive, imaginatively ideal for people to live in (Hes & Bush, 2018). By applying the biocultural approach, we can reach the ideal urban model of the eco-city, where nature and technology are fully integrated and human creativity and productivity reach their peak, while providing a quality standard of living (Hu & Xi, 2023). The eco-city, according to Liansheng Hu and Yongqin Xi (2023), is a city in which nature interacts with urban systems and is a suitable alternative that can bring the city closer to sustainable development.

Considering that the climate crisis has many factors that play a role in the exacerbation of the environment, there are some that warrant more attention than others, according to Earth Organization. The most important ecological problems we will be forced to face as humans are (i) pollution, (ii) climate change, (iii) water scarcity, (iv) deforestation, (v) loss of biodiversity, and (vi) soil erosion and degradation, but also food waste and fast fashion. As people and as societies we owe to adopt a new ethos with attitudes and values in harmony with humanity's place in the biosphere. To achieve this city model we should create social, economic, and physical ecology in order to move as Li and Zhuang (2023) mention to overcome urban disease we need to turn from gray civilization to a green one. The coordination of these three components will lay the foundations for the creation of a healthy sustainable environment, where cities will make proper use of their resources using environmental and ecological rules.

The eco-city seems today to be the only cure for the disease of excessive urbanization where it will enable the redesign of urban space making sure it is healthy, safe, cared for, alive, imaginatively ideal for people to live in (Hes & Bush, 2018). According to Wilkes-Allemann et al. (2023), to support the design of greener, cooler, more inclusive cities from numerous challenges, fields as urban development, social and human environment, and biodiversity need transdisciplinary collaboration, engaging stakeholders in transforming urban engagement and functioning of cities. Eco–cities, healthy cities, or biophilic cities are the models for future cities, which find nature-based solutions for their problems (Wilkes-Allemann et al., 2023).

Dr. Bibri (2021) identifies the importance of the following factors in smart eco-cities: Data-driven decision-making as it is important for the use of data in urban planning and development decisions, and the need for effective data management and analysis systems. Additionally, the use of smart technologies, such as the Internet of Things (IoT) and artificial intelligence (AI), to improve the efficiency and effectiveness of urban systems and services is equally important. Moreover, a holistic approach to urban development is needed, and the importance of integrating environmental, social, and economic considerations into all aspects of urban planning and development. Bibri also identifies several challenges and opportunities in the development of data-driven smart eco-cities and sustainable integrated districts, including the need for effective governance and stakeholder engagement, the need for innovative financing models, and the need for effective communication and collaboration between different stakeholders (Bibri, 2021).

Eco-cities are urban areas that aim to integrate environmental, social, and economic sustainability principles into their design, development, and operation according to Tang (2022). The trends he identifies are environmental sustainability, social sustainability, and economic sustainability, same as the sustainable approach scholars. He argues that there are several challenges in the implementation of eco-cities, including limited financial resources, institutional barriers, and lack of public participation and suggests strategies for addressing these challenges such as public-private partnerships, community participation, and capacity building (Tang et al., 2022). Khalid and Abaas (2021) are in the same page with the above scholars and additionally, they identify several theoretical frameworks that have been used to guide the development of eco-cities, including systems thinking participatory approaches and urban metabolism. Their study suggest that by adopting a multi-dimensional and interdisciplinary approach to eco-city development, policymakers and development practitioners can promote more sustainable and resilient cities that promote social equity, environmental protection, and economic growth (Khalid & Abaas, 2021). Hergül and Göker (2021) contribute in the discussion the human-nature interaction by providing access to green spaces, parks, and natural areas, and promoting a sense of connection and responsibility to the natural world. They argue that green infrastructure, community participation, and sustainable transportation can contribute to promote sustainable development (Hergül & Göker, 2021). The importance of working with natural systems and processes in urban planning and development, and the need to integrate ecological considerations into all aspects of the planning process is stressed out by Douglas (2019) as well. He also emphasizes the importance of promoting resilience in urban areas, through the use of green infrastructure, sustainable transportation systems, and climate adaptation strategies. He also identifies several challenges and opportunities in the field of ecological planning, including the need for a more comprehensive and integrated approach to urban planning, the need for innovative financing models, and the need for effective stakeholder engagement and collaboration. The study suggests that by adopting a Design with Nature approach to ecological planning and by engaging communities and stakeholders in the urban planning and development process, policymakers and development practitioners can promote more sustainable and resilient cities that protect the natural environment while promoting social equity, environmental protection, and economic growth (Douglas, 2019). The importance of environmental sustainability in promoting energy-efficient and low-carbon urban development is the focus of Marotta et al. (2021). Renewable energy, energy efficiency, and circular economy are key factors for them. They also identify several key features of positive energy districts, including the integration of renewable energy sources, the use of energy-efficient buildings and infrastructure, and the adoption of smart technologies to manage energy consumption. The study suggests that positive energy districts are an effective approach to promoting environmental sustainability in urban areas, but also highlights the need for further research and collaboration among stakeholders to develop more effective and comprehensive strategies for implementing positive energy districts. As Marotta et al. suggested, circular economy is an approach to economic development that aims to minimize waste and maximize the use of resources through the design and implementation of closed-loop systems (Ngan et al., 2019). In the same spirit, Foster and Saleh (2021) identify circular economy principles as one of the key themes and strategies used by cities and communities. They also refer to cultural heritage preservation and the need to balance their preservation with their adaptive reuse for contemporary uses. In their study, they also focus on engaging communities in the adaptive reuse of cultural heritage, to ensure that their voices and perspectives are heard, and to promote social equity and inclusion. But they also identify several challenges and opportunities in the adaptive reuse of cultural heritage in European cities, including the need for a more comprehensive and integrated approach to urban planning, the need for innovative financing models, and the need for effective stakeholder engagement and collaboration. The study suggests that by adopting a circular economy approach to the adaptive reuse of cultural heritage in European cities, and by engaging communities and stakeholders in the urban planning and development process, policymakers and development practitioners can promote more sustainable and resilient cities that preserve cultural heritage while promoting social equity, environmental protection, and economic growth (Foster & Saleh, 2021).

4.7.7 Creative Cities

A creative city is an urban area that utilizes creativity and cultural industries as principal drivers for sustainable development. The "creative city" policy approach is ambiguous due to a lack of clarity in its definition. Landry and Bianchini provided a comprehensive definition in 1995, defining a creative city as a location where citizens use innovation to address social issues like discontent, social isolation, and mobility inequity. As Landry (2008) mentions, the Creative City represents an urgent and compelling call for imaginative action in the development and management of urban life. It demonstrates the application of creative thinking, planning, and action in addressing urban issues, and emphasizes the role of creativity in urban renewal policy formulation. Since the early 1990s, the term "creative" gained prominence in 1998, reflecting a shift toward cultural and creative products (Alsayel et al., 2022) with Florida. Florida's theory focuses on the rise of the "creative class" in Western societies, arguing that cities must compete to attract and retain creative talent. According to Florida (2002), economies are increasingly driven by knowledge, cities, and provinces with creative talent thrive. The 2002 Florida study suggests that cities with a significant creative class are more prosperous, as they foster innovation and attract investors.

Cities prioritize urban environment quality, including local amenities such as authentic neighborhoods, art galleries, a vibrant music scene, outdoor spaces, cafés, and historic architecture, and a tolerant spirit, to attract human capital and encourage unconventional thinking, fostering economic prosperity (Florida, 2002, 2005; Alsayel et al., 2022). Creativity provided much of post-industrial society with the impetus needed to revive urban and economic development, responding to the stagnation caused by the collapse of industrial society (Oliveira & Paulino, 2017). It became a key part of urban development for politicians, planners, and advocates of art and culture, influenced by progressive ideals (Whiting et al., 2022).

The last decade's major conferences, books, journal themes, and journal articles have broadened the geographical and thematic focus of this field (Borén & Young, 2013). Cities designated as "Creative" have a particular focus on a range of creative domains, including literature, film, music, crafts and folk art, design, media arts, as well as gastronomy. The integration of these creative sectors into development plans is intended to enhance social cohesion, economic growth, and cultural diversity (UNESCO).

The creative city's conceptualization was influenced by post-industrial era economic and political restructuring, driven by globalization. The era saw a shift in reality and future vision, with Western countries transforming their productive models toward a globally competitive service economy (Segovia & Hervé, 2022). The UNESCO Creative Cities Network (UCCN), established in 2004, including over 350 cities worldwide, aims to promote collaboration among cities recognizing creativity as a strategic factor for urban development. Member cities engage in various creative sectors, integrating them into urban development plans for social cohesion and economic growth (UNESCO). UCCN cities integrate creative sectors like literature, film, music, crafts, design, media arts, and gastronomy into urban development plans to promote social cohesion, economic growth, and cultural diversity.

There is substantial evidence to suggest that creative cities exert a considerable beneficial influence on local economies in a multi-faceted manner. It has been demonstrated that cities which are characterized by a high level of creativity tend to demonstrate higher levels of innovation and economic growth, as well as being perceived as vibrant places to live (Alsayel et al., 2022).

More specifically, the creative industries that encompass design, media, and the arts, serve as pivotal drivers of **economic growth**. Such enterprises generate employment opportunities and contribute to the overall economic output of the city. The correlation between economic growth and employment represents a topic of significant interest within the field of economics. Several studies have confirmed that creativity is a driver of sustainability and a wealth creator for cities, supporting the claim that creativity generates economic value (Oliveira & Paulino, 2017). Cities with a creative profile frequently prove an attractive destination for tourists seeking cultural experiences, festivals, and creative events. Such an increase in visitors has a beneficial effect on local businesses, including hotels, restaurants, and shops. As a result, we have a **revenue generation through tourism**.

The existence of a robust creative sector serves to **stimulate innovation** and **enhance the city's competitive standing**. Innovation and creativity are important for economic opportunities, social solutions, new understandings, and new ideas (Pratt, 2008). In this city environment, urban industries benefit from shared knowledge, specialized customers, suppliers, and workers, allowing firms to create new products and benefit from the diversity of urban environments (Lee & Rodríguez-Pose, 2013). Investing in creative sectors can **revitalize urban areas** by transforming abandoned or underutilized spaces into vibrant cultural hubs, attracting residents and visitors. According to Sasaki, *"it is believed that such an approach can foster comprehensive urban regeneration" (Sasaki,* 2010*)*. Urban regeneration and the creative economy are linked.

Creative cities play an important role in promoting **social cohesion and civic engagement.** Evidence shows that democracy and social cohesion are strengthened through participation in cultural activities because creative cities use culture and creativity to drive innovation and economic growth. Also, creative cities often focus on creating inclusive urban spaces that meet the diverse needs of their residents. This inclusivity promotes dialogue, mutual understanding, and a stronger sense of community, that lead to more cohesive and resilient societies as residents become more involved in local governance and community initiatives (UNESCO, 2021). Cities can promote **sustainable development** by integrating creativity into urban planning. Creative cities often emphasize environmental sustainability and cultural preservation, contributing to long-term resilience. Many of the urban changes taking place in contemporary cities have been articulated around the concepts of "cultural renewal" and "creative cities" (Casals-Alsina, 2023).

Attracting the creative class alone does not guarantee a city will become truly creative. To put this another way, for creative industries, which are "driven by the creativity, skills and talents of people," to form a cluster, there must exist an "environment" that allows creativity to develop and flourish.

The importance of multilevel cooperation and knowledge-sharing in leveraging creativity for urban development and conceiving new solutions to common challenges at the local level around the globe cannot be overstated. In this regard, the UNESCO Creative Cities Network provides an unparalleled opportunity for cities to engage in peer learning and collaborative initiatives, thereby fully capitalizing on their cultural assets and utilizing them as foundational elements for sustainable, inclusive, and balanced development across economic, social, cultural and environmental dimensions (UNESCO). As George (2020) mentions, "*…Creative cities can be accelerators of a creative economy, if these cities understand how to maximise the potential of a creative city designation as a strategic policy and organising device for sustainable development…..*"

4.7.8 Co-City Model

In his 2016 article, entitled "The CO-City: Sharing, Collaborating, Cooperating, and Commoning in the City," Christian Iaione introduces an innovative approach to urban governance through the concept of the "CO-City." Iaione proposes a transition from conventional top-down governance structures to a collaborative, polycentric framework that empowers citizens, institutions, and communities to collectively manage urban resources as commons. The proposal of a new regulatory system is put forward, one which transforms individuals into distributed nodes of collective action, thus fostering shared responsibility and problem-solving in the public interest. The article delineates design principles and methodologies for implementing this urban co-governance matrix, emphasizing the necessity for adaptive, iterative approaches to urban policy-making. Iaione's work concludes with a call for a novel research methodology to be employed in the investigation of ongoing processes of state transformation and institutional genesis at the urban level. The importance of collaborative governance in addressing contemporary urban challenges is also highlighted.

4.8 A Critical Analysis of Contemporary Cities' Models

The above typology of cities created on the basis of the EU's public policies of sustainable development demonstrates that they have a clear ideological political direction within the ideology of urban neoliberalism. Neoliberal urbanism is a dominant paradigm shaping contemporary urban development. Drawing from critical urban theory, political economy, and postcolonial studies, this chapter explores how neoliberal logic restructures cities, emphasizing market-led governance, privatization, and entrepreneurialism. It interrogates the contradictions, exclusions, and socio-spatial consequences of this model, especially within the context of sustainability discourses. Through a comparative global lens, the chapter highlights how neoliberal urbanism affects cities in both the Global North and South and concludes by exploring alternative frameworks aimed at post-neoliberal urban futures centered on justice, democracy, and ecological sustainability.

The advent of neoliberal urbanism has redefined the political economy of cities across the globe. Originating from broader neoliberal reforms in the late twentieth century, this paradigm advocates for privatization, deregulation, austerity, and the entrepreneurial transformation of the city. Cities are no longer governed primarily as sites of collective welfare but are increasingly managed as competitive, investment-friendly spaces. This paragraph critically unpacks the ideological underpinnings, operational mechanisms, and spatial consequences of neoliberal urbanism. Neoliberalism, as described by Harvey (2005), is a political-economic philosophy promoting free markets, limited government, and individual entrepreneurship. It emerged as a response to Keynesian welfare-state models and found global prominence through the policies of Thatcher and Reagan in the 1980s. In cities, this ideology translates into policies that commodify urban space and transfer planning responsibilities from the public to the private sector. Harvey (1989) noted the shift from urban managerialism to entrepreneurialism. Cities now compete for investment, branding themselves through megaprojects, cultural events, and public-private partnerships. This entrepreneurialism reshapes governance, emphasizing economic growth over redistribution or equity (Peck, 2001). Neoliberal urbanism, despite its promises of efficiency and innovation, this 2024 review synthesizes smart city research, revealing gaps in evaluating inclusivity and equity in urban technology implementations. Sternberg (2023) delves into how neoliberal urban governance shapes urban ideologies and discourses, embedding capitalist values into spatial practices. In addition, Marcuse (2021) explores the role of public investment in enabling gentrification, emphasizing structural and anticipatory forms of displacement. Cocola-Gant (2019) analyzes how neoliberal urban policies catalyze gentrification and displacement, especially through public-private redevelopment projects.

In the same point of view, Clark (2020) highlights how smart city technologies, while often touted as innovative, tend to reinforce existing socio-economic disparities and governance hierarchies. Also, Anguelovski et al. (2023) critique smart city and greening initiatives, arguing that they often favor elite interests while marginalizing vulnerable groups. The authors call for the integration of social equity into urban sustainability policies. In this context, urban land and housing have increasingly become financial assets. The rise of real estate investment trusts (REITs) and speculative housing markets has led to displacement, gentrification, and housing crises (Aalbers, 2016). Housing is treated as a commodity, not a right. Public services and spaces are increasingly privatized or managed under corporate logics. Urban parks, transit systems, and utilities are operated through private entities. This erodes public accountability and reinforces inequalities in access. Neoliberal urbanism often involves governance rescaling—shifting powers to regional development authorities or special economic zones, which often lack democratic oversight. These bodies prioritize market logic over local needs (Brenner, 2004). Environmental sustainability is often co-opted into market-friendly initiatives such as eco-districts, green buildings, and carbon markets. These strategies serve elite consumption patterns and fail to address the root causes of ecological degradation (Checker, 2011).

4.8.1 Socio-Spatial Consequences

Urban sustainability initiatives have unintended social consequences. Revitalization projects in inner-city areas, framed under the guise of environmental sustainability, have accelerated gentrification. These developments tend to displace long-term, low-income residents, as wealthier populations move in, altering communities and exacerbating inequalities (Smith, 1996). Austerity urbanism is a further issue. Under fiscal pressure, municipal governments have reduced spending on essential services. These cuts have led to the deterioration of public infrastructure and have disproportionately impacted on marginalized groups who rely heavily on these services (Peck, 2012). At the same time, urban design has become increasingly exclusionary. Privatized, securitized public spaces—often referred to as "fortress urbanism"—restrict access for non-consumers, especially the homeless and racialized individuals, reinforcing patterns of socio-spatial segregation (Davis, 2006). Globally, similar patterns emerge, shaped by postcolonial dynamics. In cities such as Nairobi, Mumbai, and Jakarta, neoliberal development policies manifest through slum demolitions, mega-infrastructure projects, and smart city plans that prioritize elite interests over equitable urban development. These interventions often reproduce colonial-era spatial hierarchies, privileging mobility and access for a select few while displacing others (Roy, 2009). Informal settlements are treated ambivalently—criminalized when they obstruct capital accumulation, but tolerated or exploited when they serve economic functions. In response, insurgent planning has emerged, as grassroots movements resist top-down models and assert alternative urban visions (Miraftab, 2009). Neoliberal urban sustainability has generated crises. While projects like urban greening or transit-oriented development are environmentally beneficial, they have also fueled speculation. This has led to "green gentrification" (Anguelovski, 2016). The emphasis on climate resilience is often technocratic, sidelining socio-political causes. Disaster recovery has become an opportunity for privatization and land grabs, justified by narratives of "building back better" (Klein, 2007).

4.8.2 Case Studies

Bologna, Italy: The Co-City in Practice
Bologna is widely recognized for operating the Co-City model through its "Regulation on the Collaboration between Citizens and the City for the Care and Regeneration of Urban Commons" (2014). This framework enables citizens, civil society organizations, and the municipality to enter formal pacts of collaboration to manage and co-produce public services and spaces. The initiative reflects a shift from a consumption-based to a commons-based urban paradigm, embodying principles of inclusion, sustainability, and shared governance (Foster & Iaione, 2016).

Vancouver, Canada: Eco-City Leadership and Challenges
Vancouver's Greenest City Action Plan (GCAP), launched in 2011, positions it as a global leader in ecological urbanism. The plan includes strategies for reducing carbon emissions, enhancing green spaces, and improving urban biodiversity (City of Vancouver, 2012). While Vancouver has made measurable progress on environmental indicators, critics argue that these initiatives have also contributed to rising property values and affordability crises—demonstrating the risk of green gentrification (Anguelovski, 2016). Vancouver's case highlights the tension between ecological ambition and social equity in the implementation of eco-city frameworks.

Amsterdam, Netherlands: Circular City and Integrated Sustainability
Amsterdam has positioned itself at the forefront of circular urbanism. Through its Circular Strategy 2020–2025, the city aims to shift from a linear "take-make-waste" model to a circular economy by focusing on sustainable construction, material reuse, and local production (City of Amsterdam, 2020). The city integrates ecological goals with strong social programs, including affordable housing, participatory budgeting, and co-housing initiatives. However, challenges remain in scaling these practices across all districts and balancing global investment with community-led priorities.

Together, these models and case studies underscore the potential and limitations of post-neoliberal urban imaginaries that prioritize both environmental and social sustainability.

Neoliberal urbanism, despite its promises of efficiency and innovation, has deepened urban inequalities, eroded democratic governance, and commodified essential public goods. Its contradictions are most evident in the context of sustainability, where market-based solutions often exacerbate the very problems, they purport to solve. Moving toward post-neoliberal futures requires bold reimagining of urban governance—rooted in justice, solidarity, and ecological stewardship.

4.8.3 *Neoliberal Urbanism and the Biopolitics of the City Models: A Foucauldian Critique*

4.8.3.1 Clarification on the Selective Analytical Focus within Neoliberal Urbanism

While this study introduces several city models, including the Compact City, Eco-City, Co-City, and Fifteen-Minute City, the critical analysis within the context of neoliberal urbanism concentrates on the Smart City, Creative City, and Green City frameworks. The selection is grounded on explicit theoretical, empirical, and ideological criteria that justify their analytical relevance within contemporary urban discourse.

(a) ***Prevalence and Global Influence***

The Smart, Creative, and Green City models have emerged as the most dominant paradigms of twenty-first-century urban policy. These models are consistently promoted through global institutional frameworks such as the European Union's Urban Agenda, the OECD's Smart Cities Programme, the UNESCO Creative Cities Network (UCCN), and the World Bank's Global Urban Resilience and Land Initiatives. Their prevalence within international policy discourse makes them critical sites for examining how neoliberal logic is embedded in global urban governance (Harvey, 2005; Peck, 2012; Sternberg, 2023).

(b) ***Alignment with Neoliberal Rationalities***

Each model represents a distinct but interconnected mechanism of neoliberal urban governance:

The Smart City (OECD, 2020; Mohanty et al., 2016) advances technological managerialism and data-driven governance, privileging efficiency, control, and market-led innovation.

The Creative City (Landry, 2008; Florida, 2002) translates culture and creativity into economic capital, valorizing entrepreneurship and aesthetic productivity.

The Green City (Anguelovski et al., 2018; de Godoy et al., 2024) reflects market-friendly ecological modernization, where sustainability efforts are often tied to real estate valorization and exclusionary development.

Together, these models encapsulate the ideological triad of neoliberal urbanism—technology, creativity, and sustainability—through which market rationalities penetrate urban policy and planning (Harvey, 1989; Brenner, 2004; Checker, 2011).

(c) ***Theoretical Suitability for Foucauldian Biopolitical Analysis***

This selective approach is also justified by the models' compatibility with Michel Foucault's (1978, 2007, 2008) framework of biopolitics and governmentality. These paradigms exemplify the subtle mechanisms of biopolitical regulation, the management of life, conduct, and desire through governance rather than coercion:

The Smart City functions through algorithmic surveillance and behavioral modulation (Kitchin, 2014; Clark, 2020).

The Creative City operates through aesthetic and cultural governance, shaping subjects as entrepreneurial and innovative citizens (Florida, 2002; Landry, 2008).

The Green City imposes ecological discipline via environmental metrics and sustainable lifestyle norms (Luke, 2013; Anguelovski et al., 2023).

By contrast, models such as Compact City or Co-City emphasize spatial density and cooperative governance, providing less theoretical resonance with Foucauldian notions of population management and neoliberal rationalities.

(d) ***Empirical and Comparative Relevance***

The chosen models correspond directly to the case studies discussed in Sect. 4.5.2:

Barcelona—a leading example of Green Urbanism, illustrating both environmental innovation and green gentrification (Anguelovski, 2016).

Toronto—representing Smart Urbanism, where digital infrastructure reveals tensions between optimization and surveillance (Clark, 2020; Kitchin, 2014).

Cape Town—a Creative City case, exposing cultural commodification and sociospatial displacement (Marcuse, 2021; Cocola-Gant, 2019).

These global cases provide empirical grounding for analyzing the biopolitical dynamics of neoliberal urban models, demonstrating how global ideologies translate into local spatial and social consequences.

The Smart, Creative, and Green City models were selectively analyzed because they are (a) globally influential, (b) emblematic of neoliberal urban logic (marketization, competitiveness, privatization), and (c) theoretically relevant to a Foucauldian critique of biopolitical governance.

Other city models were introduced for typological completeness, but these three best expose how neoliberal urbanism reshapes urban life through technology, creativity, and sustainability discourses.

In summary, the Smart, Creative, and Green City models were selected for critical analysis due to their: Global institutional dominance and policy relevance, direct embodiment of neoliberal governance rationalities, theoretical alignment with Foucauldian concepts of biopower and governmentality and strong empirical grounding in comparative urban case studies.

Other models (e.g., Compact, Eco-, or Co-City) are acknowledged for typological completeness but do not exhibit the same level of neoliberal entanglement or biopolitical sophistication. The chosen triad therefore provides the most analytically productive lens through which to interrogate how neoliberal urbanism restructures urban life through innovation, creativity, and sustainability discourses.

This part of the study offers a critical analysis of the models of the Smart City, Creative City (the most popular models), and Green City within the context of neoliberal urbanism, using Michel Foucault's concept of biopolitics and governmentality. These models, while framed as progressive responses to urban challenges, serve as technologies of control—structuring urban life through metrics, aesthetics, and behavior. Through the lens of biopolitical theory, this part explores how cities become laboratories for optimizing and regulating populations, revealing the ideological underpinnings and disciplinary power embedded in modern urbanism. In the wake of global urban transformations, models such as the Smart City, Creative City, and Green City have become dominant visions for urban futures. While framed as responses to crises of sustainability and governance, these models are not ideologically neutral. Rather, they function within the paradigm of neoliberal urbanism, where market rationality and competitiveness are valorized. This part critically interrogates these urban models using Michel Foucault's biopolitics, arguing that they represent sophisticated forms of urban governmentality (Foucault, 2008).

Foucault (1978) distinguished between sovereign power and biopower—a form of modern power that regulates life through mechanisms such as health, productivity, and spatial organization. In his lectures on governmentality, he described the rise of liberalism and neoliberalism as rationalities of rule that operate through shaping desires and capacities (Foucault, 2007). Biopolitics thus offers a lens for understanding how cities govern populations not by force, but through the management of life itself (Rose, 1999).

On the other hand, neoliberal urbanism is not just a political-economic strategy but a biopolitical one. It constructs the city as a site of population regulation through data, design, and policy. Urban planning becomes a mechanism of life management, where citizens are transformed into entrepreneurial and ecological subjects (Foucault, 2008; Lemke, 2011).

a. The Smart City: Data, Discipline, and Algorithmic Governance

   Smart Cities promise efficiency through sensors, big data, and predictive analytics. These technologies represent infrastructures of biopolitical governance—modulating behaviors in real time (Kitchin, 2014). Citizens are reconstituted as data subjects who must self-regulate to align with urban metrics (Clark, 2020; Han & Kim, 2024).

b. The Creative City: Aesthetic Governance and Human Capital

   The Creative City paradigm reconfigures urban governance around innovation and cultural capital (Landry, 2008). It governs through aesthetic biopolitics, valuing lives based on creative output and entrepreneurialism (Florida, 2002). Those outside this framework—migrants, racialized bodies—are often excluded or displaced (Cocola-Gant, 2019; Marcuse, 2021).

c. The Green City: Ecological Governmentality and Sustainable Discipline

Green City strategies appear progressive but function as ecological governmentality—disciplining citizens through environmental metrics and behavioral nudging (Luke, 2013; Anguelovski et al., 2023). These projects can trigger green gentrification, where environmental upgrades lead to displacement (Checker, 2011).

Each model contributes uniquely to neoliberal governance (Sternberg, 2023). Despite their pervasiveness, these models are not unchallenged. Insurgent urbanism, counter-conducts, and grassroots movements offer alternative logics of urban life (Miraftab, 2009; Escobar, 2008). These movements emphasize care, commoning, and ecological pluralism over control and optimization.

To ground the theoretical critique in empirical reality, this paragraph examines three urban case studies that exemplify the biopolitical dynamics of the Smart, Creative, and Green City models. These cases highlight how Foucauldian mechanisms of surveillance, optimization, and exclusion manifest spatially and socially in contemporary urban contexts.

i. **Barcelona and the Biopolitics of Green Urbanism**

Barcelona has positioned itself as a leader in sustainable urban planning through initiatives like the "superblocks" (superilles), which reorganize traffic to create pedestrian-friendly zones. While these projects reduce pollution and encourage public life, studies (Anguelovski et al., 2023) show they have also contributed to green gentrification-raising property values and displacing low-income residents. Environmental upgrades are thus coupled with biopolitical mechanisms of spatial cleansing and behavioral regulation.

ii. **Toronto's Sidewalk Labs and Smart Surveillance**

In 2017, Sidewalk Labs (a Google subsidiary) proposed a data-driven smart neighborhood on Toronto's waterfront. The project promised technological innovation, but faced massive backlash over data privacy, governance, and corporate control. The backlash revealed how Smart City infrastructure not only enables urban optimization but also raises questions about algorithmic governance, surveillance, and digital sovereignty (Clark, 2020; Kitchin, 2014).

iii. **Cape Town's Creative District and Cultural Displacement**

Cape Town's Woodstock neighborhood has become a creative hub, with art galleries, co-working spaces, and boutique developments. However, this transformation has led to cultural erasure and forced evictions of long-standing communities, particularly black and working-class residents. The Creative City model here operates through aesthetic governance that privileges "innovative" bodies while rendering others disposable (Cocola-Gant, 2019; Marcuse, 2021).

These case studies affirm that Smart, Creative, and green urbanism do not exist in isolation from biopolitical rationalities. Instead, they represent modes through which neoliberal power is spatialized, disciplining urban populations through data, aesthetics, and ecological norms.

Clearly, Smart, Creative, and Green Cities are not neutral responses to urban crises, they are biopolitical technologies rooted in neoliberal rationalities. By using Foucault's concepts of governmentality and biopower, we see how these models regulate life under the guise of sustainability, creativity, and innovation. Critical urban futures must reclaim space for justice, plurality, and autonomy.

The expansion of megacities within a globalized society during the twenty-first century, compounded by the advent of the Fourth Industrial Revolution, has given rise to unprecedented challenges. In contemporary times, the global city has evolved into a pivotal platform for a myriad of functions, encompassing politics, economics, society, and culture (Sassen, 2005; Pooch, 2016). In the context of population fluctuations and technological advancements, material resources are subject to adaptation, ideologies and values acquire novel characteristics, and institutional arrangements undergo restructuring.

Significant challenges emerge during the process of reformulating developmental plans and objectives. These challenges are not only related to the context of these plans but also to their scope, which is influenced by the phenomenon of "globalization." In his 2003 argument, Habermas posits that the concept of globalization ought to be considered as a dynamic rather than a static phenomenon (Habermas, 2003). It facilitates the extent and intensity of movement, communication, and exchange beyond a country's geographical boundaries. In the contemporary era, it is no longer feasible for any city to adopt a legislative stance in complete isolation; as such, it is imperative that cities do not neglect their financial policies or vulnerabilities when confronting the monumental challenges arising within the context of the prevailing economic crisis, recession, and rampant globalization (Bitsani, 2008; Leontidou, 2010).

Bauman characterizes the contemporary era as a mutation in the process of transitioning from an ageing world that is fading away. The advent of globalization has precipitated the formation of a new world order. The fall of the Berlin Wall was a seminal moment in global history, precipitating a paradigm shift in which previously insular countries became open to external influences (Bauman, 1998). Economic globalization and global colonialization remain blind toward democratic deficiencies, social exclusions (Stiglitz, 2002 in Ferri, 2003; Kovel, 2007), financial inequalities, and the environmental destruction (Beck, 1992) that it creates.

The public policies of sustainable urban development so far, as presented above, focus only on the environmental and economic dimensions, secondarily on the social, so under this perspective, the existing city models that urban policies embrace are examined below.

In this framework, the concept of sustainable cities encompasses the ability to integrate environmental sustainability with economic growth. These cities are characterized by a reduction in their ecological footprint, alongside the promotion of sustainable consumption and production patterns. This integration is influenced by a range of geographical, social, economic, and cultural factors.

> The quality of life of the city's inhabitants is a key characteristic, with high levels of satisfaction being a prerequisite for the successful implementation of sustainable urban development strategies. Key factors in achieving this include the establishment of a comprehensive healthcare system, access to educational opportunities, the implementation of environmentally friendly and efficient transport systems, and the creation of comfortable and accessible green spaces for the enjoyment of all residents. Moreover, industries, businesses, and households that implement sustainability and circular economic standards—such as the reduction, reuse, and recycling of materials—are also responsible for a significant proportion of their electricity needs being met by renewable sources.

World Bank's Global Practice for Urban, Disaster Risk Management, Resilience and Land (GPURL) concept of sustainable cities and communities includes four key dimensions: First, sustainable communities are environmentally sustainable in terms of cleanliness and efficiency. Second, sustainable communities are resilient to social, economic, and natural shocks. They are well prepared for natural disasters, which are increasing in intensity and frequency due to climate change. Third, sustainable communities are inclusive communities. They involve all dimensions of society and all groups of people—including the marginalized and vulnerable—in their markets, services, and development. Finally, Sustainable Communities are competitive communities that can remain productive and create jobs for community members.

In this sense, the city of the future must have environmentally conscious residents, rehabilitation and preservation of natural spaces, local economy of the city, and renewable energy techniques.

4.8.4 Literary Urbanism and the Poetics of the City: A Critical Analysis of Italo Calvino's Invisible Cities

While the above part of the study has explored neoliberal urbanism and its manifestations through biopolitical models like the Smart, Creative, and Green City, it is important to recognize the role of literature and poetics in shaping alternative urban imaginaries. Italo Calvino's "Invisible Cities" (1972) offers a profound literary critique of the rational, technocratic, and commodified visions of the city. Through a fictional dialogue between Marco Polo and Kublai Khan, Calvino constructs 55 cities—each a metaphorical exploration of memory, desire, signs, and space.

Authors in world literature transport us through their works to cities both familiar and unfamiliar, cities and locations both exotic and mysterious. Novels of classical literature transport us to cities that no longer exist in the form described to us in literary work. Charles Dickens in his novel Great Expectations transports his readers to a foggy London, Thomas Hardy depicts his own Essex in The Mayor of Casterbridge, Charles Baudelaire transports us to his own Paris with the poem "The

Swan," Stratis Cirkas in his trilogy The Ungoverned States brings to the surface a Jerusalem, a Cairo and an Alexandria that have been lost, George Ioannou and George Seferis record the gradual transformation of Athens into the megalopolis that we all know.

We see, then, that literary texts can give us not only descriptions and representations of cities that have been lost in depths of time, but also convey to us the atmosphere, the experience of inhabiting a city. Modern geographers and sociologists often use literary texts to collect information about the relationship between man and space in a specific area.

Urban theory has long grappled with the material and political dimensions of city life. Yet literature—especially poetic and speculative fiction—offers a unique vantage point from which to reimagine urban experience. Italo Calvino's *Invisible Cities* (1972) stands as a landmark in literary urbanism, offering a series of meditations that challenge the dominant epistemologies of the city.

Calvino in Invisible Cities, as another Marco Polo, as another Scheherazade, tells us stories from exotic and distant cities. In reality, however, Calvino, with his narratives, tells us that every city is a cauldron of stories that never end. Every city is an ideal destination for the traveler and explorer who knows how to use their senses. Invisible Cities is a call to learn to read the city we live in like a book—Calvino's text is a travel guide for the nomads of the city.

"Invisible Cities" presents the urban not as a fixed entity but as a poetic and conceptual device. Cities like Zaira, Tamara, and Eusapia are layered with historical traces, symbolic meaning, and unconscious projections. These cities defy Cartesian logic and echo Henri Lefebvre's (1991) notion that space is socially produced and inherently political. Calvino's cities are palimpsests-poetic disruptions to the dominant urban discourse of order, planning, and control.

By foregrounding perception, memory, and emotion, Calvino critiques the modernist and neoliberal obsession with urban efficiency. His cities resist totality, mirroring Deleuze and Guattari's (1988) idea of the "rhizome"—a non-linear, non-hierarchical mode of thought. In this sense, "Invisible Cities" functions as a counter-cartography, challenging the epistemological violence of technocratic urbanism.

Moreover, Calvino's literary urbanism aligns with Foucauldian critiques of surveillance and normalization. Kublai Khan's imperial desire to map and control is continually frustrated by Marco Polo's elusive narratives, suggesting that urban life eludes domination. In cities like Tamara, where over-signification renders meaning opaque, Calvino prefigures Baudrillard's (1983) concept of the hyperreal.

Ultimately, "Invisible Cities" is a poetic meditation on urban life, offering an ethical and imaginative stance toward the city. It invites planners, architects, and citizens alike to approach urbanism with openness, humility, and creativity. In an age where data, aesthetics, and sustainability are instrumentalized by neoliberal governance, Calvino's cities remind us of the irreducible complexity and beauty of urban experience.

Urban theory has long grappled with the material and political dimensions of city life. Yet literature—especially poetic and speculative fiction—offers a unique vantage point from which to reimagine urban experience. Italo Calvino's "Invisible Cities" (1972) stands as a landmark in literary urbanism, offering a series of meditations that challenge the dominant epistemologies of the city. This paper situates Calvino's work within contemporary urban theory, exploring its relevance to debates on space, power, and representation.

4.8.4.1 The Structure and Themes of "Invisible Cities"

"Invisible Cities" consists of 55 brief prose descriptions of imaginary cities, embedded within a fictional dialogue between Marco Polo and Kublai Khan. These cities are organized thematically—Memory, Desire, Signs, the Dead—and serve as allegories for urban complexity. Each city becomes a poetic device, revealing the layered, affective, and symbolic dimensions of urban life. Calvino's cities resist the rationalist, master-planned vision of the modern city championed by figures like Le Corbusier. Instead, they evoke a sense of contingency and multiplicity. Cities like Zaira or Baucis cannot be understood through metrics or blueprints; they demand interpretation. This aligns with Henri Lefebvre's (1991) argument that space is socially produced and inherently political.

In cities like Tamara and Zemrude, Calvino explores the semiotic saturation of urban environments. These narratives prefigure Baudrillard's (1983) theory of hyperreality, where signs no longer refer to reality but to other signs. The city becomes a space of simulation—an urban spectacle—echoing contemporary concerns with branding, image, and media-driven identity in cities.

Although Marco Polo opens himself to cities all over the world, and not just the Mediterranean, he nevertheless has Italy at his heart, and indeed his own city with alternating identities, his homeland, Venice, which is constantly recreated in the geographical imaginations of people and cultures. The dialogue is revealing:

> "There's one left that you never talk about."
> Marco Polo bowed his head.
> "Venice," said Khan.
> Marco smiled. "And what else did you think I was talking to you about?"
> The emperor didn't even bat an eyelid. "And yet, I never heard you say her name."
> And Polo: "Every time I describe a city, I say something about Venice." (Calvino, 1972, p. 112).

Calvino's "invisible cities" is a magnet. Zenobia (pp. 57–8), Valdrada (pp. 75–6), little Venices where something is always missing from the charms of this city. One on stilts without water, the other on the water with mirror lives haunting it. Thus, the enigmatic Marco Polo seems to narrate his impressions of many cities, but he revolves around one city, as Kublai Khan slowly realizes:

> Kublai Khan had realized that Marco Polo's cities resembled each other, as if passing from one to another did not imply a journey but simply a change of their elements. (Calvino, 1971/2009, p. 65)

Calvino's treatment of memory and desire recalls the phenomenological insights of Gaston Bachelard and the psychoanalytic geographies of Freud. Cities like Zora (perfectly remembered) and Anastasia (the city of desire) represent urban space as a projection of the unconscious. The affective and subjective dimensions of the city thus take precedence over its infrastructural realities.

The framing device of Kublai Khan attempting to map and control his empire contrasts with Marco Polo's elusive and poetic descriptions. This reflects Foucault's (2007) critique of governmentality: the limits of power's ability to know and regulate life. The city, in Calvino's vision, defies totalizing logics.

Calvino's work is ultimately a call for alternative urban imaginaries, ones that resist commodification, surveillance, and rationalization. His cities suggest that the future of urban life depends not only on policy and design but on "imagination, ethics, and language." They call us to dwell poetically in the world, to embrace the unknowable, and to read the city as text and texture.

"Invisible Cities" challenges urban scholars, planners, and citizens to think differently about the city—not as a fixed object to be managed, but as a living process to be interpreted, co-created, and felt. Literature, in this light, becomes a vital component of critical urban theory, capable of revealing the contradictions, dreams, and ghostly traces that haunt every street and skyline.

While Italo Calvino's "Invisible Cities" offers a unique poetic engagement with urban space, it belongs to a broader tradition of literary works that explore the philosophical, political, and affective dimensions of the city. This section expands the discussion by examining other writers who have approached the city as a site of fragmentation, memory, resistance, and dream.

4.9 Anthropocentric Models of Cities from Another Perspective

4.9.1 The First Landmark and City Model in History Knossos: The City of Knowledge and Memory

At the heart of this reflection stands the silver stater of Knossos (Fig. 4.4), a coin that transcends its material value to become a profound symbol of civic identity and philosophical inquiry. Minted around 300–270 BC, its intricate depiction of the labyrinth—rendered through geometric precision and encircled by the inscription "ΚΝΩΣΙΟΣ"—captures not merely a mythic motif but a metaphor for the human condition. The labyrinth's meandering paths evoke the complexity of self-knowledge, civic memory, and the existential journey toward understanding. In this sense, Knossos may be seen not only as a physical city but as an early anthropocentric model of urban consciousness, where architecture, myth, and identity converge to mirror the inner labyrinth of human thought. This intersection between material culture and symbolic meaning sets the foundation for reinterpreting ancient cities as

Fig. 4.4 Knossos stater with labyrinth (silver coin) minted in Knossos, Crete around 300–270 BC. Photograph taken at TEFAF 2020. Wikipedia common his file is licensed under the Creative Commons Attribution-Share Alike 4.0 International license

mirrors of human cognition, anticipation, and collective memory—a perspective that continues to inform later models such as the Human Intercultural City.

The labyrinth is depicted with two basic geometric motifs: a square frame with concave meander patterns and a circular/uni-cursal (single-course). The inscription "ΚΝΟΣΙΟΣ"/ΚΝΟΣΙΟΝ indicates the origin of the coins and functions as a sign of civic pride and political autonomy.

Could it be that Knossos was not merely a city of stones and kings, but rather a site of inner contemplation, where the human being encountered the very limits of their nature? Perhaps the Labyrinth was not an architectural structure alone, but the mind itself, where the soul wandered, searching for an exit toward knowledge.

The resemblance of the word Knossos to the Greek gnōsis (knowledge)—although etymologically ungrounded (Beekes, 2010)—reveals, in symbolic terms, a deeper truth: this city stood as a beacon of civilization, a place of early wisdom, a locus of memory and existential inquiry.

It was here that the myth of the Minotaur was born—yet the monster may not have lived in the physical Labyrinth, but in the unknown depths of the self. And Theseus, courageous with Ariadne's thread, may not have slain a beast, but the darkness of ignorance (Jung, 1968).

The Knossians, "those of Knossos," may thus be interpreted—not linguistically but allegorically—as "those of knowledge": individuals who dared to map the unknown, whether through the design of the Labyrinth, the invention of writing, or the silent testimony of stone (Marinatos, 1993).

Knossos is no longer a mere city; it has become a symbol: of a civilization whose depth transcends surface; of memory that is not past, but illumination for the future; of knowledge that can only be attained through the prior experience of being lost.

From Knossos as the archetype of the "City of Knowledge and Memory," we move from mythic introspection to social praxis. The ancient labyrinth, once a symbol of humanity's inner quest, becomes in the modern context a metaphor for **interconnected human relations** within complex urban systems. Where the Cretan city embodied the individual's encounter with self and truth, the **Human Intercultural Model** reinterprets that encounter at the collective level—within the living fabric of contemporary multicultural cities. Here, the city is no longer a monument of stone and myth but a dynamic organism of people, cultures, and shared values. The anthropocentric essence of Knossos thus finds its modern echo in a model that seeks harmony not through isolation, but through **dialogue, recognition, and coexistence**.

4.9.2 Human Intercultural Model

Building upon the anthropocentric vision embodied by Knossos—the city as a reflection of human consciousness and collective identity—the Human Intercultural Model extends this paradigm into the realm of contemporary urban theory, where cities are understood as living systems shaped by interaction, diversity, and shared human values.

To understand this model, it is first necessary to define the meaning of the "identity of a city" from an anthropological, political, administrative, and sociological perspective.

The concept of identity, according to Bitsani (2008, 2013b) and other researchers, relates to three fundamental ideas: unity (meaning that everyone should be able to recognize the same and the same in the same), stability (recognizing space), and imagination (memories and collective memory). However, each of these three ideas presents significant challenges. In actuality, stability does not exist, unity is never easy to achieve, and it is not always evident when similarities are recognized (Bitsani et al., 2003; Althusser, 2006; Elias, 2009).

The city's intercultural identity is based on intercultural coexistence, social solidarity, intercultural relations with "interactive dialogue" and socio-political participation and active civil society. Rousseau was among the first to observe that dependence on others within the framework of a structured state does not imply dependence on political power, nor does it imply any other kind of dependence that presupposes the survival or achievement of a specific purpose, though it may represent the desire one has to be respected by others (Sakellaropoulos, 2004; Chomsky, 2013a). According to Alex Honneth's theory of recognition, the perception of man's intellectual development is closely linked with the perception of social participation and with the changes man undergoes (Alex Honneth, 2007). Francis Fukuyama notes, referring to Plato, that anger "as a proud aspect of the human personality" demands that our worth and dignity is recognized by others. (Fukuyama, 1995).

In this sense, the city's human identity derives from the Universal Declaration of Human Rights, from its inherent ontological characteristics (humanity), and from its cultural values and dignity, forming a notion of symbiosis—that is, the human being in harmony with the environment.

The city is recognized as "the greatest symbiotic community of all," which means it is home to multiple kinds of symbiotic relationships. It is the best in terms of both quality and content. Furthermore, the "common good" is its primary goal.

In constructing the human-intercultural model, we can recognize, at the first level, the basic realities and conceptions of space. With that in mind, a conceptual model was created. This level corresponds to the ontology of space, and the model formed is the ontological model that describes the basic realities and definitions of space, as well as how they interact.

At the second level, which concerns logical modelling, the focus shifts to concepts, their relations, and their interactions within the relevant domain.

Though its description is more specialized and is based on the systemic methodology and Parson's theory, specifically in relation to social relationships and the initial recognition of individuals, the definition of fundamental realities and conceptions of space is considered. This involves conceptual modeling entities and logical modeling at the second level.

At the third level, political relationships are examined. The political level is based on the logical level of policies, laws, measures, and associations. Therefore, the logical model enhances, enriches, and progresses the political level. The political level is grounded in the logical level, and it transforms the logical model, which portrays concepts and definitions of logical organization, into a model that addresses the institutionalized organization of a governmental and administrative system in practice.

As Bitsani (2013a) notes, "…the systemic ontological approach emphasises the city's multifactorial and multidimensional character. The system is more than the sum of its parts; it is an organic entity that is qualitatively distinct from the constituent parts. As a result, the city is a system, a collection of interconnected elements that influence and are influenced by one another. When a particular element within the system changes, the entire system changes. This system is defined as an aggregate of: (a) the members/entities, (b) the entities' characteristics (e.g., political, claims, actions, etc.), and (c) the members'/entities' relations…" City's intercultural identity allows us to reap the benefits (or learn from the drawbacks) of coexistence and shared historical experience, which cannot be replaced by any other product in the global cultural superstore (Bitsani, 2013a).

Finally, the requirement in this city model is to define "excellence" in the analysis of concepts and all that they imply. The specific properties and qualities that a concept may have are determined by the needs that a specific model must address, as well as global environmental conditions. The goal is to create a flourishing environment in which a system of harmonious coexistence operates in such a way that all needs are met. This duty/achievement was rightfully attributed to the Classical period's open, human-centered Greek city. People and "active citizenship" are inextricably linked to the city.

The Human Intercultural City Model grounds its identity and policy in the recognition that diversity can serve as a resource for societal advancement. Adopting (and putting into practice) policies that support constructive cross-cultural interactions and exchanges as well as equitable and proactive community and resident participation in city development are the first steps toward meeting the demands of a varied populace.

The concept of synergy proposes that control over ecosystems ultimately eludes us. Furthermore, the close link emphasized between human perception and behavior stipulates that a transformation of the entire cultural context is required for changes in human behavior to occur (Chomsky, 2013a). It is within this cultural context that the system of values governing relations and behavior arises (Chomsky, 2013a, 2013b). The question of how much control we have over the ecological and cultural systems as a society remains open.

In order to link our approach with a general anthropological theoretical discussion, we are evidently talking about an incidence of contested space. We may also recall Foucault (2002), who expounds on *heterotopias*, "Spaces where cultural attractions are represented become objects of negotiation." Historical villages are included in these heterotopias, according to M. Foucault (2002). Not only do practices like this perpetuate economic exploitation and allow dominant groups to exploit "urban nostalgia" in pursuit of profit from what is seen as genuine (whether natural or cultural), but they also conceal the ideologies that underlie management and organization (Foucault, 2002; Chomsky, 2013a, 2013b; Touraine, 2012; Habermas, 2003).

The Human Intercultural City Model represents a paradigm shift in urban governance and diversity management. The model emphasizes diversity as an asset and intercultural dialogue as a cornerstone of social cohesion. At this point, this paragraph critically evaluates the strengths and limitations of the model, comparing it with traditional assimilationist and multiculturalist approaches. It concludes that while the model offers a promising framework for inclusive urban development, its implementation requires strong political will, structural reform, and contextual adaptation. Contemporary cities are increasingly diverse due to global migration, economic interdependence, and political displacement. In this context, the Human Intercultural City Model emerges as a response to the limitations of both assimilationist and multiculturalist approaches. The model envisions cities as agents of change that actively foster equality, diversity, and interaction among all residents. This paper provides a critical analysis of the model, highlighting its practical tools, normative foundations, and challenges in implementation. The Human Intercultural City Model is underpinned by three interrelated principles: equality, diversity, and interaction (as Council of Europe, 2020 adopted). Unlike assimilationist models that require conformity to a dominant culture, or multiculturalist models that may tolerate but segregate cultural differences, the intercultural model promotes active engagement across cultural lines. It seeks to build a shared civic identity through dialogue and mutual recognition.

4.9.2.1 Strengths of the Model

Holistic and Human-Centered Approach: The model places individuals and communities at the heart of urban policymaking. It emphasizes not just legal equality but also cultural belonging, civic participation, and interpersonal trust.

Practical Implementation Tools: The Intercultural Cities Index and self-assessment tools provide cities with benchmarks and guidance for policy development (Council of Europe, 2020). These tools translate abstract principles into actionable strategies.

Promotion of Social Innovation: By framing diversity as a resource, the model fosters innovation in education, governance, and public services. It supports inclusive entrepreneurship and culturally responsive public spaces.

Conflict Prevention: Intercultural policies facilitate dialogue, reduce prejudice, and mitigate intergroup tensions. The model anticipates and addresses potential sources of conflict through proactive engagement.

4.9.2.2 Weaknesses

Idealism Versus Political Reality: The model's aspirational tone may not align with political and economic constraints faced by municipalities. Populist rhetoric, austerity measures, and restrictive immigration policies can undermine its implementation.

Implementation Gaps: There is a risk of superficial adoption where policies are implemented symbolically rather than substantively. Without genuine institutional commitment, the model may become a checkbox exercise.

Local-National Disconnection: The model concentrates on municipal governance but often overlooks the influence of national or supranational frameworks that shape local possibilities.

The Human Intercultural City Model is a progressive and holistic framework for managing diversity in urban settings. It transcends tolerance to promote meaningful interaction and shared citizenship. In addition, Bitsani said (2013a) identified the lack of connection between policies and the environment, also identified the immediate necessity of incorporating ecological and environmental parameters into its model, and left the issue open for further research, a fact that is now being realized with the present study and research.

Nevertheless, its success depends on the willingness of city leaders, institutional support, and alignment with broader policy environments. Future research should explore how the model can be adapted to varying political and socio-economic contexts.

In conclusion, the Human Intercultural City Model stands as a progressive and integrative framework for managing diversity in urban environments. It moves beyond tolerance to cultivate active citizenship, dialogue, and shared belonging. As Bitsani (2013a) identified, earlier models lacked integration of ecological and environmental dimensions—an omission now being addressed in current research.

The model's success ultimately depends on political will, institutional support, and the alignment of local and global policy frameworks. Future inquiry should focus on its contextual adaptability, exploring how intercultural principles can be effectively embedded across varying political, economic, and cultural settings. Through such continued evolution, the intercultural city may fully embody humanity's enduring pursuit of coexistence, dignity, and collective flourishing.

4.10 Conclusion

In summarizing the issues discussed in this chapter, it becomes evident that the interdependence and complexity of sustainable development priorities—combined with their value-laden nature—situate them within the realm of polycrisis and wicked policy problems. This is particularly apparent in the way proposed solutions often collide with competing economic and social interests. The diversity and distinct characteristics of large urban centers further intensify these tensions, rendering uniform, one-size-fits-all policy approaches ineffective.

Within this context, the typology of contemporary cities, encompassing models such as the Smart City, Creative City, and Green City, illustrates distinct governance strategies for managing complexity in the polycrisis era. While these paradigms aim to advance sustainability, innovation, and cultural vitality, they frequently operate within neoliberal frameworks that prioritize efficiency and market competitiveness over inclusivity and equity. Consequently, they risk reproducing socio-spatial inequalities under the guise of progress. In contrast, emerging alternatives such as the Co-City, Eco-City, and Human Intercultural City advocate for participatory, human-centered, and ecologically grounded approaches, reframing resilience as a collective and ethical endeavor rather than a technocratic one.

Enhancing the design of public policies for sustainable urban development therefore requires innovative and integrative strategies that address wicked challenges across interconnected domains. This includes the coordinated planning of green and digital transitions, education for digital literacy, and cross-sectoral collaboration involving both societal and market actors. Addressing complex policy issues also demands interdisciplinary cooperation—particularly between the social and economic sciences—to develop context-specific and adaptive solutions.

Moreover, because many of these challenges stem from conflicting values and interests rather than purely technical issues, systematic public dialogue remains essential. Awareness-raising, inclusive participation, and transparent communication can help bridge divisions, although policymakers must recognize that deep structural conflicts often limit the potential for full consensus.

Ultimately, the governance of sustainable urban futures depends on moving beyond narrow economic logics toward integrated, participatory, and justice-oriented frameworks. Cities that embrace ethical, intercultural, and collaborative models of development will be better equipped to navigate the uncertainties of the polycrisis while fostering equitable and resilient communities.

Overall, Chap. 4 transforms the philosophical and contextual insights of the previous chapters into a pragmatic framework for empirical investigation. It reframes the city as a site of overlapping crises and opportunities that can only be understood through adaptive, participatory governance. By articulating how global and European policy agendas translate into local action, the chapter bridges the conceptual architecture of Part I with the methodological framework of Part II. The subsequent chapters apply this integrated governance lens to the case of Ixelles, examining how biocultural sustainability is enacted, negotiated, and experienced within a real urban setting.

Conflicts of Interest The authors have no conflicts of interest to declare that are relevant to the content of this chapter.

References

Aalbers, M. B. (2016). *The financialization of housing: A political economy approach* (1st ed.). Routledge. https://doi.org/10.4324/9781315668666

Adelman, S. (2018). Chapter 2: The sustainable development goals, anthropocentrism and neoliberalism. In *Sustainable development goals*. Edward Elgar. Retrieved July 28, 2025, from https://doi.org/10.4337/9781786438768.00008

Agora Energiewende. (2023). *How Europe can make its power market more resilient: Recommendations for a short-term reform*. https://www.agora-energiewende.org/publications/how-europe-can-make-its-power-market-more-resilient

Ahmed, I. (2020). The Basel convention on the control of transboundary movements of hazardous wastes and their disposal: A legal misfit in global ship recycling jurisprudence. *Washington International Law Journal, 29*(2), 411–446. https://digitalcommons.law.uw.edu/wilj/vol29/iss2/7

Alexandris Polomarkakis, K. (2019). The European pillar of social rights and the quest for EU social sustainability. *Social & Legal Studies, 29*(2), 183–200. https://doi.org/10.1177/0964663919829199. (Original work published 2020).

Allam, Z., Bibri, S. E., Chabaud, D., & Moreno, C. (2022a). The theoretical, practical, and technological foundations of the 15-minute City model: Proximity and its environmental, social and economic benefits for sustainability. *Energies, 15*(16), 6042. https://doi.org/10.3390/en15166042

Allam, Z., Jones, D. S., & Sharifi, A. (2022b). Smart cities and the SDGs: From concept to practice. *Sustainable Cities and Society, 87*, 104–164. https://doi.org/10.1016/j.scs.2022.104164

Alsayel, A., Jong, W. M., & Fransen, J. (2022). Can creative cities be inclusive too? How do Dubai, Amsterdam and Toronto navigate the tensions between creativity and inclusiveness in their adoption of city brands and policy initiatives? *Cities, 128*, 103786. https://doi.org/10.1016/j.cities.2022.103786

Althusser, L. (2006). *Philosophy of the encounter: Later writings, 1978–1987* (G. M. Goshgarian, Trans. & Ed.). Verso.

Ameli, N., Chenet, H., Falkenberg, M., Kothari, S., Rickman, J., & Lamperti, F. (2025). Driving sustainability transitions through financial tipping points. *Earth System Dynamics, 16*, 333–345. https://doi.org/10.5194/esd-16-333-2025

Anguelovski, I. (2016). From toxic sites to parks as (green) LULUs? New challenges of inequity, privilege, gentrification, and exclusion for urban environmental justice. *Journal of Planning Literature, 31*(1), 23–36. https://doi.org/10.1177/0885412215610491

Anguelovski, I., Connolly, J., & Brand, A. L. (2018). From landscapes of utopia to the margins of the green urban life: For whom is the new green city? *City, 22*(3), 417–436. https://doi.org/10.1080/13604813.2018.1473126

Anguelovski, I., Honey-Rosés, J., & Marquet, O. (2023). Equity concerns transformative planning: Barcelona's superblocks are under scrutiny. *Cities & Health, 7*(6), 950–958. https://doi.org/10.1080/23748834.2023.2207929

Arauzo-Carod, J. M., Kostakis, I., & Tsagarakis, K. P. (2022). Policies for supporting the regional circular economy and sustainability. *The Annals of Regional Science, 68*, 255–262. https://doi.org/10.1007/s00168-022-01124-y

Banister, D. (2008). The sustainable mobility paradigm. *Transport Policy, 15*(2), 73–80. https://doi.org/10.1016/j.tranpol.2007.10.005

Barton, H., Grant, M., & Guise, R. (2003). *Shaping neighbourhoods: A guide for health, sustainability and vitality*. Routledge.

Baudrillard, J. (1983). *Simulacra and simulation* (S. F. Glaser, Trans.). University of Michigan Press. (Original work published 1981)

Bauman, Z. (1998). On glocalization: Or globalization for some, localization for some others. *Thesis Eleven, 54*(1), 37–49. https://doi.org/10.1177/0725513698054000004. (Original work published 1998).

Beck, U. (1992). *Risk society: Towards a new modernity*. Sage.

Beekes, R. S. P. (2010). *Etymological dictionary of Greek* (Vol. 1).

Bibri, S. E. (2021). Data-driven smart eco-cities and sustainable integrated districts: A best-evidence synthesis approach to an extensive literature review. *European Journal of Futures Research, 9*, 16. https://doi.org/10.1186/s40309-021-00181-4

Birnie, P., Boyle, A., & Redgwell, C. (2009). *International law and the environment* (3rd ed.). Oxford University Press.

Bithas, K. (2001). The word 'society' and its roots in ancient Greek: 'Socius' as 'to share'. In *Proceedings of the 3rd International Conference on Urban Studies, Athens, Greece, 15–17 November 2001*.

Bitsani, E. (2008). *Social services and management in multicultural societies: Contemporary approach to intercultural relations and intercultural education. The case of Trieste in Italy*. Annual Report, University of L'Aquila.

Bitsani E. (2013a). *"Intercultural city identity" and "human intercultural cities" (H.I.C). An ontological/dynamic model for the social cohesion of modern and postmodern cities*. The research example of Trieste Italy" Nova Publishers.

Bitsani, E. (2013b). *"Intercultural city identity" and "Human Intercultural Cities (H.I.C.): A dynamic ontological model" for the social co-existence and social cohesion of modern and post-modern cities*. https://www.researchgate.net/publication/290364147_Intercultural_city_identity_and_Human_Intercultural_Cities_HIC_A_dynamic_ontological_model_for_the_social_co-existence_and_social_cohesion_of_modern_and_post-modern_cities

Bitsani, E., Bulles, I., & Panagou, V. (2003). Greece a modern multicultural society and the role of local government in the new conditions. In T. Papatheodosiou (Ed.), *Essays in memory* (pp. 361–393). Interbooks.

Bodansky, D. (2016). The legal character of the Paris Agreement. *RECIEL, 25*, 142–150. https://doi.org/10.1111/reel.12154

Bogers, M., Biermann, F., Kalfagianni, A., & Kim, R. E. (2022). Sustainable development goals fail to advance policy integration: A large-n text analysis of 159 international organizations. *Environmental Science & Policy, 138*, 134–145. https://doi.org/10.1016/j.envsci.2022.10.002

Borén, T., & Young, C. (2013). Getting creative with the 'Creative City'? Towards new perspectives on creativity in urban policy. *International Journal of Urban and Regional Research, 37*, 1799–1815. https://doi.org/10.1111/j.1468-2427.2012.01132.x

Bosch, M. (2019). *Urban resilience in the age of crisis*. Routledge.

Brenner, N. (2004). *New state spaces: Urban governance and the rescaling of statehood*. Oxford University Press.

Brundtland Commission. (1987). *Our common future*. Oxford University Press.

Brusselaers, J., Breemersch, K., Geerken, T., et al. (2022). Macroeconomic and environmental consequences of circular economy measures in a small open economy. *The Annals of Regional Science, 68*, 283–306. https://doi.org/10.1007/s00168-021-01079-6

Burton, E., Jenks, M., & Williams, K. (2000). *Achieving sustainable urban form* (1st ed.). Routledge. https://doi.org/10.4324/9780203827925

C40 Cities Knowledge Hub. (2020). C40 City climate action planning monitoring, evaluation and reporting (MER) guidance. C40 Cities. https://www.c40.org/wpcontent/uploads/2021/08/C40-MER-Guidance.pdf

Calvino, I. (1972). *Le città invisibili*. Einaudi.

Calvino, I. (2009). *Invisible cities* (W. Weaver, Trans.). Vintage Books. (Original work published 1971)

Campbell, S. D., & Zellner, M. (2020). Wicked problems, foolish decisions: Promoting sustainability through urban governance in a complex world. *Vanderbilt Law Review, 73*(6), 1643–1690. https://scholarship.law.vanderbilt.edu/vlr/vol73/iss6/3

Casals-Alsina, È. (2023). Can the 'creative city' be sustainable? Lessons from the Sant Martí district (Barcelona). *International Journal of Cultural Policy, 30*(4), 511–528. https://doi.org/10.1080/10286632.2023.2222770

Cavada, M., Hunt, D.V.L. & Rogers C.D.F. (2014). Smart cities: Contradicting definitions and unclear measures. In *Conference: 4th world sustainability forum*. https://doi.org/10.13140/2.1.1756.5120

Cervero, R. (1996, September). Mixed land-uses and commuting: Evidence from the American Housing Survey. Transportation Research Part A: Policy and Practice, Elsevier, vol. 30(5), pages 361-377.

Checker, M. (2011). Wiped out by the "Greenwave": Environmental gentrification and the paradoxical politics of urban sustainability. *City & Society, 23*(2), 210–229. https://doi.org/10.1111/j.1548-744X.2011.01063.x

Chomsky, N. (2013a, June 4). *Humanity imperiled: The path to disaster*. TomDispatch.com.

Chomsky, N., (2013b, March 5). *Can civilization survive capitalism?* Alternet.

City of Amsterdam. (2020). *Amsterdam circular strategy 2020–2025*. https://www.amsterdam.nl/en/policy/sustainability/circular-economy/

City of Vancouver. (2012). *Greenest City 2020 Action Plan*. https://vancouver.ca/green-vancouver/greenest-city-action-plan.aspx

Claringbould, D., Koch, M., & Owen, P. (2019). Sustainable finance: The European Union's approach to increasing sustainable investments and growth—Opportunities and challenges. *Vierteljahrshefte zur Wirtschaftsforschung, 88*(2), 11–27. https://doi.org/10.3790/vjh.88.2.11

Clark, J. (2020). *Uneven innovation: The work of smart cities*. Columbia University Press.

Cocola-Gant, A. (2019). Gentrification and displacement: Urban inequality in cities of late capitalism. In *Handbook of gentrification studies* (pp. 49–63). Edward Elgar. https://doi.org/10.4337/9781785364600.00030

Conklin, J. (2006). *Wicked problems & social complexity*. CogNexus Institute.

Convention on Biological Diversity. (2022). *COP 15 decision 04: Kunming-Montreal global biodiversity framework*. https://www.cbd.int/doc/decisions/cop-15/cop-15-dec-04-en.pdf

Council of Europe. *Intercultural cities: Good practice examples*. https://www.coe.int/en/web/interculturalcities/good-pratice/-/asset_publisher/lA8MBwPd2Unw/content/intercultural-public-space

Council of the European Union. (2020). Council recommendation of 30 October 2020 on A Bridge to Jobs – Reinforcing the Youth Guarantee and replacing the Council Recommendation of 22 April 2013 on establishing a Youth Guarantee (2020/C 372/01). Official Journal of the European Union. https://eur-lex.europa.eu/legalcontent/EN/TXT/?uri=CELEX%3A32020H1204%2801%29

Davis, M. (2006). *City of quartz: Excavating the future in Los Angeles*. Verso Books.

de Godoy, J. A. R., Benini, S. M., & da Silva, A. L. C. (2024). Green cities, gray realities: The rhetoric of sustainability and urban segregation. *RGSA-Revista de Gestão Social e Ambiental, 18*(12), e010455. https://doi.org/10.24857/rgsa.v18n12-157

DeFries, R., & Nagendra, H. (2017). Ecosystem management as a wicked problem. *Science, 356*(6335), 265–270. https://doi.org/10.1126/science.aal1950

Deleuze, G., & Guattari, F. (1988). *A thousand plateaus: Capitalism and schizophrenia*. Bloomsbury Publishing.

Douglas, I. (2019). A critical review 50 years change in urban land use and ecological planning globally in the era of design with nature. *Ecosystem Health and Sustainability, 5*(1). https://doi.org/10.1080/20964129.2019.1656111

Drexhage, J., & Murphy, D. (2010). Sustainable development: From Brundtland to Rio 2012. In *Background paper prepared for consideration by the high level panel on global sustainability at its first meeting, 19 September 2010*. UN Headquarters.

Earle, A. G., & Leyva-de la Hiz, D. I. (2020). The wicked problem of teaching about wicked problems: Design thinking and emerging technologies in sustainability education. *Management Learning, 52*(5), 581–603. https://doi.org/10.1177/1350507620974857. (Original work published 2021).

Ekardt, F., Günther, P., Hagemann, K., Helmers, L., & Schmidt, S. (2023). Legally binding and ambitious biodiversity protection under the CBD, the global biodiversity framework, and human rights law. *Environmental Sciences Europe, 35*(80). https://doi.org/10.1186/s12302-023-00786-5

Elias, N. (2009). The creed of a nominalist: Observations on popper's the logic of scientific discovery. In R. Kilminster & S. Mennell (Eds.), *Essays I: On the sociology of knowledge and the sciences* (Elias collected works) (Vol. 14, pp. 161–191). University College Dublin Press.

Enrich, B., Garcíα-Romero, M., & Patel, S. (2024). *Renewable energy development: Current landscape, impacts, and goals*. Scipedia. https://www.scipedia.com/public/Enrich_et_al_2024b

Escobar, A. (2008). *Territories of difference: Place, movements, life, Redes*. Duke University Press.

European Commission. (2019). *The European Green Deal*. https://ec.europa.eu/stories/european-green-deal/

European Commission. (2021a). Better regulation: Joining forces to make better laws (COM(2021) 219 final). Publications Office of the European Union. https://eurlex. europa.eu/legal-content/EN/TXT/?uri=COM%3A2021%3A219%3AFIN

European Commission. (2021b). *A new European Bauhaus: Shaping more beautiful, sustainable and inclusive forms of living together*. https://europa.eu/new-european-bauhaus/

European Commission. (2022). *Recovery and resilience facility: Implementation report*. https://ec.europa.eu/info/business-economy-euro/recovery-coronavirus/recovery-and-resilience-facility_en

European Commission. (2023). *A green Deal industrial plan for the net-zero age (COM(2023) 62 final)*. https://eur-lex.europa.eu/legal-content/EN/TXT/?uri=CELEX%3A52023DC0062

European Environment Agency. (2024). *Trends and projections in Europe 2024 (EEA Report No. 11/2024)*. https://www.eea.europa.eu/publications/trends-and-projections-in-europe-2024

European Fiscal Board. (2022b). Annual report 2022. Publications Office of the European Union. https://economy-finance.ec.europa.eu/system/files/2022-10/EFB_Annual_Report_2022.pdf

European Parliament & Council of the European Union. (2024). Regulation (EU) 2024/1735 of the European Parliament and of the Council of 13 June 2024 on establishing a framework of measures for strengthening Europe's net-zero technology manufacturing ecosystem and amending Regulation (EU) 2018/1724 (Text with EEA relevance). *Official Journal of the European Union, L, 202*, 1–36. https://eur-lex.europa.eu/legal-content/EN/TXT/?uri=CELEX:32024R1735

Fabius L. (2016). *Laurent Fabius'Reflections on the Paris Conference and implications for 2016*. COP21/CMP11 Bureau. https://unfccc.int/news/laurent-fabius-reflections-on-the-paris-conference-and-implications-for-2016

Farman, J. C., Gardiner, B. G., & Shanklin, J. D. (1985). Large losses of total ozone in Antarctica reveal seasonal ClOx/NOx interaction. *Nature, 315*(6016), 207–210. https://doi.org/10.1038/315207a0

Ferri, G. (2003). Review of globalization and its discontent, by J. E. Stiglitz. *The Journal of Socio-- Economics, 32*(6), 669–673. https://doi.org/10.1046/j.0391-5026.2003.00107.x

Florida, R. (2002). *The rise of the creative class*. Basic Books.

Florida, R. (2005). *Cities and the creative class* (1st ed.). Routledge. https://doi.org/10.4324/9780203997673

Foster, S. R., & Iaione, C. (2016). The city as a commons. *Yale Law & Policy Review, 34*(2), 281–349.

Foster, G. J., & Saleh, R. (2021). The circular city and adaptive reuse of cultural heritage index: Measuring the investment opportunity in Europe. *Resources, Conservation and Recycling, 175*(10588). https://doi.org/10.1016/j.resconrec.2021.105880

Foucault, M. (1978). *The history of sexuality, Volume 1: An introduction*. Vintage Books.

Foucault, M. (2002). *The archaeology of knowledge*. Trans of 1969 original by A. M. Sherridan Smith. Routledge. Ice, action, and policy. Teachers College Press.

Foucault, M. (2007). *Security, territory, population: Lectures at the Collège de France 1977–78*. Palgrave Macmillan.

Foucault, M. (2008). *The birth of biopolitics: Lectures at the Collège de France 1978–79*. Palgrave Macmillan.

Fukuyama, F. (1995). *Trust: The social virtues and the creation of prosperity*. Penguin Books.

Gehl, J. (2010). *Cities for people*. Island Press.

George, H. (2020). Creative cities and sustainable development: A framework. In X. Gu, M. K. Lim, & J. O'Connor (Eds.), *Re-imagining creative cities in twenty-first century Asia*. Palgrave Macmillan. https://doi.org/10.1007/978-3-030-46291-8_18

Grek, S., Tichenor, M., & Bandola-Gill, J. (2023). Numbers as utopia: Sustainable development goals and the making of quantified futures. *The British Journal of Politics and International Relations, 26*(3), 742–758. https://doi.org/10.1177/13691481231210385. (Original work published 2024).

Habermas, J. (2003). *The future of human nature* (W. Rehg, Trans.). Polity Press.

Hale, T. (2016). "All hands on deck": The Paris Agreement and nonstate climate action. *Global Environmental Politics, 16*(3), 12–22. https://doi.org/10.1162/GLEP_a_00362

Han, M. J. N., & Kim, M. J. (2024). A systematic review of smart city research from an urban context perspective. *Cities, 150*, 105027. https://doi.org/10.1016/j.cities.2024.105027

Harvey, D. (1989). From managerialism to entrepreneurialism: The transformation in urban governance in late capitalism. *Geografiska Annaler: Series B, Human Geography, 71*(1), 3–17. https://doi.org/10.1080/04353684.1989.11879583

Harvey, D. (2005). *A brief history of neoliberalism*. Oxford University Press. https://doi.org/10.1093/oso/9780199283262.001.0001

Harvey, D. (2008). The right to the city. *New Left Review, 53*, 23–40. https://newleftreview.org/issues/ii53/articles/david-harvey-the-right-to-the-city

Head, B. W., & Alford, J. (2013). Wicked problems: Implications for public policy and management. *Administration & Society, 47*(6), 711–739. https://doi.org/10.1177/0095399713481601. (Original work published 2015).

Hennicke, P., Rasch, J., Schröder, J., & Lorberg, D. (2019). *The energy transition in Europe: A vision of progress (Wuppertal Spezial No. 54)*. Wuppertal Institute for Climate, Environment and Energy. https://epub.wupperinst.org/frontdoor/deliver/index/docId/7336/file/WS54.pdf

Hergül, Ö. C., & Göker, P. (2021). *Evaluating eco-cities with a sustainable perspective in human-nature interaction*. https://doi.org/10.31590/ejosat.822062

Hes, D., & Bush, J. (2018). *Enabling eco-cities: Defining, planning and creating a thriving future*. Springer. https://doi.org/10.1007/978-981-10-7320-5

Höft, R. (2008). *Convention on biological diversity*. Wiley Online Library. https://doi.org/10.1002/9780470015902.a0020474

Honneth, A. (2007). Recognition as ideology. Recognition and power: Axel Honneth and the tradition of critical social theory, 323–347. Contention in contemporary social science. *The Sociological Quarterly, 48*(4), 641–659.

Hoyer, D., Bennett, J., Reddish, J., Holder, S., Howard, R., Benam, M., Levine, J., Ludlow, F., Feinman, G., & Turchin, P. (2023). Navigating polycrisis: Long-run socio-cultural factors shape response to changing climate. *Philosophical Transactions of the Royal Society B: Biological Sciences, 378*, 20220402. https://doi.org/10.1098/rstb.2022.0402

Hu, L., & Xi, Y. (2023). Evaluation of effective factors in achieving ecological city in the direction of sustainable development. *Sustainability, 15*(8), 7006. https://doi.org/10.3390/su15087006

Hughes, A., & Grumbine, R. (2023). The Kunming-Montreal global biodiversity framework: What it does and does not do, and how to improve it. *Frontiers in Environmental Science, 11*. https://doi.org/10.3389/fenvs.2023.1281536

Iaione, C. (2016). The CO-City: Sharing, collaborating, cooperating, and commoning in the city. *American Journal of Economics and Sociology, 75*, 415–455. https://doi.org/10.1111/ajes.12145

IPCC. (2014). *AR5 synthesis report: Climate change 2014*. https://www.ipcc.ch/report/ar5/syr/

Jenks, M., & Burgess, R. (2000). *Compact cities: Sustainable urban forms for developing countries*. Spon Press.

Jia, N., Li, Y., Cheng, J., Jiang, Z., Ma, T., Zheng, L., Xia, Z., Wang, R., Zhang, Z., Yin, C., Chen, R., & Liu, J. (2024, October 26). Polycrisis: A framework for understanding multiple crises in the metacoupled world. SSRN. https://doi.org/10.2139/ssrn.5003139

Johansson, N. (2021). Circular agreements—Exploring the role of agreements and deals as a political tool for a circular economy. *Circular Economy and Sustainability, 1*, 499–505. https://doi.org/10.1007/s43615-021-00004-5

Jung, C. G. (1968). *The archetypes and the collective unconscious* (R. F. C. Hull, Trans., 2nd ed.). Princeton University Press. (Original work published 1959).

Karybekova, G. (2018). International instruments in the field of climate change. *Regional Problems, 21*, 108–111. https://doi.org/10.31433/1605-220X-2018-21-3(1)-108-111

Keohane, R., & Oppenheimer, M. (2016). Paris: Beyond the climate dead end through pledge and review? *Politics and Governance, 4*(3), 142–151. https://doi.org/10.17645/pag.v4i3.634

Khalid, Z., & Abaas, R. Z. (2021). Defining the aspects of the local urban sustainability: Eco-cities as a model. *IOP Conference Series: Earth and Environmental Science., 754*, 012005. https://doi.org/10.1088/1755-1315/754/1/012005

Khavarian-Garmsir, A. R., Sharifi, A., & Sadeghi, A. (2023). The 15-minute city: Urban planning and design efforts toward creating sustainable neighborhoods. *Cities, 132*, 104101. https://doi.org/10.1016/j.cities.2022.104101

Kitchin, R. (2014). The real-time city? Big data and smart urbanism. *GeoJournal, 79*(1), 1–14. https://doi.org/10.1007/s10708-013-9516-8

Klein, N. (2007). *The shock doctrine: The rise of disaster capitalism*. Metropolitan Books.

Klein, L., Buckle, P., Nguyen, N., Preiser, R., & Ison, R. (2023). Navigating the polycrisis—Governing for transformation: The 2024 agenda for the systems community. *Systems Research and Behavioral Science, 40*(6), 973–977.

Kolb, D. A. (1984). *Experiential learning: Experience as the source of learning and development* (Vol. 1). Prentice-Hall.

Kovel, J. (2007). *The enemy of nature—The end of capitalism or the end of the world*. Zed Books.

Landry, C. (2008). *The creative city: A toolkit for urban innovators* (2nd ed.). Routledge. https://doi.org/10.4324/9781849772945

Lawrence, M., Janzwood, S., & Homer-Dixon, T.(2022). *'What is a global polycrisis?' Version 2.0. Discussion paper 2022-4*. Cascade Institute. https://cascadeinstitute.org/technical-paper/what-is-a-global polycrisis/

Lawrence, M., Homer-Dixon, T., Janzwood, S., Rockström, J., Renn, O., & Donges, J. (2024). Global polycrisis: The causal mechanisms of crisis entanglement. *Global Sustainability, 7*, 1–36. https://doi.org/10.1017/sus.2024.1

Lee, N., & Rodríguez-Pose, A. (2013). Original innovation, learnt innovation and cities: Evidence from UK SMEs. *Urban Studies, 50*(9), 1742–1759. https://doi.org/10.1177/0042098012470395. (Original work published 2013).
Lefebvre, H. (1991). *The production of space* (D. Nicholson-Smith, trans.). Basil Blackwell. (Original work published 1974).
Lemke, T. (2011). *Biopolitics: An advanced introduction*. NYU Press.
Lencucha, R., Kulenova, A., & Thow, A. M. (2023, January 23). Framing policy objectives in the sustainable development goals: Hierarchy, balance, or transformation? *Global Health, 19*(1), 5. https://doi.org/10.1186/s12992-023-00909-w
Leontidou, L. (2010). Urban social movements in 'weak' civil societies: The right to the city and cosmopolitan activism in southern Europe. *Urban Studies, 47*(6), 1179–1203. https://doi.org/10.1177/0042098009360239. (Original work published 2010).
Lepawsky, J. (2015). Are we living in a post-Basel world? *Area, 47*, 7–15. https://doi.org/10.1111/area.12144
Li, X., & Zhuang, X. (2023). Eco-City problems: Industry–City–ecology, urbanization development assessment in resource-exhausted cities. *Sustainability, 15*(1), 166. https://doi.org/10.3390/su15010166
Li, Q., Ge, Y., & Sayer, J. A. (2023). Challenges to implementing the Kunming-Montreal global biodiversity framework. *Land, 12*(12), 2166. https://doi.org/10.3390/land12122166
Lindblom, C. E. (1959). The science of 'muddling through'. *Public Administration Review, 19*(2), 79–88.
Lippi, A., & Tsekos, T. N. (2024a). Introduction: Governing sustainable development from theory to practice: Wicked problems, expected capacities, and design in uncertain environments. In A. Lippi & T. Tsekos (Eds.), *Policy capacity, design and the sustainable development goals: Wicked problems in uncertain environments* (pp. 1–23). Emerald Publishing.
Lippi, A., & Tsekos, T. N. (2024b). Conclusions: Taking seriously complexity, wickedness, and capacity for sustainable development policymaking. In A. Lippi & T. Tsekos (Eds.), *Policy capacity, design and the sustainable development goals: Wicked problems in uncertain environments* (pp. 281–315). Emerald Publishing.
Litman, T. (2012). Current mobility trends – Implications for sustainability. In *Keep moving, towards sustainable mobility* (pp. 23–44). Eleven International Publishing.
Litman, T. (2015). *Evaluating transportation land use impacts: Considering the impacts, benefits and costs of different land use development patterns*. Victoria Transport Policy Institute.
Luke, T. W. (2013). Corporate social responsibility: An uneasy merger of sustainability and development. *Sustainable Development, 21*(2), 83–91.
Macnaughton, G., & Frey, D.. (2018, December). Challenging neoliberalism: ILO, human rights, and public health frameworks on decent work. *Health and Human Rights, 20*(2), 43–55. PMID: 30568401; PMCID: PMC6293342.
Małkowska, A., Urbaniec, M., & Kosała, M. (2021). The impact of digital transformation on European countries: Insights from a comparative analysis. *Equilibrium. Quarterly Journal of Economics and Economic Policy, 16*(2), 325–355. https://doi.org/10.24136/eq.2021.012
Marcuse, P. (2021). Gentrification, displacement, and the role of public investment. *Journal of Planning Literature, 33*(1).
Marinatos, N. (1993). *Minoan religion: Ritual, image, and symbol*. University of South Carolina Press.
Marotta, I., Guarino, F., Longo, S., & Cellura, M. (2021). Environmental sustainability approaches and positive energy districts: A literature review. *Sustainability, 13*(23), 13063. https://doi.org/10.3390/su132313063
Meier, H.-R., Petzet, M., & Will, T. (2007). *Cultural heritage and sustainable development*. ICOMOS.
Miraftab, F. (2009). Insurgent planning: Situating radical planning in the global south. *Planning Theory, 8*(1), 32–50.

Mohanty, S. P., Choppali, U., & Kougianos, E. (2016, July). Everything you wanted to know about smart cities: The internet of things is the backbone. *IEEE Consumer Electronics Magazine, 5*(3), 60–70. https://doi.org/10.1109/MCE.2016.2556879

Moreno, C. (2016, October 5). *La ville du quart d'heure: pour un nouveau chrono-urbanisme*. La Tribune.

Morin, E., & Kern, A. B. (1999). *Homeland earth: A manifesto for the new millennium*. Hampton Press.

National Aeronautics and Space Administration. (n.d.). *Ozone watch: Latest status of ozone*. NASA Goddard Space Flight Center. https://ozonewatch.gsfc.nasa.gov/

Neuman, M. (2005). The compact city fallacy. *Journal of Planning Education and Research, 25*(1), 11–26. https://doi.org/10.1177/0739456X04270466. (Original work published 2005).

Newman, P., & Kenworthy, J. (1999). *Sustainability and cities: Overcoming automobile dependence*. Island Press.

Newman-Storen, R. (2014). Leadership in sustainability: Creating an interface between creativity and leadership theory in dealing with "wicked problems". *Sustainability, 6*(9), 5955–5967. https://doi.org/10.3390/su6095955

Ngan, S. L., How, B. S., Teng, S. Y., Promentilla, M. A. B., Yatim, P., Er, A. C., & Lam, H. L. (2019). Prioritization of sustainability indicators for promoting the circular economy: The case of developing countries. *Renewable and Sustainable Energy Reviews, 11*, 314–331. https://doi.org/10.1016/j.rser.2019.05.001

OECD. (2020). *Beyond growth: Towards a new economic approach*. https://www.oecd.org

OECD. (2025). *EU carbon border adjustment mechanism: What is it, how does it work and what are the effects?* https://www.oecd.org/en/blogs/2025/03/eu-carbon-border-adjustment-mechanism-what-is-it-how-does-it-work-and-what-are-the-effects.html?utm_source=chatgpt.com

Oliveira, A. F., & Paulino, F. F. (2017). European creativity and urban regeneration. *The Journal of Public Space, 2*(2), 127. https://doi.org/10.5204/jps.v2i2.98

Paiho, S., Tuominen, P., Rökman, J., Ylikerälä, M., Pajula, J., & Siikavirta, H. (2022). Opportunities of collected city data for smart cities. *IET Smart Cities, 4*(4), 275–291. https://doi.org/10.1049/smc2.12044

Papadopoulos, E., Sdoukopoulos, A., & Politis, I. (2023). Measuring compliance with the 15-minute city concept: State-of-the-art, major components and further requirements. *Sustainable Cities and Society, 99*, 104875. https://doi.org/10.1016/j.scs.2023.104875

Peck, J. (2001). *Workfare states*. Guilford Press.

Peck, J. (2012). Austerity urbanism: American cities under extreme economy. *City, 16*(6), 626–655.

Perez, M. E., Scholten, D., & Stegen-Smith, K. (2019). The multi-speed energy transition in Europe: Opportunities and challenges for EU energy security. *Energy Strategy Reviews, 26*. https://doi.org/10.1016/j.esr.2019.100415

Pooch, M. U. (2016). *DiverCity - Global cities as a literary phenomenon: Toronto, New York, and Los Angeles in a globalizing age*. (Lettre). transcript Verlag. https://doi.org/10.14361/9783839435410

Poorthuis, A., & Zook, M. (2023). Moving the 15-minute city beyond the urban core: The role of accessibility and public transport in The Netherlands. *Journal of Transport Geography, 110*, 103629. https://doi.org/10.1016/j.jtrangeo.2023.103629

Pratt, A. C. (2008). Innovation and creativity. Originally published in Hall, T., Hubbard, P., and Short, J. R.; *The SAGE companion to the city* (pp. 138–153). Sage.

Rajamani, L., & Bodansky, D. (2019). The Paris rulebook: Balancing international prescriptiveness with national discretion. *International and Comparative Law Quarterly, 68*(4), 1023–1040. https://doi.org/10.1017/S0020589319000320

Redclift, M. (2005). Sustainable development (1987–2005): An oxymoron comes of age. *Sustainable Development, 13*, 212–227. https://doi.org/10.1002/sd.281

Riordon, J. R. (2024, October 30). *NASA, NOAA rank 2024 ozone hole as 7th-smallest since recovery began*. NASA. https://www.nasa.gov/earth/climate-change/ozone-layer/nasa-noaa-rank-2024-ozone-hole-as-7th-smallest-since-recovery-began/

Rittel, H. W. J., & Webber, M. M. (1973). Dilemmas in a general theory of planning. *Policy Sciences, 4*(2), 155–169. https://doi.org/10.1007/BF01405730

Ritzer, G., & Stepnisky, J. (2020). *Classical sociological theory*. Sage.

Rodwell, D. (2007). *Conservation and sustainability in historic cities*. Blackwell Publishing.

Rogelj, J., den Elzen, M., Höhne, N., Fransen, T., Fekete, H., Winkler, H., et al. (2016). Paris Agreement climate proposals need a boost to keep warming well below 2 °C. *Nature, 534*, 631–639. https://doi.org/10.1038/nature18307

Rose, N. (1999). *Powers of freedom: Reframing political thought*. Cambridge University Press.

Roseland, M. (2012). *Toward sustainable communities: Solutions for citizens and their governments* (4th ed.). New Society Publishers.

Roy, A. (2009). Why India cannot plan its cities: Informality, insurgence and the idiom of urbanization. *Planning Theory, 8*(1), 76–87.

Sachs, J. D. (2012). From millennium development goals to sustainable development goals. *The Lancet, 379*(9832), 2206–2211. https://doi.org/10.1016/S0140-6736(12)60685-0

Sakellaropoulos, T. (2004). The open method of coordination. A sound instrument for the modernisation of the European social model. In T. Sakellaropoulos & J. Berghman (Eds.), *Connecting welfare diversity within the European social model* (pp. 55–92). Intersentia.

Saleem, M., Abbas, S., Ghazal, T. M., Khan, M. A., Sahawneh, N., & Ahmad, M. (2022). Smart cities: Fusion-based intelligent traffic congestion control system for vehicular networks using machine learning techniques. *Egyptian Informatics Journal, 23*(3), 417–426. https://doi.org/10.1016/j.eij.2022.03.003

Sasaki, M. (2010). Urban regeneration through cultural creativity and social inclusion: Rethinking creative city theory through a Japanese case study. *Cities, 27*. https://doi.org/10.1016/j.cities.2010.03.002

Sassen, S. (2005). The global city: Introducing a concept. *The Brown Journal of World Affairs, 11*(2), 27–43. http://www.jstor.org/stable/24590544

Segovia, C., & Hervé, J. (2022). The creative city approach: Origins, construction and prospects in a scenario of transition. *City Territory and Architecture, 9*, 29. https://doi.org/10.1186/s40410-022-00178-x

Serhan, Y. (2023, January 17). *'Polycrisis' is the buzzword of the day at Davos*. Time. Retrieved May19, 2024, from https://time.com/6247799/polycrisis-in-davos-wef-2023/

Seshadri, A. K., Gambhir, A., & Debnath, R. (2025). Navigating systemic risks in low-carbon energy transitions in an era of global polycrisis. *Global Sustainability, 8*, e6. https://doi.org/10.1017/sus.2025.7

Seto, K. C., Güneralp, B., & Hutyra, L. R. (2012). Global forecasts of urban expansion to 2030 and direct impacts on biodiversity and carbon pools. *Proceedings of the National Academy of Sciences, 109*(40), 16083–16088. https://doi.org/10.1073/pnas.1211658109

Siirilä, J., & Salonen, A. O. (2024). Towards a sustainable future in the age of polycrisis. *Frontiers in Sustainability, 5*. https://doi.org/10.3389/frsus.2024.1436740

Simon, H. A. (1976). *Administrative behavior: A study of decision-making processes in administrative organizations* (3rd ed.). Free Press.

Simpson, C., & Rodrigues, F. (2025). JRC highlights report 2024 – From science to impact (I. Bonjean, J. Bray, I. Čeliković, M. Fornara, K. James, et al., Eds.). Publications Office of the European Union. https://doi.org/10.2760/6280845

Sjoberg, G. (1965). *The preindustrial city: Past and present*. Simon & Schuster.

Smith, N. (1996). *The new urban frontier: Gentrification and the revanchist city*. Routledge.

Smith, D. A., Shen, Y., Barros, J., Zhong, C., Batty, M., & Giannotti, M. (2020). A compact city for the wealthy? Employment accessibility inequalities between occupational classes in the London metropolitan region, 2011. *Journal of Transport Geography, 86*, 102767. https://doi.org/10.1016/j.jtrangeo.2020.102767

Sookhak, M., Tang, H., He, Y., & Yu, F. R. (2019). Security and privacy of smart cities: A survey, research issues and challenges. *IEEE Communications Surveys & Tutorials, 21*(2), 1718–1743. https://doi.org/10.1109/COMST.2018.2867288

Stephens, T. (2023). The Kunming–Montreal global biodiversity framework. *International Legal Materials, 62*(5), 868–887. https://doi.org/10.1017/ilm.2023.16
Sternberg, C. (2023). *Neoliberal urban governance: Spaces, culture and discourses in Buenos Aires and Chicago*. Springer. https://doi.org/10.1007/978-3-031-21718-0
Stevenson, D. (2003). *The city*. Polity Press.
Stiglitz, J. E. (2002). *Globalization and its discontents*. W. W. Norton & Company.
Takkinen, P., & Heikkurinen, P. (2024). Peripheral sustainability expertise on technology: An autoethnography amidst the polycrisis. *Acta Borealia, 41*(1), 16–30. https://doi.org/10.1080/08003831.2024.2334176
Tang, W., Niu, Z., Wei, Z., & Zhu, L. (2022). Sustainable development of eco-cities: A bibliometric review. *Sustainability, 14*(17), 10502. https://doi.org/10.3390/su141710502
Tooze, A. (2022) *Welcome to the world of the polycrisis*. FT-opinion global economy 28/10/2022.
Touraine, A. (2012). After the crisis. For a future without marginalization. *RIMCIS - International and Multidisciplinary Journal of Social Sciences, 1*(2).
Tsavachidis, M., & Petit, L. Y. (2022). Re-shaping urban mobility – Key to Europe's green transition. *Journal of Urban Mobility, 2*. https://doi.org/10.1016/j.urbmob.2022.100014
Tyas, W. P., Nugroho, P., Sariffuddin, S., Purba, N. G., Riswandha, Y., & Sitorus, G. H. I. (2018, October 25–26). Applying smart economy of smart cities in developing world: Learnt from Indonesia's home based enterprises. In *Proceedings of the 1st international conference on Smart City innovation (ICSCI 2018)* (Vol. 248, p. 012078). IOP Conference Series: Earth and Environmental Science https://doi.org/10.1088/1755-1315/248/1/012078
UNESCO (2021). *Cities, culture, creativity: Leveraging culture and creativity for sustainable urban development and inclusive growth*.
United Nations. (1972). *Report of the United Nations conference on the human environment (Stockholm, 5–16 June 1972)*. United Nations. https://www.un.org/en/conferences/environment/stockholm1972
United Nations. (1992a). *Convention on biological diversity*. https://www.cbd.int/convention/text/ and https://treaties.un.org/doc/Treaties/1992/06/19920605%2008-44%20PM/Ch_XXVII_08p.pdf
United Nations. (1992b). *Report of the United Nations Conference on Environment and Development*. https://www.un.org/esa/documents/ga/conf151/aconf15126-1.htm
United Nations. (1992c). *United Nations framework convention on climate change (A/RES/48/189)*. https://unfccc.int/resource/docs/convkp/conveng.pdf
United Nations. (1998). *Kyoto protocol to the United Nations framework convention on climate change*. https://unfccc.int/resource/docs/convkp/kpeng.pdf
United Nations. (2000). *United Nations millennium declaration*. https://www.ohchr.org/en/instruments-mechanisms/instruments/united-nations-millennium-declaration
United Nations. (2015a). *Paris Agreement*. https://unfccc.int/sites/default/files/english_paris_agreement.pdf
United Nations. (2015b). *Transforming our world: The 2030 agenda for sustainable development (A/RES/70/1)*. United Nations General Assembly. https://sdgs.un.org/publications/transforming-our-world-2030-agenda-sustainable-development-17981
United Nations, Department of Economic and Social Affairs. (2025). *World social report 2025: A new policy consensus to accelerate social Progress*. UNDESA. 118 pp.
United Nations Environment Programme. (1987). *Montreal protocol on substances that deplete the ozone layer*. https://ozone.unep.org/treaties/montreal-protocol
United Nations Environment Programme. (2020). *Handbook for the Montreal protocol on substances that deplete the ozone layer* (2020th ed.). United Nations Environment Programme. https://ozone.unep.org/sites/default/files/Handbooks/MP-Handbook-2020-English.pdf
Valasaki, M. (2020). The mask during the modern pandemic: Sociological approaches and issues of interpretation. *The Greek Review of Social Research, 154*, 149–159. https://doi.org/10.12681/grsr.25129

Velders, G. J. M., Andersen, S. O., Daniel, J. S., Fahey, D. W., & McFarland, M. (2007). The importance of the Montreal protocol in protecting climate. *Proceedings of the National Academy of Sciences, 104*(12), 4814–4819. https://doi.org/10.1073/pnas.0610328104

Vîrjan, D., Popescu, C. R., Pop, I., & Popescu, D. (2023). Energy transition and sustainable development at the level of the European Union. *Amfiteatru Economic, 25*(63), 429–446. https://doi.org/10.24818/EA/2023/63/429

Voskoboinikova, J. (2023). *Green Deal industrial plan: The EU support for green technologies*. Linklaters Sustainable Futures. https://sustainablefutures.linklaters.com/post/102i8hr/green-deal-industrial-plan-the-eu-support-for-green-technologies

Wesseler, J. (2022). The EU's farm-to-fork strategy: An assessment from the perspective of agricultural economics. *Applied Economic Perspectives and Policy, 44*(4), 1826–1843. https://doi.org/10.1002/aepp.13239

Whiting, S., Barnett, T., & O'Connor, J. (2022). 'Creative City' R.I.P.? *M/C Journal, 25*(3). https://doi.org/10.5204/mcj.2901. (Original work published June 27, 2022).

Wilkes-Allemann, J., Kopp, M., van der Velde, R., Karaca, E., Čepić, S., Tomićević-Dubljević, J., Bauer, N., Petit-Boix, A., Brantschen, E. C., et al. (2023). Envisioning the future—Creating sustainable, healthy, and resilient BioCities. *Urban Forestry & Urban Greening, 84*, 127935. https://doi.org/10.1016/j.ufug.2023.127935

World Commission on Environment and Development. (1987). *Our common future*. Oxford University Press. https://sustainabledevelopment.un.org/content/documents/5987our-common-future.pdf

World Meteorological Organization. (2022). *Scientific assessment of ozone depletion: 2022 – Executive summary (GAW Report No. 278)*. United Nations Environment Programme. Retrieved October 3, 2024, from https://sustainabledevelopment.un.org/content/documents/25418belgium2.pd

Yearworth, M. (2015). Sustainability as a 'super-wicked' problem; opportunities and limits for engineering methodology. *Intelligent Buildings International, 8*(1), 37–47. https://doi.org/10.1080/17508975.2015.1109789

Zaki, B. L., Pattyn, V., & Wayenberg, E. (2024). Policymaking in an age of polycrises: Emerging perspectives. *Policy Design and Practice, 7*(4), 377–389. https://doi.org/10.1080/25741292.2024.2432048

Zhang, S., Zhen, F., Kong, Y., Lobsang, T., & Zou, S. (2022). Towards a 15-minute city: A network-based evaluation framework. *Environment and Planning B: Urban Analytics and City Science, 50*(2), 500–514. https://doi.org/10.1177/23998083221118570. (Original work published 2023).

Zhu, Y., et al. (2022). Building better cities: Evaluating the effect of circular economy city construction on air quality via a quasi-natural experiment. *Journal of Environmental and Public Health, 2022*, 3151072. https://doi.org/10.1155/2022/3151072

Part II
Modeling the Biocultural City: An In-Depth Ixelles Case Study

Transition from Part I to Part II

The theoretical frameworks and literature discussed in Part I form the conceptual foundation of the methodological approach developed in Part II. The review of biocultural theory, polycrisis governance, and wicked policy problems, directly informed the design of the empirical study by shaping its research questions, analytical dimensions, and choice of mixed-methods strategy. Specifically, the five analytical axes identified in Part I, environmental, social, cultural, economic, and technological, serve as the guiding dimensions for the field investigation in Ixelles. Likewise, the emphasis on complexity, interdisciplinarity, and reflexivity discussed in the theoretical chapters is operationalized here through abductive reasoning, triangulation, and participatory observation. Thus, Part II translates the conceptual insights of Part I into a practical research framework, demonstrating how theory can inform the design, implementation, and interpretation of an urban biocultural case study.

Chapter 5
Methodology of Field Research

5.1 Introduction: Theoretical and Methodological Framework

This chapter establishes the theoretical and methodological foundations of the study, situating the biocultural approach within the broader landscape of cultural theory, sustainability studies, and contemporary urban research. The objective is to explore how interdisciplinary frameworks can offer new perspectives for understanding the interaction between culture, environment, and governance in European cities.

The theoretical context is grounded in the notion that cities are not static entities but dynamic cultural and ecological systems. Urban environments are continuously produced and re-produced through the interplay of social practices, collective imaginaries, and material infrastructures. As such, the city may be interpreted as a living archive of human–environment interactions—a spatial expression of collective identity and adaptation.

The research draws upon multiple epistemological traditions: from cultural geography and environmental anthropology to sociology, urban semiotics, and heritage studies. Each contributes conceptual tools for analyzing how local communities negotiate meaning and value within their environments. This interdisciplinary dialogue forms the basis of the biocultural paradigm adopted here.

Within this framework, the biocultural city functions as both an analytical and a normative concept. Analytically, it reveals how cultural expressions—architecture, art, rituals, and social behavior—are embedded in ecological contexts. Normatively, it suggests that cultural awareness and environmental responsibility must be integrated in urban policy and governance if sustainable transformation is to be achieved.

Methodologically, the study employs a mixed qualitative approach combining discourse analysis, case-study research, and comparative synthesis. The principal empirical focus lies on the City of Ixelles, selected as a representative urban context

---

Supplementary Information The online version contains supplementary material available at https://doi.org/10.1007/978-3-032-17584-7_5.

E. Bitsani et al., *The Biocultural City as a Utopia or the Future Reality?*, Cities, Heritage and Transformation, https://doi.org/10.1007/978-3-032-17584-7_5

due to its cultural diversity, historical layering, and active engagement with sustainability policies. The analysis proceeds through three interconnected levels:

1. Discursive level: examining how sustainability and culture are framed in official and civic narratives.
2. Spatial–material level: analyzing urban form, heritage sites, and cultural infrastructures.
3. Participatory level: assessing community initiatives and local governance practices.

Together, these dimensions allow for a holistic understanding of how biocultural values are articulated, negotiated, and implemented within contemporary European urbanism.

5.1.1 Linking Theory with the Case Study

This chapter presents the methodological framework adopted for the present study, which investigates the interrelations between culture, sustainability, and social equity in the Municipality of Ixelles through the lens of a biocultural approach. The research assumes that urban sustainability cannot be fully understood or achieved through environmental or economic measures alone; rather, it must integrate cultural values, practices, and meanings as dynamic forces that shape both human behavior and structural transformation.

The study's central premise is that culture constitutes the connective and generative element within the biocultural paradigm, influencing how communities interpret, respond to, and co-create sustainable futures. Therefore, the methodology is designed to capture the complex interplay between biological–environmental dimensions and cultural–social dynamics in an urban context.

Grounded in constructivist ontology and interpretivist epistemology (Guba & Lincoln, 1994), but also incorporating positivist elements to ensure empirical validity, the research follows a pragmatic, abductive strategy (Blaikie, 2010). This allows the investigation to move iteratively between theoretical insights and empirical evidence, generating both explanatory depth and generalizable understanding.

The chapter proceeds as follows: Section 5.2 outlines the core research claim and guiding questions, establishing the conceptual and strategic continuity between the central role of culture and the biocultural framework. Section 5.3 describes the research design, including the philosophical assumptions, data collection methods, and analytical procedures. Together, these sections articulate a coherent methodological pathway for exploring how a culture-centered biocultural strategy can contribute to addressing urban inequalities and promoting sustainable development in Ixelles.

When examining the challenges and opportunities in large urban centers, it becomes clear how complex they are. These cities serve as hubs for various productive, consuming, social, and residential activities involving large groups of people who represent different cultures, values, and lifestyles. All of this takes place in a limited space and defined landscape that is continuously evolving and expanding. Large cities epitomize both complication and complexity.

In Chap. 4 of the first part, we discussed innovative methods in policy science that pertain to the reception, analysis, understanding, and management of social, economic, and environmental complexity. We acknowledge that studying a city like Ixelles, which is part of the larger Brussels metropolitan area, can effectively demonstrate the relevance of these concepts and the usefulness of the associated policy tools.

Ixelles is home to residents, employees, and workers from diverse ethnic, religious, and cultural backgrounds. This diversity adds to the complexities related to various urban issues, such as environmental impact, energy use, transportation, employment, and technological transition, along with the associated challenges that emerge from these interactions.

On the other hand, this book adopts a biocultural analytical perspective to approach large urban centers, a perspective that will also be applied to the Ixelles case study. This method is an example of a complex analytical framework that integrates the study of both biological and cultural diversity, along with their interconnected relationship. Consequently, it acts as a tool with interdisciplinary characteristics, akin to those proposed by the Polycrisis and Wicked Policy Problems frameworks discussed in Part I.

Specifically, the connection and sequence of the chapters of the theoretical part with the empirical part as well as between them is analyzed below:

Chapters 1–4 of Part I form a coherent theoretical continuum that progresses from conceptual foundations to policy application. Chapter 1 establishes the overarching paradigm of the biocultural city, positioning culture and ecology as mutually constitutive pillars of sustainable urban development. It defines the study's epistemological basis and raises the central research question, whether biocultural approaches can reconcile issues of inequality, environmental degradation, and governance. This initial chapter sets the intellectual tone and provides the normative direction for the entire work, insisting that any model of urban sustainability must rest on cultural values, human creativity, and ecological interdependence.

Building upon this foundation, Chap. 2 introduces the utopia–dystopia dialectic as a theoretical lens for interpreting urban transformation. By tracing philosophical and political traditions from Plato to Foucault, it explores the tension between idealized visions of social harmony and the dystopian realities of exclusion, environmental crisis, and biopolitical control. This chapter deepens the argument of Chap. 1 by situating the biocultural paradigm within a long lineage of critical thought on governance, power, and the human condition, offering a conceptual vocabulary for analyzing cities as spaces of both aspiration and anxiety.

Chapter 3 extends this reflection into the concrete spatial and environmental context of Europe. It conceptualizes European cities as cultural landscapes, dynamic arenas where cultural heritage, social diversity, and ecological systems intersect. By mapping the material and symbolic dimensions of sustainability, Chap. 3 translates the abstract debates of the first two chapters into the socio-spatial realities of urban Europe. It thus creates the bridge between philosophical reflection and empirical observation, highlighting how the global forces of globalization, migration, and climate change reshape the lived experience of the European city.

Finally, Chap. 4 consolidates the theoretical groundwork by articulating how these cultural and ecological insights inform contemporary urban governance. It examines policymaking through the frameworks of wicked problems and polycrisis, arguing for participatory, adaptive, and multi-actor governance models capable of managing complexity. In doing so, Chap. 4 operationalizes the ideas introduced earlier: the holistic vision of Chap. 1, the critical dialectic of Chap. 2, and the contextual depth of Chap. 3 converge into a practical framework for action.

Collectively, these four chapters provide a seamless intellectual trajectory.

From paradigm to philosophy, from context to policy laying the conceptual foundation for the methodological and empirical exploration that follows in Part II (Table 5.1).

.

The methodology, defined here as the systematic analysis of the methods employed in this investigation, was devised by combining various approaches based on key principles to create a comprehensive framework. The framework is inherently broad and subject to continuous review in light of ongoing research and empirical findings.

The initial assumption underlying our hypothesis is that society is not a mechanistic system established to fulfil a specific objective, but rather a constellation of agents pursuing a diverse range of aims. However, this does not necessarily constitute an integrated or coordinated system, nor does it exclude autonomous action or conflict arising from the correlation between perceptions and behaviors. It is important to note that this correlation is significant: in uncertain situations that may result from unforeseen variations in individuals' perceptions and practices, this theoretical framework may prove particularly beneficial in specific contexts.

Our methodological framework draws on the principles of abductive reasoning, combining deductive and inductive logics (Blaikie, 2010). The research assumes a constructivist ontology and an interpretivist epistemology (Guba & Lincoln, 1994), acknowledging that reality is socially constructed and that meanings are shaped through interaction and context. However, the research also incorporates elements of positivism to ensure empirical robustness and generalizability in its quantitative components. Within this framework, society is conceived not as a mechanistic structure but as an emergent system composed of diverse actors pursuing distinct goals (Bitsani & Kavoura, 2011). These goals are historically situated and influenced both by structural forces—such as policies and institutions—and by cultural perceptions and lived experiences. This dual lens of structure and agency informs our understanding of sustainability and inequality in Ixelles.

Data collection was designed within a multimodal framework of interpretation and communication. First, a content analysis of the websites of the Municipality of Ixelles and its institutional structures, as well as those of social, cultural, and environmental organizations operating in the city, was undertaken. Subsequently, a mixed-methods approach was implemented, combining qualitative interviews with citizens of Ixelles and Brussels, and quantitative analysis based on a structured

Table 5.1 Comprehensive summary table: Theoretical and empirical connection of all chapters

Chapter	Core focus/theoretical function	Key concepts	Connection to empirical Part II (Case Study of Ixelles)
Chapter 1 Introduction: Conceptual Foundations of the Biocultural City	Establishes the overall research question and theoretical framework for biocultural sustainability. Defines the epistemological basis of the study and situates it in global urban and environmental debates.	Biocultural city, sustainability, cultural shift, holistic urbanism, interdisciplinarity	Shapes the research objectives, defines analytical dimensions (environmental, social, cultural, economic, technological), and justifies the mixed-methods design used in Ixelles.
Chapter 2, Utopia and Dystopia: Theoretical Lenses for Imagining Urban Futures	Explores philosophical paradigms of utopia and dystopia to interpret the dualities of modern urban life. Frames the city as both ideal and site of crisis.	Utopia/dystopia, governance, biopolitics, Foucault, imagination, social inequality	Provides interpretive tools for qualitative analysis; frames residents' narratives and local governance practices as expressions of utopian aspirations and dystopian constraints.
Chapter 3, The Context of Culture, Environment, and Sustainability: European Cities as cultural landscapes	Contextualizes European urban transformations, connecting environmental sustainability, cultural heritage, and governance in a unified analytical framework.	Cultural landscapes, resilience, sustainability, circular economy, urban diversity	Justifies case selection (Ixelles as representative of European diversity); provides empirical variables such as governance, participation, and cultural-environmental interaction.
Chapter 4, Contemporary Approaches to Policymaking: Wicked Problems, Polycrisis, and Urban Governance	Translates philosophical and contextual insights into governance theory. Examines how policy frameworks manage complexity and interdependence in cities.	Wicked problems, polycrisis, adaptive governance, UN 2030 Agenda, EU Urban Agenda, resilient/smart cities	Bridges theory and methodology; justifies participatory, cross-sectoral, and multi-actor approaches; provides policy indicators for evaluating Ixelles' biocultural performance.
Transition to Part II - +Methodology and Case Study of Ixelles	Operationalizes the theoretical constructs into a research design combining quantitative and qualitative data collection and analysis.	Triangulation, participatory observation, mixed methods, abductive reasoning	Implements the theoretical model empirically; tests how biocultural sustainability manifests in local governance, everyday life, and spatial practices.

questionnaire administered to a random sample of residents and members of immigrant communities.

In parallel, since 2018, on-site participatory observation and documentation of urban changes have been carried out. The data and research results collected through these complementary methods are cross-referenced through the technique of triangulation, thereby enhancing the validity and depth of the findings.

5.2 Basic Research Question

Building on the conceptual foundation outlined above, this section articulates the core research claim and guiding questions that direct the study's investigation. The research proceeds from the understanding that sustainability in contemporary urban contexts is inherently biocultural; that is, ecological systems, social structures, and cultural practices are interdependent and co-constitutive. Within this paradigm, culture is not treated as an ancillary or decorative dimension of urban life, but rather as a structuring and transformative force—one that mediates perceptions, behaviors, and collective action toward sustainable development.

The study explores how a biocultural strategy centered on culture can inform and enhance urban planning, environmental policy, and social inclusion in Ixelles. The following main research claim and guiding questions are designed to reflect this integrated perspective, linking the cultural dimension of human experience to the biocultural processes underpinning urban sustainability and equality.

In this framework, the investigation contends that culture, when placed at the center of sustainability discourse, constitutes a transformative axis capable of reshaping urban development within a biocultural paradigm. Culture is understood here as the dynamic medium through which ecological, social, and technological processes acquire human significance, ethical direction, and civic resonance.

In order to examine this proposition empirically, the study adopts a mixed-methods strategy grounded in the biocultural perspective yet guided principally by the cultural dimension of sustainability. The ensuing research questions translate this conceptual stance into an operational framework for empirical enquiry.

Our central research claim is that embedding culture as a core component of a biocultural approach to urban sustainability can provide more equitable and inclusive solutions to the challenges of inequality, social cohesion, and environmental transformation in European cities. In other words, a biocultural strategy that foregrounds culture, its values, practices, and institutions, constitutes a crucial mechanism for achieving sustainable and socially just urban development.

Accordingly, the study is guided by the following central research questions:

1. To what extent does a culture-centered biocultural approach to sustainability address structural inequalities related to power, minority inclusion, and gender equity? Specially, to what extent can culture-centered strategies advance the objectives of biocultural sustainability within the urban context of Ixelles?

2. How does the integration of cultural values and practices within the biocultural framework influence urban responses to climate change, adaptation strategies, and sustainable development? Specially, how does the prioritization of cultural values and practices influence local responses to environmental, economic, and social challenges?
3. How do citizens, particularly those from migrant and minority backgrounds, perceive, experience, and respond to the Municipality of Ixelles's cultural and environmental policies when interpreted through a biocultural lens? In what ways do residents and municipal authorities perceive culture as a resource for biocultural transformation and for the advancement of inclusive urban governance?

Collectively, these questions establish the analytical pathway through which the subsequent sections of this chapter delineate the methodological design, data-collection instruments, and interpretative strategies employed in the field research.

These questions align with global policy priorities, including the United Nations Sustainable Development Goals (SDGs), particularly Goals 5 (Gender Equality), 10 (Reduced Inequalities), 11 (Sustainable Cities and Communities), and 13 (Climate Action). Together, they reinforce the strategic continuity between culture as a transformative force and the biocultural paradigm as the theoretical and methodological framework guiding this research.

5.3 Research Design

The research design translates the theoretical orientation of the study—the biocultural approach centered on culture as a transformative force—into a coherent methodological framework capable of addressing the central research questions. It is structured to capture the interdependence between cultural meanings, social dynamics, and environmental processes in the context of urban sustainability and inequality in Ixelles. To achieve this, the design combines qualitative and quantitative methods within a mixed-methods strategy, reflecting the study's abductive and pragmatic orientation (Saunders et al., 2007).

The investigation was developed progressively between 2018 and 2025, combining longitudinal and cross-sectional analyses to align empirical evidence with theoretical reflection. The design follows the "research onion" model (Saunders et al., 2007), which provides a structured pathway for aligning philosophical assumptions, methodological choices, and research techniques. Accordingly, the study integrates constructivist and interpretivist paradigms—to capture the socially constructed and context-dependent nature of knowledge—with positivist elements that ensure analytical rigor and generalizability.

This blended framework enables the study to examine how cultural practices and values influence sustainability at both structural and experiential levels. The following sections detail the philosophical underpinnings, methodological strategies, data collection procedures, and analytical techniques that together operationalize the culture-centered biocultural approach in the empirical investigation of Ixelles.

From 2018 to 2022, a thorough literature review and critical analysis were conducted, with particular emphasis on recent and interdisciplinary research. This process helped to define both the general philosophical perspective and the theoretical framework of the research within which it is situated. The literature review continued until the conclusion of the research in 2025, ensuring a dynamic engagement with emerging studies relevant to its evolving scope.

Given the complexity of the issues addressed (Bryman & Bell, 2019), the authors followed the "research onion" design to provide comprehensive contextualization. This framework supported the alignment of research philosophy, strategy, and methods (Bitsani & Anselmi, 2021) (Fig. 5.1).

Following the "research onion" model (Saunders et al., 2007), the research design was developed in progressive layers that align philosophical assumptions with methodological and empirical choices. Each layer of the design reflects the overarching biocultural framework, positioning culture as a mediating dimension connecting environmental, social, and economic systems. This layered structure ensures that the research remains theoretically grounded while also flexible enough to accommodate the complex, evolving realities of urban sustainability and inequality in Ixelles.

The principal layers of the design are as follows:

- **Philosophical Layer—Pragmatic Paradigm Integrating Constructivism and Positivism:**

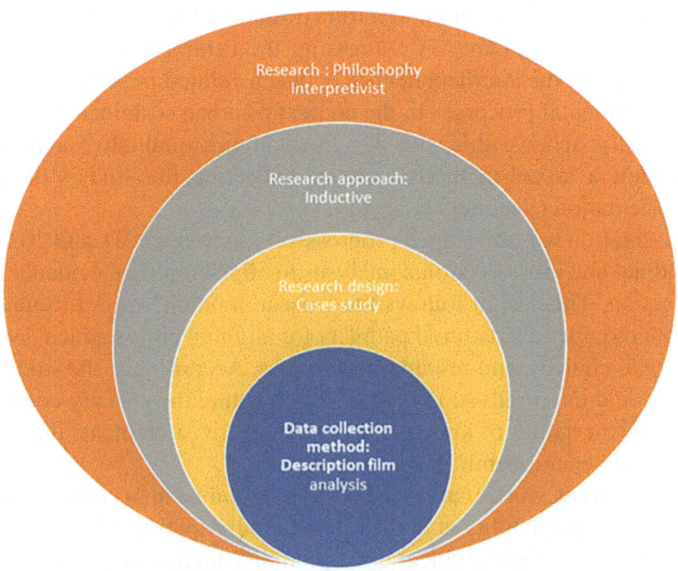

Fig. 5.1 The "research onion" (Saunders et al., 2007)

The study adopts a pragmatic worldview that accommodates both interpretive understanding and empirical validation. It combines constructivist ontology and interpretivist epistemology—acknowledging that realities are socially constructed through culture and interaction—with positivist elements that support measurable and generalizable insights. This philosophical integration reflects the biocultural notion that cultural meanings and material conditions are mutually constitutive (Guba & Lincoln, 1994; Blaikie, 2010).

- **Approach Layer—Abductive Reasoning:**
  The investigation follows an abductive logic, moving iteratively between theory and data to refine conceptual understanding. This cyclical process allows for the continuous reinterpretation of empirical evidence through the lens of the biocultural framework, ensuring that theoretical assumptions evolve alongside observed realities.
- **Strategy Layer—Case Study, Action Research, and Ethnography:**
  The case study of Ixelles serves as the focal point for examining how cultural and environmental factors interact in real-life contexts. The use of action research and ethnographic elements reflects the study's engagement with local communities and participatory observation, both onsite and online. These strategies operationalize the biocultural approach by situating culture within lived, spatial, and digital environments.
- **Methodological Choice—Mixed Methods:**
  Together, they embody the biocultural synthesis of meaning and measurement. The research employs a mixed-methods design, combining qualitative and quantitative techniques to capture both subjective meanings and objective patterns. Qualitative interviews and participatory observation explore the cultural dimensions of sustainability, while quantitative surveys provide empirical evidence of trends and correlations. Together, these methods embody the biocultural synthesis of meaning and measurement.
- **Time Horizon—Cross-Sectional and Longitudinal:**

  The design integrates cross-sectional analyses (e.g., surveys conducted in 2022–2023) with longitudinal observation (2018–2025). This dual temporal perspective reflects the evolutionary nature of the biocultural process, allowing the study to document both immediate social dynamics and long-term transformations within the city's cultural and environmental landscape.

  In sum, the research onion framework enables a coherent translation of the culture-centered biocultural paradigm into a practical methodological structure, ensuring that every stage—from philosophical grounding to data collection—reflects the study's integrative vision of sustainability as both a cultural and ecological process.

As Gruber (2009) suggests, the authors employ a combination of ontology and epistemology. Ontology concerns the authenticity of information and how one understands its existence, while epistemology refers to valid knowledge and the means by which it can be obtained (Bitsani, 2013, p. 117). The former assumes that knowledge is independent of the object of study, whereas the latter asserts that individuals construct their own perceptions and understandings of reality. Hence, positivist studies are often more scientific and focused on testing phenomena, whereas interpretivist studies are typically qualitative in nature.

In recent years, scholars have increasingly emphasized the need for literature reviews to move beyond mere description and towards analytical depth. Torraco (2005) argues that an effective review not only recounts what is known but also explains how knowledge has evolved and where it is heading. Moreover, a theoretically informed literature review provides a structured approach to evaluating the conceptual and empirical contributions of previous studies. It highlights how different theoretical models have been applied, exposes inconsistencies, and suggests pathways for theoretical refinement—an especially important endeavor in disciplines where multiple frameworks compete to explain complex phenomena such as sociology, education, or urban studies.

The literature review was guided by key themes, including—but not limited to—'sustainability," "biocultural approach," "sustainable cities," "social sustainability," "cultural development," "city models," "environmental sustainability," "cultural heritage," and "public and local policies," etc.

To fulfil the research objectives, eight databases were selected: Dimensions, SpringerLink, Semantic Scholar, ProQuest, Elsevier (ScienceDirect), Emerald Insight, Scopus, and Taylor & Francis Online. The findings of the literature review structured the theoretical foundation of this study and led to the identification of the following analytical dimensions:

a. Environmental dimension (ecological policies, climate risk management).
b. Societal dimension (multiculturalism, interculturalism, disability, poverty, education, housing, public health).
c. Cultural dimension (cultural heritage).
d. Economic dimension (economic development, cultural industries).
e. Technological dimension (creative sector).

Given these dimensions, the researchers employed the case study as their primary strategy (Babbie, 2011). Despite ongoing debates surrounding the term "case study" in the social sciences, it was chosen not simply because it is widely used (Priya, 2021) but because it is particularly suited to achieving in-depth understanding and exploration of complex issues in real-world settings (Crowe et al., 2011; Baxter & Jack, 2010; Farquhar et al., 2020). It enables a detailed and nuanced understanding of contemporary issues within defined systems (Coombs, 2022) and facilitates the investigation of causal links, pathways, and implementation strategies while addressing the "how," "what," and "why" of social phenomena (Babbie, 2011).

The time framing was determined by the complexity of the research purpose; the researchers employed cross-sectional, longitudinal, and consecutive independent samples (Bryman & Bell, 2015). Observations were gathered both at single points in time and across extended periods to capture change dynamics. Historical research on Ixelles also informed the study, recognizing that historical knowledge influences contemporary perceptions (Christmann & Heimann, 2017).

Both primary and secondary data were collected using qualitative and quantitative techniques. The use of mixed methods and diverse data sources established a dialectical relationship between past and present, providing a basis for documenting social behaviors and interpreting their cultural significance through cross-sectional surveys. Adopting a historical perspective and situating it within a defined timeframe further enabled examination of societal continuities and discontinuities, enriching understanding of social change (Bitsani & Anselmi, 2021).

Findings were cross-checked through triangulation to ensure reliability, objectivity, and validity (Noble & Heale, 2019; Bitsani & Kavoura, 2008; Farquhar et al., 2020; Bans-Akutey & Tiimub, 2021). Claims about the biocultural city model were corroborated by data triangulation, with findings substantiated by multiple sources. Quantitative data validated qualitative insights, ensuring robustness.

Methodological triangulation was also employed: a systematic review of academic literature was supplemented with expert interviews and surveys. Multiple approaches—including both "across-method" and "within-method" techniques—were used, combining closed-ended (quantitative) and open-ended (qualitative) questions to capture a comprehensive picture of sustainability in Ixelles. Investigator triangulation further enhanced credibility, as researchers from diverse academic and professional backgrounds contributed to the analysis and interpretation of results. Environmental triangulation was also applied, comparing findings across different neighborhoods to assess sustainability efforts in areas such as green infrastructure, air quality, water management, and biodiversity conservation.

The research possesses a hybrid and polymetric character, operating simultaneously within the tangible world and the immaterial sphere of the internet. As computing power increases, the immaterial domain expands beyond individualized digital experiences to encompass collective spaces of communication, commerce, and community (Fraser, 2006). Alternative and digital media foster new forms of interaction and belonging, contributing to the evolving biocultural fabric of contemporary urban life (Bitsani & Anselmi, 2021).

The case study of Ixelles thus functions as the empirical locus where these methodological dimensions converge. Through its hybrid, multimodal, and participatory design, the research examines both the tangible and symbolic layers of urban life—on-site, within communities, and across digital environments. This enables a dynamic understanding of how cultural expressions, policy frameworks, and ecological transformations co-evolve within the city.

The following sections elaborate on the specific data collection methods, analytical procedures, and triangulation techniques employed to ensure methodological rigor and interpretive validity. Together, these processes translate the culture-centered biocultural paradigm into a systematic empirical investigation, providing robust insights into the interconnections between culture, sustainability, and social equity in Ixelles.

5.4 Data Collection and Analytical Framework: Case Study Context: Ixelles

To operationalize the culture-centered biocultural strategy, the research employed a multi-source and multimethod data collection framework, designed to capture the complex interactions between cultural, social, and environmental dimensions of urban sustainability in Ixelles. Data were gathered from primary and secondary sources through a combination of content analysis, participatory observation, semi-structured interviews, and structured questionnaires. This design ensured both empirical breadth and interpretive depth, in line with the study's mixed-methods orientation.

The content analysis involved a systematic examination of the official websites of the Municipality of Ixelles, its affiliated institutions, and various cultural, social, and environmental organizations operating within the city. This analysis focused on representations of sustainability, cultural initiatives, and public communication strategies, thereby bridging the material and digital layers of the urban ecosystem.

The participatory observation was conducted from 2018 to 2025 and had a hybrid and multimodal character. Researchers engaged in on-site fieldwork, attending public events, local meetings, and community initiatives, while also conducting online ethnographic observation of digital forums and municipal communication channels. This dual perspective allowed for a biocultural reading of urban change—observing how cultural meanings, practices, and narratives evolve across both physical and virtual spaces.

The qualitative component consisted of semi-structured interviews with citizens, municipal officials, and members of migrant and minority communities, conducted both in person and via videoconference to ensure inclusivity and accessibility. The analysis of qualitative data followed the principles of Grounded Theory, allowing theoretical insights to emerge inductively from participants' lived experiences and narratives. Coding and thematic development were performed iteratively, identifying recurring concepts and patterns that illuminate how cultural values influence perceptions of sustainability, inclusion, and inequality. This process ensured that the biocultural perspective remained embedded in the empirical interpretation of urban life.

In parallel, the quantitative component consisted of a structured questionnaire administered to a random sample of residents and members of immigrant communities. The survey measured attitudes, perceptions, and practices related to culture, environment, and social inclusion. Statistical analyses identified correlations between these variables, providing empirical validation of the qualitative findings and contributing to methodological triangulation (Noble & Heale, 2019).

The integration of these methods—qualitative, quantitative, and digital—reflects the study's biocultural methodology, which views knowledge as both socially constructed and empirically observable. Through the use of Grounded Theory, cross-referenced with quantitative results and digital observations, the research establishes a dialectical link between meaning and measurement, ensuring that the cultural dimension remains central to the analysis of sustainability and urban transformation.

Ixelles, a municipality in the Brussels-Capital Region, was selected for its multicultural demographic profile, visible urban regeneration efforts, participatory governance structures, and rich cultural infrastructure. The population includes a significant number of immigrants from Morocco, Greece, Turkey, Congo, and other countries, making it an ideal site for exploring intercultural urban sustainability.

The study examined various city zones, such as Matongé, Flagey, and Cimetière, to understand how environmental policies and cultural activities intersect with social cohesion and public space use.

The city of Ixelles can demonstrate how the implementation of sustainable development strategy plans can lead to addressing cohesion policy challenges in the urban fabric and enhance sustainable and inclusive urbanization (Medeiros & Van Der Zwet, 2020).

Using an interdisciplinary approach, we combine both qualitative and quantitative methods.

5.5 Qualitative Methods and Techniques

The qualitative research design is thus regarded as one in which the researcher typically assigns knowledge rights in accordance with constructivist perspectives (Creswell, 2009). The research design encompasses a range of methods, including the use of narratives, phenomenology, ethnography, grounded theory studies, and case studies (Abuhamda et al., 2021). This approach allows for the utilization of scientific methodologies for the collection and interpretation of data, thereby facilitating the process of generalization (Abuhamda et al., 2021).

In qualitative research, an interpretive and naturalistic methodology is employed to gain insight into phenomena within their contextual settings. The objective is to elucidate the meaning individuals ascribe to these phenomena (Pandey, 2019; Denzin & Lincoln, 2005). Lune and Berg (2017) define qualitative research as a process of uncovering meanings, establishing a term, defining a concept, identifying metaphors, symbols, and descriptions of phenomena. Qualitative data methods, such as evaluation, open-ended questions, in-depth interviews, and field notes, are used to gather data from participants in their natural environments (Leedy & Ormrod, 2014). These techniques provide a comprehensive overview of the study, allowing for a nuanced interpretation of behavior and providing insight into individuals and their real-life contexts (Abuhamda et al., 2021).

5.5.1 Interviews

We also used interviews with city's key stakeholders, as "the knowledge of local experts is important for local construction of knowledge" (Baines et al., 2000). The researchers, before conducting any interview, defined the objectives, researched the stakeholder's role, background, and interests and prepared the questions on an open-ended basis that aligned with their objectives.

In this framework, semi-structured interviews formed the cornerstone of the qualitative data collection strategy, allowing for both guided inquiry and flexibility to explore emergent themes (Kallio et al., 2016). Between 2022 and 2024, 27 interviews were conducted with key informants, including:

- Mr. Christos Doulkeridis, Mayor of Ixelles (across three sessions)
- Professor of ULB, European Employees, characteristic representatives of Ixelles and representative of immigrant communities, stakeholders of the city
- Local business owners, artists, educators, and migrant community leaders

Interviews lasted between 45 and 90 minutes and were conducted in person or via video call. A semi-structured protocol was used to ensure consistency across themes, including:

- Governance and sustainability priorities
- Perceptions of cultural inclusion and urban identity
- Observed changes in public space and citizen engagement

All interviews were audio-recorded, transcribed verbatim, and analyzed using QDA Miner software. Analysis followed an iterative coding process with both deductive themes (based on literature) and inductive categories (emerging from responses) (Nowell et al., 2017).

Reflexive memos were maintained by each interviewer to document positionality and ensure transparency. Interview data provided critical depth to triangulate findings from focus groups, surveys, and Delphi panels.

During the Interview, the researchers made the stakeholder comfortable, by active listening and showing empathy. They acknowledged their points and kept the conversation focused on the objectives. Wherever it was necessary, they asked for confirmation, and they summarized the key points, keeping at the same time notes on personal reflections and on the stakeholders' body language. After the interviews, they summarized and reviewed the key points to capture any additional thoughts and used the insights gained to inform the research design if it was considered necessary. The acquired information was evaluated in conjunction with data obtained from fieldwork, archives, and the researchers' own field observations.

The stakeholders were chosen, among a broad basis from public policymakers, academia, representatives of political parties, entrepreneurship community, local community, civil sector, media, cultural practitioners, social scientists, etc. The choice was made by the researchers by using tools as old-fashioned brainstorming, brainwriting in a given timeframe, historical information, lessons learned, etc. One crucial point on stakeholders' engagement was if there was a prior migration experience.

One of the key stakeholders was the Mayor of Ixelles, Mr. Doulkerides, with whom the researchers conducted interviews at regular intervals, as time variable was given to demonstrate patterns of change. On this basis, the researchers were able to validate patterns of relationships and social interactions among people who share a common past in addition to their migration experience.

The initial session occurred in 2022, and the subsequent one took place in 2023 and in 2024. The basic axes of the interview were about sustainability, climate change, culture governance, etc.

We conducted interviews with over 25 key stakeholders between 2022 and 2024, including:

- Mr. Christos Doulkeridis, Mayor of Ixelles (multiple sessions)
- Local councilors and cultural officers
- Civil society actors, NGO leaders, and community members

Interviews followed a thematic guide and were transcribed, coded, and analyzed using QDA software. Topics included sustainability strategies, governance, migration, cultural identity, and participatory planning.

5.5.2 *Participatory Observation*

From 2018 to 2025, researchers engaged in fieldwork in public spaces, community festivals (e.g., Matongé and Chatelain), local council meetings, and parks. Researchers maintained diaries, conducted informal interviews, and collected photographic documentation.

The researchers, after defining the research goals, selected the key locations such as parks, festivals, and community meetings. They obtained necessary permissions to observe in certain areas and participated in community activities. During observation there was engagement with the community, and the writers took detailed notes and kept a research diary. Additionally, informal interviews, surveys, and photography were used to enrich the data.[1]

Participant observation was necessary when exploring complex social dynamics, especially in communities that were not completely familiar to the researcher. But mainly for the development of grounded theory, since the purpose of our research is to formulate theory that will arise inductively from the field data (Bernard, 2017).

The type of participation was "Participant-as-observer": Participants are aware of the research, and the researcher is active in the group (Bernard, 2017).

Following the observation, the researchers analyzed the data by reviewing the notes taken and categorizing the findings based on themes related to sustainability.

---

[1] Participatory observation is a qualitative research method in which the researcher becomes actively involved in the daily activities of the group or community being studied. Unlike traditional observation where the researcher remains detached, participatory observation blends participation with observation, allowing for deeper immersion and understanding. Participatory observation is a powerful tool for understanding lived experience from the inside out. It bridges the gap between etic (outsider) and emic (insider) perspectives, making it invaluable for studying social phenomena in context. However, it demands careful ethical navigation, critical reflexivity, and a substantial time commitment (Angrosino, 2011).

They reflected then on any potential biases by considering how their presence and perspective might have influenced the observations.

Ethical considerations were followed in the present research: Transparency and consent: Especially crucial if full participation involves deception, Anonymity and confidentiality: Must be protected when publishing sensitive data. Researcher safety and emotional toll: Immersion can expose researchers to stress or danger (Angrosino, 2011).

During these participating sessions, the researchers followed the basic principles of participant observation like the principle of informed consent, which requires that participants were made aware of the nature of the research and gave their consent to being observed and the principle of confidentiality which requires that the identity of participants and any sensitive information they may provide be protected.

5.5.3 Delphi Panels

To deepen the understanding of sustainability challenges and opportunities in Ixelles, the Delphi technique was used to harness the expert judgment of a multidisciplinary panel. This structured method is particularly effective for addressing complex, multidimensional issues where empirical evidence is limited or fragmented (Linstone & Turoff, 1975; Bitsani, 2013).

The Delphi technique is a structured process that enables the gathering of expert opinions with the objective of reaching a consensus. Hostrop (1975) states that the "Delphi" technique offers researchers a more objective approach to evaluate varying opinions on goals and objectives. The Delphi technique is particularly well-suited to address complex issues such as sustainability, where the input of experts and the formation of a consensus are of paramount importance. Prior to commencing the study, it is essential that the researchers define the research problem. This clearly delineates the sustainability issue that is to be addressed and selecting experts. To this end, a diverse panel of experts in the field of sustainability must be identified and recruited, comprising academics, industry professionals, and policymakers. The survey employed the Delphi technique to gather information from staff members of diverse institutions representing different cultural groups and communities, including city government officials, the regional administration, and political officials, administrative bodies or local community representatives (social capital) and proprietors of public and social amenities within the city (Ferri et al., 2005; Cupta & Clark, 1996; Linstone & Turoff, 1975; Bitsani, 2013).

The Delphi process consisted of three iterative rounds conducted between mid-2022 and late 2025. A total of experts was recruited, including academics in sustainability and urban studies, municipal policymakers, environmental engineers, cultural practitioners, and leaders from civil society organizations. Selection criteria emphasized both subject matter expertise and local contextual knowledge.

Round 1 began with open-ended questions designed to elicit broad opinions on what constitutes a biocultural approach to sustainability in the context of Ixelles. Questions included:

- What are the key cultural dimensions that influence sustainable urban development in diverse communities?
- How can municipal policies incorporate cultural values and social practices in climate adaptation?
- What are perceived obstacles to inclusive urban regeneration?

Responses were anonymized and analyzed thematically. Common themes included the importance of integrating immigrant cultural heritage in public space design, fostering participatory governance, and ensuring environmental justice.

Round 2 involved a structured questionnaire based on the Round 1 themes. Experts rated the relevance and feasibility of proposed strategies (e.g., urban gardening programs, intercultural festivals, decentralized energy initiatives). Interquartile deviation scores were calculated to assess consensus levels (Okoli & Pawlowski, 2004).

Round 3 presented the aggregated group responses along with dissenting viewpoints, allowing participants to revise or confirm their judgments. This round solidified areas of agreement, including:

- The need for neighborhood-specific sustainability indicators
- A strong role for municipal arts and culture departments in environmental communication
- Institutional support for co-designed urban spaces involving minority communities

The results of the Delphi process were used to refine our interview questions and inform the thematic coding framework for the broader study. Additionally, the panel's feedback was instrumental in shaping policy recommendations.

By engaging diverse and informed voices, the Delphi method enhanced the validity of our conceptual model and ensured that multiple stakeholder perspectives were integrated into the research design. It exemplified participatory foresight planning within a local governance context (Bitsani, 2013).

Three rounds of the Delphi method were conducted with sustainability experts, urban planners, and social scientists. The goal was to refine consensus on what constitutes biocultural urban sustainability.

The research team objectively validated the conclusions drawn by streamlining the questionnaire process to the point of opinion formation. Composed of focus group representatives, **round table discussions were planned and organized to evaluate the progress** and results of the research.

The researchers participated in these meetings in the capacity of "facilitators."

The research provides a deeper understanding of cultural social groups through the compilation of invaluable information shedding light on the topics researched.

Therefore, thanks to this methodological technique, we were able to delve deeper into the relations among people residing in a specific area and relations among minorities and the majority of Ixelles's. Doing so, we have gained an insight into the issues of intercultural relations and the creation of a multicultural city and of cultural identities.

5.5.4 Focus Group

Foucault (2002, 2009a, 2009b) recognizes the importance of actors in generating, transforming, and disseminating discursive knowledge. Focus groups represent a widely utilized qualitative methodology within the social sciences, designed to explore participants' perceptions, attitudes, and experiences through the medium of moderated group interaction. In contrast to individual interviews, focus groups exploit group dynamics to reveal shared understandings or divergent viewpoints (Bitsani, 2013).

Focus groups were used to capture community-level insights, stimulate group dialogue, and assess shared perceptions of biocultural sustainability. A total of six focus groups were conducted between January and October 2023, each consisting of 6 to 10 participants from different neighborhoods and cultural communities of Ixelles. Recruitment was done using purposive sampling, ensuring representation across gender, age, migration background, and residential area.

Each session lasted approximately 90 minutes and followed a semi-structured discussion guide covering:

- Perceptions of safety, inclusiveness, and environmental quality
- Experiences with cultural programs, municipal services, and participatory platforms
- Recommendations for improving sustainability and cohesion in Ixelles

Sessions were audio-recorded, transcribed verbatim, and analyzed using thematic coding. Data was coded in NVivo and triangulated with interview and survey findings.

Limitations included variability in participant familiarity with urban policies and minor dominance issues in two groups, which were addressed through reflective facilitation. Focus group methodology was especially effective in exploring intercultural tensions and collective memory around urban regeneration.

Six focus groups with residents from different neighborhoods were held. Participants discussed perceptions of urban change, intercultural coexistence, and environmental challenges. Analysis followed thematic and discourse analysis frameworks.

As effective focus group research is dependent upon thoughtful planning, the researchers followed structured steps regarding participant selection, group size, and moderator role to limit bias and guarantee inclusivity (Stewart & Shamdasani,

2014). For data analysis the researchers used thematic coding, in the context of grounded theory and content analysis approaches (Bloor et al., 2001).

But however, the adequate planning was, the limitations of the technique must be mentioned; focus groups are not designed to capture information of a deeply personal or sensitive nature (Carey & Asbury, 2016).

5.5.5 *Multimodal Content Analysis*

To examine how Ixelles communicates its values and priorities through digital platforms, a multimodal content analysis was conducted. This method integrates visual, textual, and semiotic analysis of digital media to uncover the discursive construction of sustainability and intercultural identity (Kress & van Leeuwen, 2001).

Between November 2022 and December 2024, over 30 digital sources were analyzed, including:

- The official municipal website and departmental subpages (culture, mobility, environment, etc.)
- Social media accounts (Facebook, Instagram, Twitter/X) of the Ixelles Municipality and affiliated organizations
- Websites of community-based organizations and cultural venues (e.g., Flagey, CIVA, BiblioXL)

Using a combination of visual grammar and critical discourse analysis (CDA), we coded for:

- Use of inclusive language (e.g., multilingualism, gender-neutral terms)
- Depiction of minority groups and cultural scenes
- Visual metaphors for sustainability (e.g., trees, bicycles, recycled materials)
- Emphasis on participation, innovation, or heritage

We employed the Visual Analysis Matrix (Rose, 2016) to classify layout, gaze, color palette, and symbolism. Findings revealed a dominant visual narrative that reinforces Ixelles as both a green and multicultural city yet revealed underrepresentation of youth voices and informal migrant groups.

We analyzed over 30 official and unofficial websites, social media posts, and online archives from 2022 to 2024. The analysis applied multimodal discourse analysis, identifying how visual language, layout, and semiotics communicate municipal values such as sustainability, safety, and inclusion.

5.6 Quantitative Methods

The principal objective of a quantitative research design is to control the association between an independent variable and a dependent variable or outcome variable representative of a population (Abuhamda et al., 2021). The objective of quantitative research is to quantify the data and generalize findings from a sample of a study from a variety of perspectives. The process of quantitative research entails the collection, analysis, and interpretation of quantifiable data, with the objective of substantiating the hypothesis formulated within a particular study (Ghanad, 2023).

Quantitative research was employed to complement the qualitative findings and provide measurable insights into the perceptions, behaviors, and experiences of residents in Ixelles concerning sustainability, inclusion, and urban governance. This approach allowed for generalizable findings based on statistical analysis and aimed to validate qualitative trends through empirical evidence.

5.6.1 Questionnaire Design

A structured questionnaire consisting of 24 items was developed following Creswell and Poth (2018) guidelines for instrument construction. The tool included both closed-ended Likert scale items (5-point scales measuring agreement and frequency) and open-ended questions to allow limited qualitative elaboration. The questionnaire was divided into the following thematic sections:

- Demographic background (age, gender, education, nationality, neighborhood)
- Awareness and satisfaction with public services
- Experience of and participation in cultural and environmental programs
- Perceptions of safety, quality of life, and sustainability
- Views on inclusion, migration, and municipal governance

The survey was distributed both online (via Google Forms) and in printed form, in French, English, and Greek, ensuring accessibility for diverse linguistic groups.

5.6.2 Sampling Procedures

The sampling strategy followed a stratified random approach to ensure representation across major demographic and cultural groups within Ixelles (Lohr, 2021). The total sample included 160 respondents, with oversampling applied to underrepresented minority groups to account for potential non-responses.

The population was stratified by:

- Neighborhood clusters (e.g., Flagey, Matongé, Cimetière)
- National/ethnic background (Belgian, Moroccan, Greek, Turkish, Congolese, etc.)
- Gender and age

This method allowed the research team to draw a sample reflective of Ixelles' socio-demographic heterogeneity and was particularly effective in studying intergroup perceptions and place-based differences (Fink, 2019).

5.6.3 Data Collection and Quality Assurance

Data was collected between March 2022 and June 2023. For the digital survey, we ensured participant anonymity and applied CAPTCHA verification to prevent bot responses. For paper-based responses, data were manually entered into a coded SPSS dataset with double-checking to reduce entry errors.

Survey administration was supported by trained volunteers in community centers and municipal offices. Non-probability techniques (e.g., snowball sampling) were avoided except in cases where digital literacy or access were barriers to participation.

5.6.4 Data Analysis Techniques

Data was analyzed using IBM SPSS Statistics Version 27. Descriptive statistics (means, standard deviations, frequency distributions) were calculated for each survey item. Inferential analyses included:

- Cross-tabulation and chi-square tests to detect significant differences between demographic groups
- Independent t-tests and ANOVA for comparing means across neighborhood or ethnic subgroups
- Pearson correlation to assess the relationship between perceptions of sustainability and civic participation

Open-ended responses were analyzed using content analysis to identify recurring terms and themes (Neuman, 2022).

5.6.5 Validity and Reliability

Internal consistency reliability was tested using Cronbach's alpha, with subscales for "perception of sustainability" and "civic trust" yielding coefficients above 0.80. Face and content validity were established through expert review and pilot testing with 15 participants prior to full deployment.

Where possible, validated items were adapted from previous urban sustainability surveys, such as the Eurobarometer on Environment, ensuring comparability with broader European trends.

In the present study, a sample of 160 responses was considered adequate owing to a combination of practical constraints and methodological considerations. The challenges inherent to field-based research, namely limited time, resources, and accessibility, necessitated a reduction in the initially intended sample size. Furthermore, ethical concerns and public health considerations related to the post-COVID context further constrained the data collection process. Whilst it is acknowledged that the reduced sample size engenders elevated margins of error and the potential for diminished generalizability, these limitations are reported with transparency. In order to mitigate potential biases and enhance the reliability of the findings, stratified or quota sampling techniques were employed in order to ensure that key demographic characteristics of the population were proportionally represented. Convenience sampling was avoided unless it was explicitly warranted and justified, in order to maintain methodological rigor and support the validity of the results within the scope of the study's limitations.

Sample Limitations

A limitation is defined as a factor that may or will affect the study but is not under the control of the researcher. In studies utilizing questionnaires, a common limitation is the willingness of the individuals to respond at all, to respond in a timely fashion, and to respond accurately.

The writers identified five major limitations as a basis for this study.

Sample Size: The study was conducted with a sample size of about 10%. Future research with a larger and more diverse sample could provide more robust results.

Geographical Scope: The research was limited to a specific geographical area, which may not reflect the experiences or behaviors of individuals in other regions. Expanding the study to include multiple locations could enhance the applicability of the results.

Data Collection Method: The reliance on self-reported data may introduce bias, as participants might not accurately recall or may choose to present themselves in a favorable light. Utilizing multiple data collection methods, such as observations, can mitigate this issue.

Time Constraints: The study was conducted over two years' time.

External Factors: Uncontrolled external factors, such as economic conditions or policy changes, may influence the results in the future. Future research could control these variables.

5.7 Data Analysis

Quantitative data were coded and analyzed using SPSS. Descriptive and inferential statistics were applied, including frequency analysis, cross-tabulations, and correlation tests to explore relationships among variables.

5.8 Triangulation and Validity (Expanded)

To ensure the robustness of findings, triangulation was employed on multiple levels (Denzin, 1978):

- **Data triangulation:** Interviews, focus groups, surveys, web content, and fieldnotes
- **Methodological triangulation:** Combining qualitative and quantitative tools
- **Investigator triangulation:** Multidisciplinary research team from sociology, urban planning, and communication
- **Environmental triangulation:** Multiple neighborhoods and settings within Ixelles

Triangulation was not only a methodological choice but a foundational principle underpinning the entire research design. It strengthened credibility, dependability, and confirmability, especially important for a longitudinal and multi-perspective case study such as this.

Data triangulation involved combining diverse data source, semi-structured interviews, focus groups, participatory observation, visual and multimodal content, and a community-wide quantitative survey. This ensured that we could capture phenomena from both institutional and grassroots perspectives, across time and settings.

Methodological triangulation merged qualitative and quantitative paradigms. For instance, resident perceptions captured in focus groups were cross-verified with statistical trends from survey responses. Insights from digital content analysis were substantiated by interviews with policymakers and stakeholders. This methodological hybridity allowed for richer insight into contradictions and convergences in community experience and policy narratives.

Investigator triangulation was realized through a multidisciplinary research team consisting of sociologists, urban planners, and sustainability scholars. This allowed for reflexive dialogue on interpretation, reduced individual researcher bias, and brought varied epistemic perspectives into the analytical process.

Environmental triangulation was key given the spatial heterogeneity of Ixelles. The research incorporated experiences from different neighborhoods (e.g., Matongé, Flagey, Cimetière), providing a geographic spread of findings. This also allowed for comparisons between areas with different demographic makeups, urban planning histories, and institutional access.

Each triangulation strategy supported the principles of transferability, transparency, and academic rigor.

5.9 Data Analysis of all the Stages of the Ixelles's Case Study

The **data** for this study were collected in two steps from both primary and secondary sources. The qualitative approach was applied through conducting interviews with prominent personalities of the city and the greater Ixelles area, while the quantitative methodology was utilized through the administration of a structured questionnaire. A random sample of residents, as well as members of immigrant communities, was surveyed. Our qualitative data included interview transcripts, documents, open-ended survey responses, diary logins, and even the interpretation of images and videos as qualitative method goes beyond text-based data and explores the softer side of human experiences.

The writers chose qualitative data analysis as a QDA tool, a streamlined and intuitive platform specifically designed for Qualitative Research. QDA design helped maintain a clear overview of analyzing interviews data, reducing the complexity often associated with qualitative research.

The **content analysis** of media and social media, including the city's website and official Facebook page was a very important source as discourses are mirrors of the typical knowledge of society (Keller, 2013 in Christmann & Heimann, 2017). Also, the researchers retrieved news reports and articles related to Ixelles city from IBSA Institute bruxellois de statistique et d' analyse, municipality of Ixelles brochure, stabelharmonia project, info Ixelles Elsene 2023, on-line journal as Taylor and Francis group, Brussels Studies, online newspapers, policy proposals leaflets from parties. Moreover, they acquired data from various sources, including original archival texts from the Municipality of Ixelles, texts from various social communities as culture ixelles.be, elsene.bibliotheek.be, weKONEKT.brussels, "Ixelles Culture/we are XL" and cultural industries. Widely used for interpreting meaning from text data, content analysis helped the researchers to identify important aspects of content. Researchers analyzed patterns, themes, and recurring ideas in the data.

The researchers also focused on understanding and interpreting stories, narratives, and personal accounts and they used narrative analysis to do so. This method helped the researchers to explore how individuals construct meaning through their narratives.

In analyzing the data the researchers used also the **Grounded Theory Analysis**, as through data collection and analysis they started with a single case to formulate a theory, which can be further tested with additional cases to see if they contribute to that theory.

This approach facilitated the way, the data obtained by the processing of files, texts, and testimonials, which matters relevant to this society have been addressed and integrated as components in our subsequent fieldwork (Bitsani & Kavoura, 2011).

5.9.1 Indicators

For the creation of the interviews and questions, indicators which are based on the biocultural approach were used. As the discussion is about cities' sustainability development, the focus is based on how we address human needs, as these are demonstrated by Marslow pyramid.[2]

[2] Maslow's theory of motivation in psychology is based on a model of human needs that has five levels. These levels are organized in a pyramid structure. Maslow's Hierarchy of Needs is well known in the fields of psychology and management: the pyramid is easily recognizable.

Maslow Pyramid (Hierarchy of Needs) (Fig. 5.2)

The bio-cultural indicators were categorized according to the following axes; their correspondence by axis with the UN Sustainable Development Goals is presented in the next table (Table 5.2).

The above indicators can help policymakers and practitioners to evaluate sustainability from a biocultural perspective, considering both cultural and biological diversity. By using these indicators, it is possible to promote sustainability that supports the interconnections between cultural and biological diversity and benefits both people and the environment.

5.10 Ethics

Volunteerism, anonymity, confidentiality, informed consent, potential for harm, and communication of results are the main ethical pillars that our research was planned, conducted, and written on.

Ethical considerations were central to the research design and execution:

- Ethical approval was secured from the relevant university board
- Participants provided informed consent, with materials available in French, English, and Greek
- The anonymity and confidentiality of participants were preserved

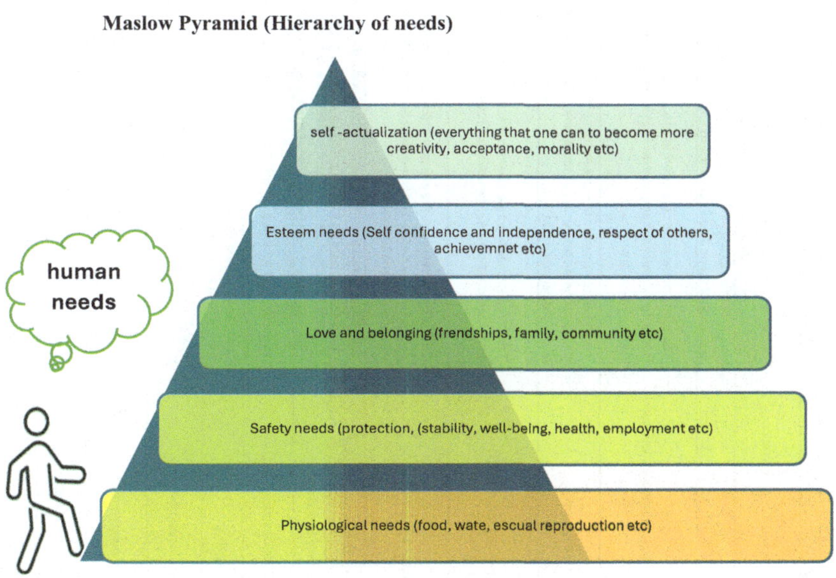

Fig. 5.2 A visual representation of Maslow's hierarchy of needs. Processing/editing

Table 5.2 Bio-cultural indicators

Title	Indicates	SDGs	TITLE	TARGET	DIVEDED FROM:
Cultural diversity preservation	This indicator measures the extent to which different cultural identities are being preserved and promoted in a given area, including language, customs, and traditions. The indicator could include the presence of cultural institutions and events, the	5, 11, 10, 11	Cultural Diversity Preservation	Target 5.1., 5.5,Target 11.3, 11.4, 11.6, 11.7,Target 10.2,10.3,10.4	Swyngedouw & Swyngedouw 2009, Bitsani, 2014
Sense of place	This indicator includes the existence of local landmarks and historical sites, the presence of public art and cultural expressions, and the level of engagement and	11	Sence of place	Target 11.3,11.4,11.6,11.7	Stevensen, 2007
Traditional knowledge preservation	This indicator measures the extent to which traditional knowledge related to natural resource management and conservation is being preserved and passed down to future generations.	3, 4, 7, 11, 13, 15	Traditional knowledge preservation	Target4.7,4.a, Target 7.1,7.a, 11.6, Target 13.3, 15.3,15.4,15.5,15.6	Rakotondrabe & Girard, 2021, Hanspach et. all, 2020; Reyen-Gard et all, 2022).
Social cohesion	The indicator includes the existence of community events and activities, the level of participation in community initiatives, and the perceived sense of safety and security within the neighborhood.	5, 16	Social cohesion	Target 5.5, Target 16.3,16.4,16.5,16.6,16.7	Medeiros & Van Der Zwet, 2020
Biodiversity conservation	This indicator measures the extent to which biodiversity is being conserved and protected in a given area, including both cultural and biological diversity	5, 8, 10, 11, 13, 15	Biodiversity conservation	Target 5.1,5.a, Target 8.3,8.8, Target 10.3,10.6, Target 11.4, Target 13.3, Target 15.1,15.6, 15.7,15.9	Batista & Andrade, 2021, Xin Li & Xinyu Zhuang (2023)
Community involvement	This indicator measures the extent to which local communities and indigenous peoples are involved in decision-making processes related to natural resource management and conservation	4, 13, 17	Community involvement	Target 4.7, Target 13.3	Reyes-Garcia et.all, 2022, Medeiros & Van Der Zwet, 2020
Access to natural resources	This indicator measures the extent to which local communities have access to natural resources and are able to use them in a sustainable manner	2, 6, 1, 3	Access to natural resources	Target 2.3,2.4,2.5,Target 6.1,6.3,6.b, Target 1.5,1.a, Target 3.8	Santos,2022, Braubach et all, 2017, Serrat, 2017, Valente,2020, Ravenscroft et all, 2013
Sustainable livelihoods	This indicator measures the extent to which local communities are able to maintain sustainable livelihoods that are based on traditional practices and natural resource				
Local food systems	This indicator measures the extent to which local food systems are being promoted and supported, including community gardens, farmers' markets, and local food processing facilities.	2, 12	Local food systems	Target 2.3, Target 12.8,12.b	Santos,2022, Diekmann et. all, 2020
Accessibility	The indicator could include the availability of public transportation, the walkability of the city, and the existence of accessible public spaces.	11	Accessibility	Target 11.1,11.2,11.7	Reyes-Garcia et.all, 2022
Green space provision	This indicator measures the extent to which urban areas provide green spaces that support both cultural and biological diversity.	3, 11	Green space provision	Target 11.3,11.4	Peters et al 2010, Santos,2022, Vierikko et all, 2015
Environmental quality	The indicator could include the level of green space and tree cover, the presence of sustainable infrastructure such as rain gardens and green roofs, and the perceived cleanliness and safety of public spaces.	11, 12	Environmental quality	Target 11.3, Target 12.2	Santos,2022, Zheng et al, 2023; Cheshmehzangi et al, 2021
Environmental justice	This indicator measures the extent to which environmental benefits and harms are being distributed equitably among different communities and social groups.	7, 10, 13	Environmental justice	Target 13.3,7.1, Target 13.3,10.4, Target 13.3,13.1,13.2,13.3	Hellgren & Zapata-Barrero 2022, Brinkley & Wagner, 2022
Climate change adaptation	This indicator measures the extent to which communities are adapting to the impacts of climate change using traditional knowledge and practices.	11, 13	Climate change adaptation	Target 11.6, Target 13.1,13.2,13.3	Lin et all, 2021 ,Akturk, 2021

- Digital data collection minimized power imbalances and facilitated access for vulnerable populations (e.g., undocumented migrants)
- Sensitive data were securely stored and used only for academic purposes

These procedures ensured compliance with the Declaration of Helsinki and EU GDPR standards.

Doing so, we also wanted to demonstrate our respect for the individual, which is expressed by recognizing that his or her autonomy and right to self-determination are based on his or her ability to make decisions and choose. The use of digital methods and the care of not exposing the identity of our informants prohibit the creation of conditions in which the respondents may be at risk of suffering psychological harm. We provided an introductory letter (Appendix III), stating the purpose of the research and all the relevant information about it, including the risks and potential benefits of getting involved, the time duration, and the contact details and approval number of the head researcher. We also let them know that their information will be kept confidential and that they are free to stop completing the survey at any time and for any reason. They can also withdraw their information by contacting you or your supervisor, at a specific given time (prior to the publication). Furthermore, they were informed where and when they can access the results of our research.

In another level, we were sincere towards our informants about the aim of our research, and we pursued to have their written consent (Appendix I) toTo do so, we designed not only the online questionnaire, but a written onewhich as the google formed one was prepared in three languages; french, english, Greek (Appendix II) to avoid discriminations.

5.11 Conclusion

In conclusion, the research methodology operationalize the biocultural approach through a coherent integration of philosophical paradigms, methodological strategies, and analytical techniques. By situating culture at the center of the biocultural framework, the design captures the reciprocal relationship between human meanings, social structures, and environmental realities. The combination of qualitative and quantitative methods, structured within a pragmatic and abductive logic, allows the study to move fluidly between interpretation and measurement, linking subjective experience with objective evidence.

Conflict of Interest The authors have no conflicts of interest to declare that are relevant to the content of this chapter.

References

Abuhamda, E., Ismail, I., & Bsharat, T. (2021). Understanding quantitative and qualitative research methods: A theoretical perspective for young researchers. *International Journal of Research, 8*, 71–87. https://doi.org/10.2501/ijmr-201-5-070

Angrosino, M. (2011). Doing ethnographic and observational research. In *Sage research methods*. Sage. https://doi.org/10.4135/9781849208932

Babbie, E. (2011). *The practice of social research* (13th ed.). Wadsworth.

Baines, J., Taylor, N., Buckenham, B., Douglas, J., & McClintock, W. (2000, June). *The importance of local knowledge*. https://www.researchgate.net/publication/237513773_The_Importance_of_Local_Knowledge

Bans-Akutey, A., & Tiimub, B. M. (2021). Triangulation in research. *Academia Letters, 2*, 1–7. https://doi.org/10.20935/AL3392

Baxter, P., & Jack, S. (2010). Qualitative case study methodology: Study design and implementation for novice researchers. *The Qualitative Report, 13*(4), 544–559. https://doi.org/10.46743/2160-3715/2008.1573

Bernard, H. R. (2017). *Research methods in anthropology: Qualitative and quantitative approaches*. Rowman & Littlefield.

Bitsani, E. (2013). *"Intercultural City identity" and "human intercultural cities" (H.I.C). An ontological/dynamic model for the social cohesion of modern and post-modern cities. The research example of Trieste Italy" research monograph.* Nova Publishers.

Bitsani, E., & Anselmi, W. (2021). From the today's dystopia to the future utopia reality: Recovering the human cultural identity of European City. In *Virtual 27th international conference of Europeanists, Europe's past, present, and future: Utopias and dystopias*. Columbia University.

Bitsani, E., & Kavoura, A. (2008). Methodological considerations for qualitative communication research. *Procedia-Social and Behavioral Sciences, 147*, 544–549. https://doi.org/10.1016/j.sbspro.2014.07.156

Bitsani, E., & Kavoura, A. (2011). Organizational networks, migration, and intercultural relations in Trieste, Italy. *International Journal of Advances in Culture, Tourism and Hospitality Research, 4*, 26–37.

Blaikie, N. (2010). *Designing social research. The logic of anticipation* (2nd ed.). Polity Press.

Bloor, M., Frankland, J., Thomas, M., & Robson, K. (2001). *Focus groups in social research*. Sage.

Bryman, A., & Bell, E. (2015). *Business research methods*. Oxford University Press.

Bryman, A., & Bell, E. (2019). *Social research methods* (5th Canadian ed.). Oxford University Press.

Carey, M. A., & Asbury, J.-E. (2016). *Focus group research*. Routledge. https://doi.org/10.4324/9781315428376

Christmann, G. B., & Heimann, T. (2017). Understanding divergent constructions of vulnerability and resilience: Climate change discourses in the German cities of Lübeck and Rostock. *International Journal of Mass Emergencies and Disasters, 35*(2), 120–143.

Coombs, H. (2022). *Case study research: Single or multiple [White paper]*. Southern Utah University. https://doi.org/10.5281/zenodo.7604301

Creswell, J. W. (2009). *Research design qualitative, quantitative and mixed methods approach* (3rd ed.). Sage.

Creswell, J. W., & Poth, C. N. (2018). *Qualitative inquiry and research design choosing among five approaches* (4th ed.). Sage.

Crowe, S., Cresswell, K., Robertson, A., et al. (2011). The case study approach. *BMC Medical Research Methodology, 11*, 100. https://doi.org/10.1186/1471-2288-11-100

Cupta, G. U., & Clark, E. R. (1996). Theory and applications of the Delphi technique: A bibliography (1975–1994). *Technological Forecasting and Social Change, 53*(2), 185–211. https://doi.org/10.1016/S0040-1625(96)00094-7

Denzin, N. K. (1978). Triangulation: A case for methodological evaluation and combination. *Sociological Methods*, 339–357.

Denzin, N. K., & Lincoln, Y. S. (2005). Introduction: The discipline and practice of qualitative research. In N. K. Denzin & Y. S. Lincoln (Eds.), *The Sage handbook of qualitative research* (3rd ed., pp. 1–32). Sage.

Farquhar, J., Michels, N., & Robson, J. (2020). Triangulation in industrial qualitative case study research: Widening the scope. *Industrial Marketing Management, 87*, 160–170. https://doi.org/10.1016/j.indmarman.2020.02.001

Ferri, C. P., Prince, M., Brayne, C., Brodaty, H., Fratiglioni, L., Ganguli, M., Hall, K., Hasegawa, K., Hendrie, H., Huang, Y., Jorm, A., Mathers, C., Menezes, P. R., Rimmer, E., Scazufca, M., & Alzheimer's Disease International. (2005). Global prevalence of dementia: A Delphi consensus study. *Lancet (London, England), 366*(9503), 2112–2117. https://doi.org/10.1016/S0140-6736(05)67889-0

Fink, A. (2019). *How to conduct surveys: A step-by-step guide*. Sage.

Foucault, M. (2002). *The archaeology of knowledge*. Trans of 1969 original by A. M. Sherridan Smith. Routledge. Ice, action, and policy. Teachers College Press.

Foucault, M. (2009a). *Security, territory, population: Lectures at the Collège de France 1977–1978*, trans. Graham Burchell. Picador.

Foucault, M. (2009b). *Security, territory, population: Lectures at the Collège de France 1977–1978* (M. Senellart, Ed.; G. Burchell, Trans.). Palgrave Macmillan.

Fraser, M. (2006). *Immatriality, virtual reality and interaction*. https://www.dhi.ac.uk/san/waysofbeing/data/health-jones-hindmarsh-2006.pdf

Ghanad, A. (2023). An overview of quantitative research methods. *International Journal of Multidisciplinary Research and Analysis, 6*(8), 52. https://doi.org/10.47191/ijmra/v6-i8-52

Gruber, T. (2009). Ontology. In L. Liu & M. T. Özsu (Eds.), *Encyclopedia of database systems* (pp. 1963–1965). Springer. https://doi.org/10.1007/978-0-387-39940-9_1318

Guba, E. G., & Lincoln, Y. S. (1994). Competing paradigms in qualitative research. In N. K. Denzin & Y. S. Lincoln (Eds.), *Handbook of qualitative research* (pp. 105–117). Sage.

Hostrop, R. W. (1975). *Managing education for results*. ETC Publications.

Kallio, H., Pietilä, A. M., Johnson, M., & Kangasniemi, M. (2016). Systematic methodological review: Developing a framework for a qualitative semi-structured interview guide. *Journal of Advanced Nursing, 72*(12), 2954–2965. https://doi.org/10.1111/jan.13031

Kress, G., & van Leeuwen, T. (2001). *Multimodal discourse: The modes and media of contemporary communication*. Arnold.

Leedy, P., & Ormrod, J. E. (2014). *Practical research planning and design* (10th ed.). Pearson Educational.

Linstone, H. A., & Turoff, M. (1975). The Delphi Method: Techniques and applications. *Journal of Marketing Research, 13*(3), 317–318. https://doi.org/10.2307/3150755

Lohr, S. L. (2021). *Sampling: Design and analysis*. Chapman and Hall/CRC. https://doi.org/10.1201/9780429298899

Lune, H., & Berg, B. L. (2017). *Qualitative research methods for the social sciences* (9th ed.). Pearson.

Medeiros, E., & Van Der Zwet, A. (2020). Evaluating integrated sustainable urban development strategies: A methodological framework applied in Portugal. *European Planning Studies, 28*(3), 563–582. https://doi.org/10.1080/09654313.2019.1606898

Neuman, W. L. (2022). *Social research methods: Qualitative and quantitative approaches* (9th ed.). Pearson.

Noble, H., & Heale, R. (2019). Triangulation in research, with examples. *Evidence-Based Nursing, 22*(3), 67–68. https://doi.org/10.1136/ebnurs-2019-103145

Nowell, L. S., Norris, J. M., White, D. E., & Moules, N. J. (2017). Thematic analysis: Striving to meet the trustworthiness criteria. *International Journal of Qualitative Methods, 16*(1). https://doi.org/10.1177/1609406917733847. (Original work published 2017).

Okoli, C., & Pawlowski, S. D. (2004). The Delphi method as a research tool: An example, design considerations and applications. *Information & Management, 42*(1), 15–29. https://doi.org/10.1016/j.im.2003.11.002

Pandey, J. (2019). Deductive approach to content analysis. In M. Gupta, M. Shaheen, & K. Reddy (Eds.), *Qualitative techniques for workplace data analysis* (pp. 145–169). IGI Global. https://doi.org/10.4018/978-1-5225-5366-3.ch007

Priya, A. (2021). Case study methodology of qualitative research: Key attributes and navigating the conundrums in its application. *Sociological Bulletin, 70*(1), 94–110. https://doi.org/10.1177/0038022920970318

Rose, G. (2016). *Visual methodologies: An introduction to researching with visual materials* (4th ed.). Sage.

Saunders, M., Lewis, P., & Thornhill, A. (2007). *Research methods for business students* (4th ed.). Financial Times Prentice Hall.

Stewart, D. W., & Shamdasani, P. N. (2014). *Focus groups: Theory and practice* (3rd ed., pp. 39–139). Sage.

Torraco, R. J. (2005). Writing integrative literature reviews: Guidelines and examples. *Human Resource Development Review, 4*(3), 356–367. https://doi.org/10.1177/1534484305278283. (Original work published 2005).

Chapter 6
Visual Content Analysis of the Municipality of Ixelles Through Its Websites

6.1 Overview of Analyzed Websites

The analysis of the websites was carried out using a multimodal approach. This method was chosen because meaning on a website can be conveyed through language (written or spoken), images, and kinesthetic actions.

Every image carries a semantic script and, according to the theory of multimodality, constitutes a system of multiple modes (Bitsani & Anselmi, 2021; Kress & Van Leeuwen, 2001).

Within this framework of multimodal interpretation, a visual, aesthetic, political, and semiotic analysis was conducted. The websites examined and analyzed include the official pages of the Ixelles municipality, the electronic press, the website of the community leader, local organizations, NGOs, and community social media platforms. These are as follows:

brusselstimes.com, The Brussels Times.
brusselslife.be.,
https://efus.eu/about-us/brussels/,
https://www.brussels.info/,
https://besustainable.brussels/fr,
https://culture.ixelles.be/fr/,
https://lowtrafficneighbourhoods.brussels/,
https://www.rtbf.be/
https://besustainable.brussels/fr,
https://ms-a.be/office, universities,
https://monument.heritage.brussels/ STABEL, IBSA.

The following list comprises the websites of the Commune of Ixelles, along with the links that were examined (Table 6.1).

E. Bitsani et al., *The Biocultural City as a Utopia or the Future Reality?*, Cities, Heritage and Transformation, https://doi.org/10.1007/978-3-032-17584-7_6

Table 6.1 Directory of websites of Services and Contacts

Organization/Service	Website	Email
Commune d'Ixelles	www.ixelles.be	info@ixelles.brussels
CPAS d'Ixelles	cpasixelles.brussels	info@cpasixelles.brussels
Service des Archives		archives@ixelles.brussels
État Civil		etat.civil@ixelles.brussels
Service Population		population@ixelles.brussels
Service Étrangers		etrangers@ixelles.brussels
Aménagement du territoire		amenagementterritoire@ ixelles.brussels
Cohésion sociale	www.ixelles.be/ site/639-Cohesion-sociale	
CSS Bruxelles Sud-Est	www.cssbse.be	
ARPAIJE asbl	www.arpaije.be	
FEMMES Prod asbl	www.codef.be/asbl/ femmes-prod	
Théâtre Varia	www.varia.be	
Petit Théâtre Mercelis	culture.ixelles.be/en/ mercelis-theatre/	
Théâtre Molière	www.muziekpublique.be	
Flagey	www.flagey.be	
CIVA	www.civa.brussels	
BiblioXL	www.biblioxl.be	bibliotheque@ixelles.brussels
Piscine d'Ixelles	www.swim.brussels/ la-piscine-dixelles	piscine@ixelles.be
Culture Ixelles	culture.ixelles.be	culture@ixelles.brussels
Accueil Temps Libre	atl.ixelles.be	
Plan Culturel Ixelles (PDF)	culture.ixelles.be/wp-content/ uploads/2023/03/Plan-Local- de-Politique-Culturelle-- Ixelles-2020-2025- Annexes-FR.pdf	
Répertoire Social Ixelles (PDF)	cpasixelles.brussels/ wp-content/uploads/2022/02/ Répertoire- CSXL_21-12-2020_Web.pdf	

Data collection and analysis took place between November 2018 and December 2024, with ongoing feedback and cross-referencing.

6.1.1 Description of Ixelles Official Websites

To support the analysis outlined above, a set of websites was selected based on their relevance to the research objectives. Table 6.2 summarizes the main features of each site, including their thematic focus and structural characteristics (Fig. 6.1).

The literature review of the municipality of Ixelles and the first analysis of its website clearly indicate the municipality's situation and the challenges it faces (Table 6.3).

6.1.2 Conclusion

The design of these websites supports the principles of multilingualism and multiculturalism by integrating modern user interfaces and participatory content structures that encourage inclusivity and civic engagement. The aesthetic choices convey key institutional values such as sustainability, community involvement, and social welfare.

The official websites of the Municipality of Ixelles employ visual identity to:

- Reinforce the urban and cultural character of the municipality through images of public spaces, parks, and cultural activities.
- Strengthen multilingualism and multiculturalism through modern interfaces and participatory content.
- Communicate messages of sustainability, community engagement, and social care.

Their visual aesthetics portray a municipality that is informed, flexible, vibrant, and open to society, while maintaining respect for its historical heritage.

6.2 Correlation and Comparison of Evaluation Indicators with Municipal Governance by Quarters

6.2.1 General Evaluation Based on the Research Indicators

Security

- Technical security: The ixelles.be website uses HTTPS, modern web design standards, and up-to-date cookie banners, indicating a commitment to privacy and data protection.

Table 6.2 Sites and key details

Thematic axles	Focus points
1. Visual Identity & Branding	– The official logo, flag, and colors (green, white, green-black, and white) reinforce the collective identity and respect for tradition. – The use of simple geometric shapes reflects a modern aesthetic, while at the same time being durable and consistent, respecting the history of the region.
2. Structure & Navigation	– The multilingual websites (French and Dutch) highlight the bilingual and multicultural reality of Ixelles. – The menu is divided into sections such as "Culture," "Events," "Citizen Services," and "FAQ." These sections reflect the emphasis on social participation and the open democratic model of administration.
3. Cultural representation	– Events such as XL Danse (folk and international dance performances in the heart of the city) illustrate an open cultural dialogue and an emphasis on everyday citizen participation. – Photos from parks, schools, and museums enhance the visual identity of the Commune. – Other cultural activities (museums, schools, universities, public art spaces) are promoted with photos that highlight the urban landscape such as parks, neo-colonial buildings, and Art Nouveau monuments.
4. Visual Language & Aesthetics	– Combination of photos and modern graphics/illustrations. – A combination of professional photos (parks, historical buildings) and modern graphics/illustrations (e.g., artistic posters, digital sketches) is used. – The photographs function as a digital projection of urban culture: Tenbosch Park, school buildings, museums, vibrant neighborhoods. – The graphics of the type "XL Danse," etc. with lively, simplistic designs in bright colors convey a feeling of openness, joy, and creativity.
5. Cultural Values & Messages	– Open, flexible, and participatory commune. – Emphasis is placed on "public space" as a cultural platform—markets, parks, squares, museums—an element that characterizes the social philosophy of the Commune. – The websites are characterized by the presence of social services and ecological actions; services such as animal welfare and social welfare highlight the values of sustainable urban life and community solidarity. – Multilingualism and international collaborations. – Multilingual information and international ties (strategic twins, etc.) reflect the linguistic and cultural diversity of the Commune.
6. Overall Impression	– Cultural identity and sustainability are strengthened. – Website element: emphasis. – Colors/signs: green and white, representing tradition and ecology. – Languages: French and Dutch, which are typical for Brussels. – Graphics vs. photos: joy, participation, history, and everyday life. – User content: cultural events, public spaces, and social welfare.

- Perception of safety: Access to local services (police, theft/fire response) is visible through a dedicated "Service public" section, though information on preventative initiatives is limited. There is no mention of topics such as community policing or crime statistics.

Fig. 6.1 Photos of Ixelles (Ixelles.be)

Table 6.3 Situation and challenges in Ixelles

Indicator	Matongé/Cimetière	Flagey/Châtelain
Safety	Low—crime, perceived danger	High—calm mobility, surveillance
Green/ environment	Limited green space, poor infrastructure	Parks, ponds, recreational paths
Interculturality	High, but lacking institutional support	Moderate, smooth coexistence
Cultural integration	Strong street culture, informal	Formal cultural centers, events
Participation	Fragmented, lack of tools	Participatory structures and platforms

Quality of Life

- Ixelles is portrayed as a "cultural municipality par excellence", with over 100 cultural venues and more than 200 events per month (source: ixelles.be). It highlights assets such as parks, lakes, Art Nouveau architecture, and major facilities such as Flagey.
- The municipality provides multilingual content (FR–NL–EN), newsletters, and social media presence (@kult.XL), supporting accessibility and public communication.
- However, the websites lack detail on subjects such as waste management, public health, housing, and sustainable mobility.

Migration

- The municipality supports and celebrates multiculturalism: the Matongé district is described as a vibrant hub of African and Asian communities.

- Cultural events (e.g., Matongé en Couleurs) and multilingual communication enhance the visibility and participation of migrant communities.
- Nonetheless, the websites do not provide information on integration programs, official support for asylum seekers, or educational and language initiatives.

Governance

- A trilingual menu and transparent access to services, office hours, newsletters, and communication channels are available.
- However, there is no dedicated section for citizen consultation (e.g., e-participation), administrative accountability, municipal council updates, or performance indicators.
- While the digital infrastructure is robust, transparency and participatory engagement beyond the cultural domain remain limited.

Cultural Activity

- Exceptionally strong: over 100 cultural venues and more than 200 events per month, including major institutions such as Flagey, the Ixelles Museums, and the Children's Museum (sources: Wikipedia; ixelles.be).
- International cultural communication (via social media and newsletters) is accessible to both residents and visitors.
- Annual festivals such as Matongé en Couleurs, theatre performances, comedy clubs, and film events contribute to a vibrant cultural scene.

Intercultural Relations

- The municipality actively promotes cultural coexistence (e.g., Matongé and the Asian/Indian community).
- Trilingual information (FR–NL–EN) supports intercultural interaction.
- However, no specific actions were identified regarding intercultural education, child integration, structured dialogue, or collaborative problem-solving (Table 6.4).

From the above, it can be concluded that the Municipality of Ixelles presents a strong cultural profile, multilingual communication, and a visible presence of migrant communities. Its digital infrastructure is relatively responsive. However, there remains considerable room for improvement in areas such as safety (both social and urban), citizen participation, transparency, and comprehensive migrant support through specific integration programs, intercultural education, and social cohesion initiatives.

Gaps were identified in public health, sustainable mobility, waste management, housing, and migrant services. Accordingly, further emphasis was placed on the Quality-of-Life Index, which encompasses these aspects.

Table 6.4 Comparison and conclusions of Ixelles websites according to research indicators

Indicator	Current status	Gaps/proposals
Safety	Strong technical infrastructure, lack of social engagement	Enrichment with info/events on safety and public order
Quality of life	Cultural environment, good communication	Additional quality indicators (health, transport, environment)
Migration	Very active visibility of migrant communities	Inclusion, education, support for asylum structures
Governance	Trilingual, well-organized public services	Direct participation, e-consultation, transparency
Cultural activity	Strong presence, multiple events, historical and contemporary art	Preservation and expansion of international collaborations
Intercultural relations	Encouraged, diversity of communities present	Targeted community contact events, festivals, educational programs

6.2.2 *In-Depth Analysis*

A more detailed analysis of subpages and banners revealed the following findings regarding the Quality-of-Life Index in relation to municipal policy and governance:

6.2.2.1 Clear Strategy and Services (Mainly Digital)

The Cellule Qualité de la Vie (Quality of Life Office) addresses issues such as vermin, stray animals, and pigeons, and provides free interventions in cases of hygiene problems.

There is also an Animal Welfare Service, with a program to sterilize stray animals and collect dead animals.

Citizens are informed about environmental issues and air quality, and about the Climate Action Plan 2030, which aims to reduce CO_2 emissions by 55 percent.

The Commune is a member of the Harmonia Network for Improved Resilience and Sustainable Urban Areas to Cope with Climate Change and Extreme Events.

The Commune of Ixelles employs new technologies consistent with the "smart city" concept, acting as both mediator and facilitator: on the one hand, empowering citizens to shape their environment, and on the other, utilizing participatory methods in urban planning (Wilson et al., 2019).

Assessment: Strong administrative framework, with a focus on cleanliness and environmental management, though with less emphasis on health or welfare issues.

6.2.2.2 Public Spaces and Mobility

The reconstruction of the Chaussée d'Ixelles (2016–19) transformed the street from façade to façade, incorporating pedestrianization, cycle lanes, and car restrictions, creating a more pleasant and safer environment.

Implementation of the Contrat Local de Mobilité Flagey–Étangs under the Good Move program aims to alleviate traffic congestion by improving public transport and facilities for pedestrians and cyclists.

Municipal and regional transport, car-sharing, and micro-mobility schemes supplement private car use. Bus frequency (line 71) has increased by approximately 66 per cent.

The Commune has invested in pedestrian and bicycle infrastructure to encourage active transport and reduce reliance on cars. This includes the construction of cycle lanes and pedestrian zones, as well as the installation of bicycle parking facilities.

Public space and mobility remain challenging issues in many contemporary cities, as noted by Ravazzoli and Torricelli (2017).

Assessment: Dynamic upgrading of public spaces, demonstrating a good balance between accessibility, walkability, and sustainability.

6.2.2.3 Green and Open Spaces

Green spaces and ponds (Ixelles Ponds, Tenbosch Park) form a green belt approximately 700 m in length within the urban fabric.

Emphasis is placed on nature conservation, biodiversity, rainwater management, and "free garden/composting" projects within the community.

The Be Sustainable initiative aligns with wider sustainable urban development projects.

One notable project is Plan Arbres, which aims to decongest the Commune and plant trees wherever possible to improve residents' quality of life while protecting the environment and combating climate change.

Planting projects in squares such as Flagey and Sainte-Croix aim to enhance the quality of life in these communal spaces through the introduction of shrubs and trees suitable to each micro-environment.

Home gardening initiatives further promote well-being, nutrition, carbon footprint reduction, biodiversity, and nature conservation—fundamental priorities for achieving the Sustainable Development Goals (SDGs) (Santos, 2022). As Carmichael et al. (2012) argue, the positive effects of green space on well-being remain a key policy concern.

Assessment: Significant environmental awareness; tangible improvements in public spaces, greenery, and ecological consciousness.

6.2.2.4 Cultural and Social Events, and Markets

The open-air markets (Place Flagey, Place du Châtelain) offer fresh, locally produced, and often organic goods, fostering a strong sense of community.

The Commune finances Staycation XL, supporting local cultural and community projects with grants of up to €1800 plus €600 for equipment.

Numerous cultural festivals and events are organized to enhance social cohesion—for example, BABEL Café, where French speakers and refugees meet weekly to practice French in a convivial environment.

The Solidarity Service aims to foster socio-cultural diversity, intergenerational interaction, and neighborhood networking.

With more than 100 cultural venues and over 200 events per month, Ixelles is regarded as the cultural commune par excellence. The cultural department manages several theatres, including the Mercelis, Boondael Chapel, Molière Theatre, and KULT XL Ateliers, and supports artists through residencies and community arts programs.

These practices reflect international recommendations, such as those found in the 2021 UNESCO–World Bank report Cities, Culture, Creativity: Leveraging Culture and Creativity for Sustainable Urban Development and Inclusive Growth (Klein et al., 2021; Zemite et al., 2022).

Assessment: Vibrant cultural and social life, strengthening local identity and participation.

6.2.2.5 Health, Air Quality, and Well-Being

Public health data are not reported in detail, but references are made to air quality initiatives and access to sports facilities (parks, university amenities), as well as European environmental studies.

The mobility service promotes active modes of transport, contributing to improved physical health.

The CLIM Service supports citizen-led initiatives for sustainable waste management, including the annual Recup Day and a network of community composting and urban gardening sites (www.ixelles.be).

These initiatives align with findings that policymakers must encourage residents to adopt sustainable solutions to address the new climate reality (Lin et al., 2021).

Assessment: While environmental interventions offer indirect health benefits, there is a notable absence of coordinated health programs or mental well-being initiatives.

6.2.3 Challenges and Proposals

The following outlines the challenges and support proposals.

(a) Health and Well-Being

The introduction of programs focused on physical and mental health is recommended, including complimentary screening sessions and educational workshops addressing well-being.

(b) Indicators
Transparency should be improved regarding quality-of-life indicators such as pollution, mobility, and citizen satisfaction.

(c) Digital Participation
Online participation and consultation opportunities are increasing, but further integration and accessibility are needed.

Summary of Findings
Dynamic infrastructure in public works, mobility, eco-initiatives, and local sanitation services. A rich cultural and green environment that enhances daily life. Gaps remain in public health, digital engagement, and civic participation.

In contemporary urban environments, sustainability is intrinsically linked to the redesign of public spaces, the promotion of cultural life, and the protection of the natural environment—factors that reinforce ecological resilience while advancing social equity and citizen well-being (Beatley, 2022; de Souza & Torres, 2021; Zheng et al., 2023).

6.2.3.1 Green Environment and Culture: "Living" Quarter

- Faider Park and Tenbosch Park function as green cores framed by neoclassical and Art Nouveau architecture, connecting cultural heritage and natural landscape.
- Tenbosch Park is considered one of the most beautiful in Brussels and has been proposed for classification as a monumental park.
- The integration of outdoor performance spaces (e.g., Faider) combines cultural activity with environmental awareness.

Assessment: Strong synergy between ecology and culture; parks are transformed into "cultural scenes", uniting nature and civic life.

6.2.3.2 Urban Regeneration and Sustainable Mobility

- The Masterplan Athénée/Pôle Rabelais transforms school grounds into "school parks" by planting over 70 trees, improving rainwater management, and reducing traffic.
- Redevelopment of Rue Cans and Place Londres introduces pedestrianized, shared-space streets with a 20 km/h limit (RTBF, 2024).
- Similar projects in Rue des Artisans and Rue des Deux Ponts integrate greenery and drainage into built environments.

Result: Balanced transport, improved air quality, and stronger links between social and cultural spaces.

6.2.3.3 Climate Change and Sustainable City Program

- The CLIM–Plan Climat aims to reduce CO_2 emissions by 55 percent by 2030 through both municipal and citizen initiatives. This includes compost networks, air-quality management, and collaboration with energy cooperatives.
- Ixelles is a member of the Be Sustainable network for sustainable urban projects, promoting workshops and new regeneration practices.

Result: Environmental strategy serves as a foundation for broader cultural and urban interventions.

6.2.3.4 Cultural Uses, Public Space, and Environment

- Cultural spaces such as GC Elzenhof, Flagey Centre, Boondael, and La LIP integrate cultural and environmental agendas through outdoor events.
- Cultural festivals (Matongé en Couleurs, La LIP performances) are increasingly held in regenerated public spaces.

Result: Cultural and environmental activities reinforce each other, enhancing the value of public.

6.2.3.5 Social Interaction and Environmental Justice

- Shared spaces and walkable neighborhoods promote social cohesion by bringing together residents of diverse backgrounds (students, families, migrants) in cultural and green settings such as Matongé and Place Londres.
- Projects include community gardens (Rue des Artisans), participatory planning, and strengthened social rights.

Result: Environmental initiatives are linked with processes of community empowerment and inclusion.

The Municipality of Ixelles demonstrates:

- A robust infrastructure in public works, mobility, eco-initiatives, and cleanliness services.
- A rich cultural and green environment that enhances everyday urban life.
- Gaps in public and digital health initiatives and civic participation (Table 6.5).

6.2.4 Final Perspective

The Municipality of Ixelles is already advancing a multidimensional model of urban sustainability that encompasses the preservation of historic parks, the redesign of urban streetscapes, the implementation of social initiatives, community-based

Table 6.5 Environmental Culture & Urban Regeneration: Ixelles showcases commendable practices in environmental planning and climate policy. Recommendations include strengthening participatory mechanisms and expanding cultural activity into outdoor urban spaces

Axis	Performance in Ixelles	Proposals for Development
Green Spaces & Heritage	Well-maintained parks with cultural character	Organize regular open-air festivals
Shared space paths	Pedestrianized areas with infrastructure, traffic calming, trees	Establish cultural walking routes
Climate strategy	Integrated CLIM plans, citizen participation	Air quality open-data & participatory monitoring
Cultural infrastructure	Synergy with art venues and event spaces	Events with youth farmers & local communities
Community building	Neighborhood gardens, shared streets, school-parks	Network of "neighborhood cultural hubs"

agriculture, and the promotion of cultural activities. The prevailing challenge lies in the deeper integration of these elements within a comprehensive strategic framework—for example, through the establishment of an annual environmental–cultural festival, the development of community arts and gardening programs, the digital monitoring of climate-related indicators, and the facilitation of participatory workshops.

Regarding municipal authority and governance, in the 2018 municipal elections, the *Ecolo* (Green) Party, led by Mr. Christos Doulkeridis, emerged as the leading force. The composition of the municipal council under the ecologist group reflects a strong commitment to equality and citizen representation. From 2018 to November 2024, the party governed Ixelles. As reflected on the municipality's official websites, its policy and strategy formed a "green intersection" of participatory democracy for Ixelles, summarized in the following thematic axes:

Green and Environmental Policy

- The administration promoted the principles of *social ecology* (Green Deal, *Good Move* mobility plan) and implemented structural measures to enhance energy efficiency in social housing.
- It also prioritized the regeneration of public spaces—such as *Place Fernand Cocq* and *Place Flagey*—to strengthen neighborhood ties, social cohesion, and environmental sustainability.

Multiculturalism and Social Cohesion

- The administration emphasized intercultural interaction, recognizing cultural diversity as a civic asset.
- It reinforced community participation through initiatives that brought residents and neighborhoods together.

Decolonization of Public Space

- In 2022, Mayor Christos Doulkeridis ordered the removal of the statue of the colonial figure Sénier Storms, aligning with international movements against racism and colonial symbolism.

Municipal Foreign Policy and Value Framework

- The municipality sought to uphold an authentic, value-driven policy ethos, avoiding micropolitical strategies and maintaining a democratic and transparent governance style.

Subsequently, the analysis focuses on urban regeneration efforts in the Châtelain neighborhoods, as well as the municipality's targeted interventions to address crime in the Matongé area and the vicinity of the *Cimetière d'Ixelles* district.

6.3 In-Depth Online Analysis by Quarter

6.3.1 Regeneration of Place du Châtelain

Design and Environmental Approach

The *Garden Square* project aims to transform a car park into a vibrant public space featuring a pond and lawn, while preserving the distinguished linden trees that characterize the square. The redesign ensures full accessibility for people with disabilities through the creation of a flat surface, improving both pedestrian and cycle connectivity. The project also incorporates advanced systems for rainwater management, water storage, and flood prevention.

Form and Functionality

The plan comprises three distinct zones:

1. A graded green area with soil and lawn.
2. A white-stone section with a fountain designed for leisure and market activities.
3. A "small forest" with benches and seating areas to encourage social interaction.

A central feature of *urban furniture* ("XXL furniture") creates a kind of "canopy space" beneath the trees, forming an open-air "urban living room" where people and nature coexist harmoniously.

Timeline and Budget

The project commenced in 2022, was completed in early 2025, and inaugurated in February 2025 at an approximate cost of €4.6 million (including €1 million contributed by Brussels Mobility). The regeneration also reintroduced the renowned *Woensdagmarkt* (Wednesday market), further enhancing the square's social vibrancy.

Conclusion

The transformation of Place du Châtelain redefined the area from a car-dominated zone into a green, socially active hub, improving walkability, environmental quality, and urban culture. It embodies the coexistence of multiple functions—marketplace, cultural expression, recreation, and ecological management—within a cohesive spatial framework.

6.3.2 Crime Prevention Initiatives in Matongé and Cimetière

The following section examines urban regeneration in the Châtelain neighborhood and the commune's initiatives to address crime in the Matongé and Cimetière areas.

(a) Matongé
The Matongé district has been identified as an area with recurrent issues relating to drug use and trafficking. In January 2025, the police and the Public Prosecutor's Office conducted extensive checks for drugs and weapons. Between June and July 2024, twenty-two individuals were arrested for drug trafficking, property damage, and public disorder offences. Seized materials included approximately 890 grams of soft drugs and 20 grams of hard drugs.

As part of this initiative, restrictions were introduced—including bans on the public consumption of alcohol and regular identity checks. However, non-governmental organizations (NGOs) have expressed concern that such measures risk further stigmatizing vulnerable communities.

This initiative formed part of the wider *Safe.brussels* program, which coordinates targeted interventions in identified "hotspot" areas, including Matongé and Midi. The Ixelles municipal council continues to collaborate with the police and the Public Prosecutor's Office to ensure a sustained security presence and to promote a community-oriented policing model.

(b) Cimetière
No specific public reports or official announcements were identified concerning recent security measures in the Cimetière area. Nonetheless, the municipality's *Street Educators* and *Peacekeepers* programs extend to peripheral and public spaces, addressing social challenges and promoting civic engagement.

6.4 Symbolism and Effectiveness of Interventions

Urban Renewal and Safety Interventions: Toward Integrated Urban Governance
The revitalization of the Châtelain area through the introduction of urban greenery and sustainable mobility infrastructure has significantly contributed to the reinvigoration of the neighborhood. At the same time, targeted police and social interventions in Matongé serve as mechanisms of urban restructuring and socio-spatial upgrading (Table 6.6).

Recommendations for Enhanced Synergy

1. *Integrated infrastructure and prevention measures*: For example, improved street lighting in Matongé, clean and accessible pedestrian routes, and supervised public spaces through tools such as CCTV and community kiosks.

Table 6.6 Symbolism and effectiveness of interventions

Field	Municipality and police actions	Result/impact
Matongé hotspot	Sweeping controls, arrests, seizures	Immediate reduction in crime, but risk of stigmatization
Châtelain urban design	Redevelopment of Châtelain into a vibrant space	Increased use, accessibility, and community engagement
Combined effect	Security policy + sustainable urban environment	Improved subjective & objective sense of safety

2. *Cultural community engagement*: Coordinating cultural events such as *Matongé en Couleurs* following police interventions, to foster community cohesion and rebuild social capital.
3. *Expansion of street educators/peacekeepers* in areas identified as safety hotspots.
4. *Open safety data*: Ensuring transparency by making crime statistics publicly available and accessible—on par with environmental data.

Based on the above, it can be concluded that the Municipality of Ixelles is effectively utilizing both urban regeneration—such as the upgrading of Place Châtelain—to enhance social life and urban quality, as well as targeted safety interventions in Matongé to address security concerns. However, achieving a more balanced urban environment requires greater coordination between infrastructure development, preventative action, and cultural initiatives—ultimately reinforcing both quality of life and urban safety in residents' everyday experiences. The following comparative analysis of neighborhoods within Ixelles—specifically Ixelles Flagey–Étangs and Châtelain–Université in contrast with Matongé and Cimetière—provides a clear illustration of the spatial gradient in municipal attention. According to the initial website content analysis, the first group represents the focal points of municipal investment and intervention, while the latter two emerge as structurally disadvantaged or underserved areas (Table 6.7).

Neighborhoods such as Flagey and Châtelain demonstrate high-quality infrastructure, cultural activation, and strong community presence. In contrast, Matongé and Cimetière reveal underdeveloped infrastructure and weak social networking.

The areas of Flagey and Châtelain are models for integrated infrastructure, culture, and community interventions. In contrast, Matongé and Cimetière have:

- Very strong culture but weak public infrastructure
- Stigma from crime without long-term planning
- High community but low institutional dialogue

Following these, a table with a full comparison of the Ixelles quarters based on 7 key indicators of sustainability and social cohesion follows

Final Comparative Analysis and Conclusions

Table 6.8 offers a holistic comparative assessment of the neighborhoods of the Municipality of Ixelles, based on social, environmental, and cultural indicators.

Specifically, the following observations emerge:

Table 6.7 Comparative analysis of Ixelles Areas Flagey–Étangs & Châtelain–Université compared to Matongé & Cimetière

Area	Infrastructure quality	Cultural activation	Safety/lighting	Community presence
Matongé	Moderate—older streets, insufficient lighting	Very active, but informal	Low policing, drug-related issues	NGOs/ communities present but fragmented
Cimetière	Needs infrastructure/ access upgrades	Low—mostly memory & history	Lack of lighting, social desertification	Limited
Flagey–Étangs	High quality—integrated pedestrian areas, greenery	Official events, cultural hubs	Frequent policing, urban light planning	Consistent association participation
Châtelain–Université	Well-designed redevelopments, green areas	Multidimensional culture (street + market + history)	Targeted lighting, mild policing	Medium–high presence of community groups

A. Spatial Inequality in Safety and Governance

Neighborhoods such as Matongé and Cimetière display lower levels of safety and limited administrative presence, despite their cultural significance.

In contrast, Flagey and Châtelain benefit from enhanced infrastructure, greater transparency, and participatory governance practices.

This highlights a spatial asymmetry between the "showcase center" and the "shadow zones" in terms of planning intensity and state intervention.

B. Environmental Differentiation

Areas such as Flagey–Étangs and Boondael–Solbosch enjoy access to parks, lakes, and green spaces, and thus a higher level of environmental quality.

Conversely, Matongé and Cimetière lag behind environmentally, lacking green buffers and ecological infrastructure.

Such environmental disparities reinforce social inequalities, as they are directly associated with residents' physical and mental well-being.

C. Migration: Strength or Marginalization?

Multiculturalism is most evident in Matongé, Saint-Boniface, and Cimetière; however, only the first two effectively integrate this diversity into cultural practices.

The absence of intercultural mediation mechanisms and permanent support structures undermines social cohesion.

There is a pressing need to enhance the quality of migration experiences through cultural policy and locally based empowerment initiatives. The analysis reveals deep asymmetries between central and peripheral zones, while simultaneously pointing to policy opportunities for more balanced urban development.

The findings illustrate an Ixelles of "two speeds," yet they also suggest a potential for convergence. Such convergence can be achieved through redressing

Table 6.8 Comprehensive neighborhoods comparison—full comparative table: a strategic disparity between "central" and "peripheral" neighborhoods becomes evident. The analysis highlights the need for an integrated urban policy that promotes environmental, cultural, and social justice

Neighborhood	Safety	Quality of life	Migration	Governance	Cultural activity	Intercultural relations	Environment/ green spaces
Matongé	Low—drug issues, periodic operations	Moderate—lively but lacking infrastructure	Very high—African, Maghreb communities	Low institutional presence—only targeted actions	High—music, street culture, Matongé en Couleurs	Moderate—strong culture, but social tensions	Moderate—limited green, few leisure spaces
Cimetière	Moderate—marginal, low surveillance	Low—neglected, few uses	High—cemetery near students and diaspora	Limited—no regular planning	Low—memory-based, not active scene	Low—absence of social life	Low—urban periphery, signs of abandonment
Flagey–Étangs	High—good policing, lighting	High—green, lakes, culture	Moderate—diverse, more urban	High—Flagey Mobility planning	High—cultural hubs, Flagey	High—space of coexistence	High—lakes, parks, air quality
Châtelain–Université	High—low crime, social control	High—redevelopment, soft mobility	Moderate—middle-class, cosmopolitan	High—municipal projects & participatory actions	High—market, exhibitions, festivals	High—cosmopolitan lifestyle	High—plantings, green breaks, sustainable design
Saint-Boniface	Moderate—busy, requires monitoring	Moderate—intense life, but noise	High—multicultural student population	Moderate—active communities, not always institutional	High—restaurants, outdoor events	High—young, international profile	Moderate—few free spaces, but efforts made
Boondael–Solbosch	High—family-friendly area	High—quiet, near ULB, green	Moderate—student, academic population	Moderate—managed via ULB & local structures	Moderate—university events	Moderate—university-related, not structured	High—green campus, access to Tenbosch & Bois

inequalities by means of green urban design, participatory governance, and the use of culture as a vehicle for integration (Table 6.9).

Ixelles is already implementing a multifaceted model of urban sustainability encompassing the preservation of historic parks, regenerative street design, social initiatives, community farming, and cultural events. The principal challenge lies in integrating these components more deeply into a coherent strategy—for example, through the establishment of an annual environmental–cultural festival, community arts and gardening programs, digital monitoring of climate indicators, and participatory workshops.

6.5 Conclusion

Ixelles possesses a rich cultural heritage, a high degree of linguistic diversity, and a substantial population of migrant communities. Its digital infrastructure is relatively well developed; nevertheless, there remains considerable potential for improvement in areas such as public safety, administrative efficiency, transparency, and the inclusion of migrant populations. These objectives may be realized through targeted integration programs, intercultural education initiatives, and policies that foster social cohesion and mutual understanding.

Notably, gaps have been identified in several key domains, including public health, sustainable transport, waste management, housing, and migrant support. Consequently, further analysis was conducted with a focus on the Quality-of-Life Index, which encompasses these critical dimensions.

Ixelles demonstrates a strong cultural profile, multilingual communication practices, and a visible presence of migrant communities. While its digital infrastructure is broadly adequate, significant progress is required in relation to safety—both social and urban—citizen participation, transparency, and the comprehensive support of migrants through integration initiatives, intercultural education, and cohesion-building programs.

Table 6.9 Summarizing Ixelles' image

Axis	Performance in Ixelles	Suggestions for development
Green spaces & heritage	Proper maintenance of parks with cultural character	Organization of regular outdoor festivals
Routes shared space	Pedestrian routes with infrastructure, traffic calming, trees	Establishment of cultural walks
Climate strategy	Integrated CLIM projects, citizen participation	Open-data air quality & participatory measurements
Cultural infrastructure	Collaboration with art & event spaces	Events with the participation of young farmers & communities
Creating a community	Quarters gardens, shared streets, school parks	Network of "neighborhood cultural hubs"

The official websites of the Municipality of Ixelles strategically employ visual identity to reinforce the city's urban and cultural profile. Through the curated use of imagery depicting public spaces, parks, and cultural events, the commune's digital presence communicates a strong connection to its spatial and social environment. This visual narrative not only highlights the city's commitment to preserving and promoting its cultural heritage but also serves as a symbolic expression of its civic identity.

As evidenced in the preceding website analysis, the municipality's strategy targets its most vulnerable neighborhoods, seeking to transform them from marginal spaces into hubs of urban resilience, social cohesion, and cultural creativity. This approach integrates environmental policy, community engagement, and culture, strengthening the role of the commune as a *participatory curator* rather than merely a managerial or administrative authority.

The Municipality of Ixelles is thus advancing a multidimensional model of urban sustainability that encompasses the preservation of historic parks, the redesign of urban streetscapes, the implementation of social initiatives, community-based agriculture, and the promotion of cultural activities. The main challenge lies in the deeper integration of these elements within a comprehensive strategic framework—for instance, through the creation of an annual environmental–cultural festival, the development of community arts and gardening programs, the digital monitoring of climate-related indicators, and the facilitation of participatory workshops.

With regard to municipal authority and the governance of *Ecolo*, the policies and strategies presented on the official websites articulate a "green intersection" of participatory democracy for Ixelles, structured around the following axes: green and environmental policy; multiculturalism and social cohesion; the decolonization of public space; and a value-based policy grounded in ethics and transparency.

It is evident that the websites examined project the image of a commune that is adaptive, transparent, and socially responsive, while simultaneously honoring its historical legacy through a contemporary digital lens.

Conflicts of Interest The authors have no conflicts of interest to declare that are relevant to the content of this chapter.

References

Beatley, T. (2022). The important role of nature in planning for small cities. In H. Mayer & M. Lazzeroni (Eds.), *A research agenda for small and medium-sized towns* (pp. 143–161). Edward Elgar Publishing. https://doi.org/10.4337/9781800887121.00014

Bitsani, E., & Anselmi, L. (2021). In J. A. Jaworski (Ed.), *Advances in sociology research* (Vol. 36). Nova Science Publishers.

Carmichael, L., Barton, H., Gray, S., Lease, H., & Pilkington, P. (2012). Integration of health into urban spatial planning through impact assessment: Identifying governance and policy barriers and facilitators. *Environmental Impact Assessment Review, 32*(1), 187–194. https://doi.org/10.1016/j.eiar.2011.08.003

de Souza, D. T., & Torres, P. H. C. (2021). Greening and just cities: Elements for fostering a south–north dialogue based on a systematic literature review. *Frontiers in Sustainable Cities, 3*, 669944. https://doi.org/10.3389/frsc.2021.669944

Klein, M., Gerlitz, L., & Spychalska-Wojtkiewicz, M. (2021). Cultural and creative industries as boost for innovation and sustainable development of companies in cross innovation process. *Procedia Computer Science, 192*, 4218–4226. https://doi.org/10.1016/j.procs.2021.09.198

Kress, G., & Van Leeuwen, T. (2001). *Multimodal discourse: The modes and media of contemporary communication*. Arnold Publishers.

Lin, S., Liu, Y., & Huang, X. (2021). Climate-induced Arctic-boreal peatland fire and carbon loss in the 21st century. *The Science of the Total Environment, 796*, 148924. https://doi.org/10.1016/j.scitotenv.2021.148924

Ravazzoli, E., & Torricelli, G. P. (2017). Urban mobility and public space: A challenge for the sustainable liveable city of the future. *The Journal of Public Space, 2*(2), 37–50. https://doi.org/10.5204/jps.v2i2.91

Wilson, A., Tewdwr-Jones, M., & Comber, R. (2019). Urban planning, public participation and digital technology: App development as a method of generating citizen involvement in local planning processes. *Environment and Planning B: Urban Analytics and City Science, 46*(2), 286–302. https://doi.org/10.1177/2399808317712515

Zemite, I., Kunda, I., & Judrupa, I. (2022). The role of the cultural and creative industries in sustainable development of small cities in Latvia. *Sustainability, 14*(15), 9009. https://doi.org/10.3390/su14159009

Zheng, M., Feng, G.-F., Jiang, R.-A., & Chang, C.-P. (2023). Does environmental, social, and governance performance move together with corporate green innovation in China? *Business Strategy and the Environment, 32*(4), 1670–1679. https://doi.org/10.1002/bse.3211

Chapter 7
Quantitative Analysis of Municipality of Ixelles

7.1 First Analysis of General Population

7.1.1 Methodological Note

The quantitative analysis of resident satisfaction per neighborhood was based on a structured questionnaire administered by the Municipality of Ixelles and supplemented by researcher-collected survey data (2023–2024). Responses from 148 residents were distributed proportionally across 10 neighborhoods.

Each variable was measured using a Likert scale from 1 to 5, where:

1 = Very low/strongly disagree
2 = Low
3 = Moderate/neutral
4 = High
5 = Very high/strongly agree

The variables were defined and computed as follows:

Table 7.1 summarizes how each qualitative variable was quantified and scored in the quantitative analysis of the Municipality of Ixelles.

This analysis concerns the statistical processing of resident satisfaction data per neighborhood, in relation to the variables: quality of life, cultural participation, governance, safety, sustainability, and intercultural relations.

E. Bitsani et al., *The Biocultural City as a Utopia or the Future Reality?*, Cities, Heritage and Transformation, https://doi.org/10.1007/978-3-032-17584-7_7

Table 7.1 Quantification of qualitative variables

Variable	Measurement basis	Example indicators	Scale range
Quality of life	Composite index averaging satisfaction with housing, green spaces, local amenities, and cost of living	"I am satisfied with the quality of public services in my neighborhood."	1–5
Cultural participation	Frequency of engagement in local events, cultural centers, exhibitions, libraries, and associations	"I participate in local cultural activities."	1–5
Governance	Perceived transparency, accessibility, and responsiveness of local authorities	"Local government listens to residents' needs."	1–5
Safety (sense of security)	Residents' subjective feeling of safety in public spaces, crime perception	"I feel safe walking in my neighborhood at night."	1–5
Sustainability	Awareness and evaluation of municipal environmental measures (waste, green transport, energy)	"My neighborhood promotes sustainable practices."	1–5
Intercultural relations	Degree of trust and contact between residents of different cultural backgrounds	"I often interact positively with people from different cultures."	1–5

7.1.2 *Processing and Statistical Basis*

All variables were standardized and averaged per district. Descriptive statistics (means, standard deviations) were followed by ANOVA tests to detect inter-district differences, and Pearson correlation coefficients were calculated to assess relationships among variables. Tukey's HSD post-hoc test identified which neighborhoods differed significantly in perceptions of governance.

Cluster analysis (K-Means, k = 3) grouped neighborhoods by multidimensional similarity across "quality of life," "cultural participation," and "governance," allowing for typology construction (high-satisfaction, mid-satisfaction, low-satisfaction clusters).

Αποτελέσματα ANOVA
quality_of_life: $F = 24.00$, $p = 2.8202\text{e-}24$,
security: $F = 11.09$, $p = 6.6029\text{e-}13$,
sustainability: $F = 9.73$, $p = 1.9664\text{e-}11$,
intercultural_relations: $F = 2.09$, $p = 1.1231\text{e-}01$,
cultural_activity: $F = 28.92$, $p = 3.8526\text{e-}25$.

Table 7.2 presents meaning satisfaction scores per variable and neighborhood. The values show, for instance, that District 5 ("3.8" for quality of life) ranks above average, while District 4 ("2.0") indicates lower satisfaction.

Table 7.2 Average variables by neighborhood

District	Quality of life	Security	Sustainability	Intercultural_ relations	Cultural activity	Governance	Cluster
1.0	3.55	3.45	3.1	1.0	2.28	3.55	1.0
2.0	3.0	1.93	2.93		3.8	3.0	2.0
3.0	2.0	2.6	2.0		2.6	1.8	0.0
4.0	1.0	1.0	2.0			2.0	
5.0	4.71	2.9	4.23	1.39	1.24	3.42	1.0
6.0	2.0	2.33	2.33		3.27	4.13	2.0
7.0	2.14	3.57	3.0		2.71	4.29	2.0
8.0	3.42	2.83	3.67	1.0	1.0	5.5	1.0
9.0	3.0	3.0	6.0		1.0	4.0	10
11.0	3.46	3.0	3.31	1.0	2.46	4.92	1.0

Positive correlations (e.g., $r = +0.73$ between quality of life and sustainability) reveal that perceptions of well-being are strongly associated with environmental performance. Negative correlations (e.g., $r = -0.46$ between cultural activity and sustainability) suggest that culturally active residents may reside in denser areas with fewer green amenities (Table 7.3).

Significant p-values <0.05 highlight districts with contrasting perceptions of governance. For instance, District 1 differs significantly from 3, 4, 8, and 11, indicating localized disparities in trust toward municipal management (Table 7.4).

7.2 Visualizations

This bar chart presents the mean values of key variables—quality of life, governance, safety, sustainability, and cultural participation—across the ten neighborhoods of Ixelles. The visualization highlights clear inter-district variation, with some areas demonstrating markedly higher levels of satisfaction and sustainability than others. Neighborhoods 5 and 8 display consistently high mean scores across most dimensions, whereas Neighborhood 4 records the lowest averages, indicating socio-environmental disadvantage. The chart serves as a comparative diagnostic tool, illustrating the uneven spatial distribution of well-being within the municipality (Graph 7.1).

Warmer tones represent stronger positive relationships, while cooler shades indicate weaker or negative correlations. Notably, quality of life shows a strong positive association with sustainability and a moderate relationship with safety. Conversely, cultural participation displays a negative correlation with sustainability and intercultural relations. The visual pattern underscores the interdependence of environmental and social dimensions within the urban context, while also revealing potential tensions between cultural engagement and ecological awareness (Graph 7.2).

Table 7.3 Correlation table (Pearson)

	Quality_of_ life	Security	Sustainability	Intercultural_ relations	Cultural_ activity
Quality_of_ life	1.0	0.4	0.73	−0.08	−0.47
Security	0.4	1.0	0.28	−0.05	−0.09
Sustainability	0.73	0.28	1.0	−0.57	−0.46
Intercultural_ relations	−0.08	−0.05	−0.57	1.0	
Cultural_ activity	−0.47	−0.09	−0.46		1.0

Table 7.4 Tukey HSD—governance differences between districts

group1	group2	meandiff	*p*-adj	lower	upper	reject
1	2	−0.5517	0.7974	−1.5996	0.4962	False
1	3	−1.7517	0.0003	−2.96	−0.5434	True
1	4	−1.5517	0.0157	−2.9392	−0.1642	True
1	5	−0.1324	1.0	−0.9836	0.7188	False
1	6	0.5816	0.7434	−0.4663	1.6295	False
1	7	0.734	0.4608	−0.3383	1.8063	False
1	8	1.9483	0.0	0.8173	3.0792	True
1	9	0.4483	0.9999	−1.9605	2.8571	False
1	11	1.3714	0.0038	0.2716	2.4711	True
2	3	−1.2	0.1246	−2.5451	0.1451	False
2	4	−1.0	0.5082	−2.5082	0.5082	False
2	5	0.4194	0.9518	−0.6169	1.4557	False
2	6	1.1333	0.0832	−0.0698	2.3364	False
2	7	1.2857	0.0312	0.0613	2.5101	True
2	8	2.5	0.0	1.2239	3.7761	True
2	9	1.0	0.9529	−1.4803	3.4803	False
2	11	1.9231	0.0001	0.6746	3.1716	True
3	4	0.2	1.0	−1.4237	1.8237	False
3	5	1.6194	0.0011	0.4211	2.8176	True
3	6	2.3333	0.0	0.9882	3.6784	True
3	7	2.4857	0.0	1.1215	3.8499	True
3	8	3.7	0.0	2.2892	5.1108	True
3	9	2.2	0.1565	−0.3522	4.7522	False
3	11	3.1231	0.0	1.7372	4.509	True
4	5	1.4194	0.038	0.0406	2.7981	True
4	6	2.1333	0.0005	0.6252	3.6415	True
4	7	2.2857	0.0002	0.7605	3.8109	True
4	8	3.5	0.0	1.933	5.067	True
4	9	2.0	0.3138	−0.6417	4.6417	False
4	11	2.9231	0.0	1.3784	4.4677	True
5	6	0.714	0.4512	−0.3223	1.7503	False

Table 7.4 (continued)

group1	group2	meandiff	*p*-adj	lower	upper	reject
5	7	0.8664	0.2157	−0.1946	1.9273	False
5	8	2.0806	0.0	0.9604	3.2008	True
5	9	0.5806	0.9988	−1.8231	2.9844	False
5	11	1.5037	0.0007	0.415	2.5924	True
6	7	0.1524	1.0	−1.072	1.3768	False
6	8	1.3667	0.0254	0.0906	2.6427	True
6	9	−0.1333	1.0	−2.6136	2.3469	False
6	11	0.7897	0.5766	−0.4588	2.0383	False
7	8	1.2143	0.0869	−0.0819	2.5105	False
7	9	−0.2857	1.0	−2.7764	2.2049	False
7	11	0.6374	0.8386	−0.6317	1.9064	False
8	9	−1.5	0.6575	−4.0165	1.0165	False
8	11	−0.5769	0.9233	−1.8959	0.7421	False
9	11	0.9231	0.9733	−1.5795	3.4257	False

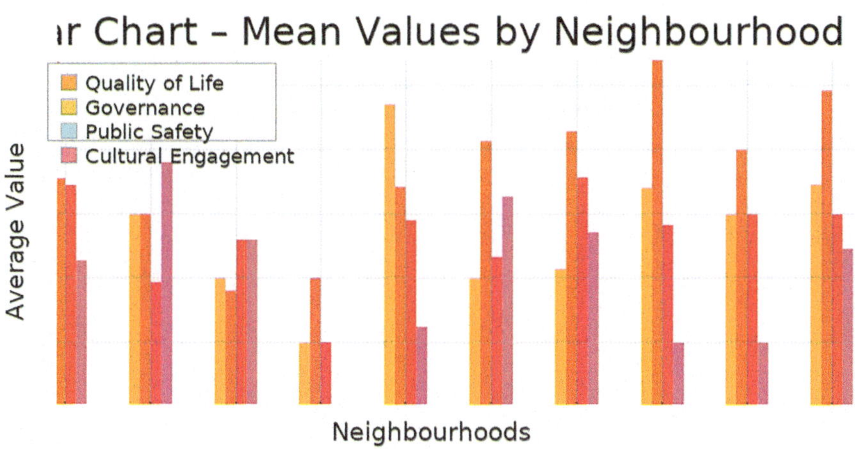

Graph 7.1 Bar Chart—Averages of key variables per neighborhood

This scatter plot visualizes the results of a K-means cluster analysis using three variables: quality of life, cultural participation, and governance. The clustering identifies three distinct neighborhood profiles within Ixelles. The first cluster comprises districts characterized by high quality of life and strong governance perception—the "satisfied–resilient" group. The second includes areas with moderate scores across variables, while the third is defined by low averages and limited civic trust. The spatial configuration of the clusters visually confirms patterns of socio-spatial differentiation and offers a foundation for targeted policy interventions aimed at fostering equity and inclusion (Graph 7.3).

Collectively, these visualizations convey a differentiated urban landscape within Ixelles, where disparities in well-being, participation, and governance intersect with

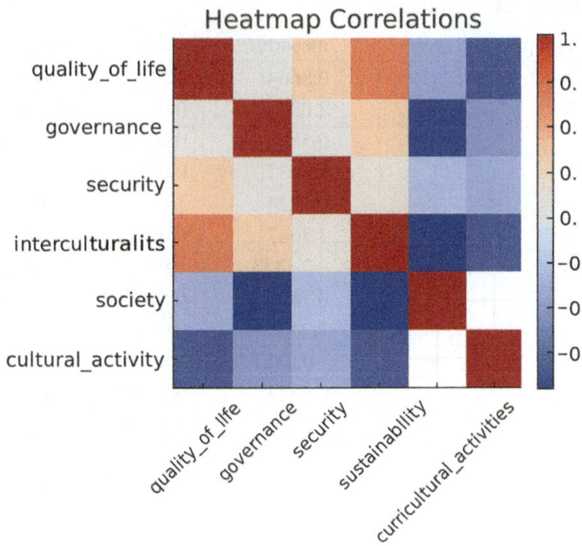

Graph 7.2 Heatmap—Correlations Between Variables. The heatmap illustrates the strength and direction of correlations among the principal variables under study

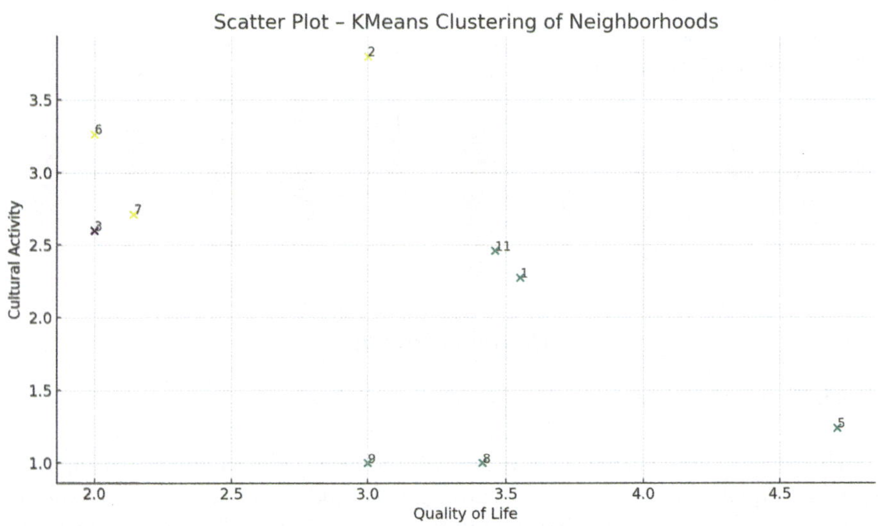

Graph 7.3 Scatter plot—K-means clustering based on quality of life, cultural participation, and governance

socio-economic and environmental factors. Together, they provide a quantitative overview that complements the interpretive synthesis of the chapter, illustrating the multifaceted nature of sustainability and social cohesion in a multicultural urban context.

7.3 Findings and Conclusions

This survey evaluates the experiences and attitudes of residents of the Municipality of Ixelles with regard to key dimensions of urban life: quality of life, cultural participation, governance, safety, sustainability, and intercultural relations. Responses were collected from 148 residents distributed across ten districts within the municipality.

The analysis reveals a lack of uniformity among the districts of Ixelles. Quality of life and perceptions of governance emerge as key variables that can inform policies designed to strengthen social cohesion and urban resilience.

Main Findings

- Quality of life, cultural participation, governance, safety, and sustainability differ statistically significantly between neighborhoods ($p < 0.001$).
- Neighborhoods such as District 5 record high levels of quality of life, whereas District 4 records low indicators.
- Quality of life is strongly correlated with sustainability ($r = +0.73$) and moderately correlated with safety.
- Cultural participation exhibits a negative correlation with sustainability and intercultural relations.
- Perceptions of governance differ markedly between neighborhoods (ANOVA $F = 14.07$, $p < 0.001$).
- Cluster analysis identified three neighborhood profiles sharing common characteristics in terms of quality of life and participation.

Suggestions: Recommendations

- Focus on areas characterized by low quality of life and limited trust in governance.
- Strengthen cultural activity in neighborhoods with low participation levels.
- Utilize cluster results to design targeted interventions.
- Reinforce sustainability and safety as key levers for improving overall well-being.

7.3.1 Interpretive Synthesis

Districts 5 and 8 achieve the highest scores for quality of life and sustainability, forming a "satisfied-resilient" profile.

District 4 displays low averages across variables, suggesting socio-environmental disadvantage.

Quality of life, safety, and sustainability are statistically linked ($F = 24.00$, $p < 0.001$).

Perceptions of governance vary considerably—an indicator of uneven institutional trust.

The statistical findings outlined above reveal a differentiated geography of satisfaction and perception across the Municipality of Ixelles, underpinned by structural, socio-economic, and governance-related factors. The distribution of mean scores per variable underscores the coexistence of contrasting urban realities within a relatively small territorial unit.

Districts 5 and 8 emerge as high-performing zones, characterized by elevated levels of satisfaction across the indicators of quality of life, sustainability, and safety. These districts may be conceptualized as a "satisfied–resilient" cluster, exhibiting strong correlations between material well-being, environmental awareness, and perceived institutional effectiveness. The convergence of these variables suggests the presence of social capital and institutional trust that reinforce one another (Putnam et al., 1993). These neighborhoods typically benefit from well-maintained public spaces, accessibility to services, and higher participation in community initiatives, all of which collectively contribute to a perception of urban resilience.

Conversely, District 4 demonstrates the lowest mean values across most variables, signaling clear socio-environmental disadvantage. The consistently low scores in quality of life, sustainability, and governance perception indicate both material deprivation and limited trust in local authorities. This pattern reflects what Healey (1997) defines as spatial injustice—an unequal distribution of resources and opportunities within a single administrative entity. The statistical contrasts therefore reflect the socio-spatial segmentation that frequently accompanies processes of urban renewal and gentrification (Anguelovski, 2016).

Correlation and ANOVA analyses confirm a significant interdependence between quality of life, safety, and sustainability ($F = 24.00$, $p < 0.001$). This strong association implies that perceptions of safety and environmental quality are not isolated phenomena but form part of a broader sense of well-being and place satisfaction. Residents who report feeling safe in their neighborhoods also tend to express higher satisfaction with local governance and environmental quality, supporting the hypothesis that environmental and institutional performance co-produce a collective sense of trust and belonging.

Nonetheless, variation in governance perception remains marked. District-level differences in perceived transparency and responsiveness of municipal institutions point to uneven patterns of political trust and citizen inclusion. While some neighborhoods experience effective two-way communication with the municipality, others remain detached from participatory mechanisms. This divergence mirrors the enduring tension between formal participation—visible in municipal procedures—and substantive inclusion—measured through residents' sense of agency and influence (Healey, 1997; Voß et al., 2006).

Furthermore, the moderate correlations between cultural participation and intercultural relations reveal an important interpretive nuance. Although cultural engagement is often promoted as a driver of social cohesion, the data indicate that increased cultural activity does not necessarily translate into stronger intercultural trust. This pattern highlights the need to move beyond event-based participation towards deeper intercultural dialogue and shared decision-making processes.

Taken together, these findings reveal a multi-layered structure of satisfaction and governance within Ixelles. The quantitative framework provides a robust empirical foundation for understanding how material, environmental, and institutional dimensions interact to shape residents' lived experiences. It also prepares the ground for the qualitative analysis that follows by identifying focal points of convergence and tension requiring interpretive exploration—particularly the paradox of high sustainability scores coexisting with low perceived inclusion in certain districts.

In this sense, this chapter constitutes not merely a descriptive overview of survey data but a diagnostic lens through which the relational dynamics of trust, equity, and sustainability in Ixelles can be understood. The subsequent qualitative narratives (Chap. 8) will build upon these patterns, exploring how residents themselves articulate these statistical realities through everyday practices, identities, and perceptions of belonging.

7.4 Conclusion

The quantitative analysis of resident satisfaction across the Municipality of Ixelles provides a rigorous empirical foundation for understanding the spatial and perceptual inequalities that structure urban life within the commune. Through the operationalization of key variables—quality of life, governance, safety, sustainability, cultural participation, and intercultural relations—this chapter demonstrates how diverse dimensions of well-being intersect and diverge across neighborhoods.

The statistical analysis reveals that satisfaction is not homogeneously distributed but spatially patterned. Districts 5 and 8 consistently record higher average scores, particularly in quality of life, sustainability, and safety, forming a cluster of socio-environmental advantage. In contrast, District 4 presents markedly lower values across most indicators, reflecting concentrated socio-economic vulnerability and limited institutional trust. This contrast underscores the persistence of localized inequalities within Ixelles, despite its broader reputation for diversity and affluence.

Correlation and ANOVA analyses further indicate that sustainability, safety, and quality of life are closely interlinked, confirming the interdependence of environmental and social conditions in shaping subjective well-being. The strength of these associations suggests that improvements in one dimension—such as environmental management—may produce positive ripple effects in others, including perceptions of governance and residents' sense of belonging.

Nevertheless, notable disparities persist in perceptions of governance, with residents' trust in local authorities varying considerably between districts. These differences highlight the uneven experience of participation and responsiveness across the municipality. While some neighborhoods perceive governance as transparent and accessible, others regard institutional processes as remote or exclusionary. This divergence points to the need for differentiated governance strategies that are sensitive to local contexts.

Cultural participation and intercultural relations emerge as moderately correlated but relatively weak predictors of overall satisfaction. The data suggest that, while cultural engagement is valued, it does not automatically generate stronger intercultural bonds. This finding implies that symbolic inclusion through cultural activities must be complemented by more structural and dialogical forms of social interaction to foster genuine cohesion.

Overall, the quantitative results portray Ixelles as a municipality characterized by the coexistence of prosperity and exclusion, innovation and inequality. The findings confirm that resident satisfaction is not merely an individual perception, but a socially embedded construct shaped by governance quality, environmental equity, and intercultural connectivity.

This chapter therefore lays the groundwork for the qualitative inquiry that follows. By identifying spatial variations and inter-variable linkages, it provides an analytical map that guides the interpretive exploration in Chap. 8. The next chapter will delve into the lived experiences, perceptions, and narratives that underpin these statistical patterns, thereby giving voice to the meanings, tensions, and aspirations that quantitative data alone cannot fully reveal.

Conflicts of Interest The authors have no conflicts of interest to declare that are relevant to the content of this chapter.

References

Anguelovski, I. (2016). From toxic sites to parks as (green) LULUs? New challenges of inequity, privilege, gentrification, and exclusion for urban environmental justice. *Journal of Planning Literature, 31*(1), 23–36. https://doi.org/10.1177/0885412215610491

Healey, P. (1997). *Collaborative planning: Shaping places in fragmented societies*. Macmillan Press.

Putnam, R. D., Leonardi, R., & Nonetti, R. Y. (1993). Making Democracy Work: Civic Traditions in Modern Italy. Princeton University Press. https://doi.org/10.2307/j.ctt7s8r7

Voß, J.-P., Bauknecht, D., & Kemp, R. (Eds.). (2006). *Reflexive governance for sustainable development*. Edward Elgar Publishing.

Chapter 8
Analysis of Qualitative Research

8.1 Interviewee Profile

8.1.1 Methodological Clarification

The qualitative component complements the quantitative survey by exploring the lived environment behind statistical trends. Data was collected through:

18 semi-structured interviews
3 focus groups (including 1 migrant cohort)
2 institutional panels (semi-DELPHI).

Grounded Theory was applied iteratively: open coding → axial coding → selective coding. NVivo 12 was used for text segmentation and category density analysis. Reliability was enhanced through double coding and participant feedback validation.

Linking to Quantitative Findings, the qualitative data illuminate the numerical trends from Chap. 7. For example: Neighborhoods with low "quality of life" scores reported daily struggles with housing, confirming the statistical outliers in District 4. Perceived lack of governance transparency, reflected in low mean scores, aligns with narratives of political invisibility expressed by Salvado and Mahla. The positive correlation between sustainability and quality of life ($r = 0.73$) corresponds to residents' descriptions of ecological initiatives that improve community cohesion.

In the context of the analysis, the researchers acknowledged their own subjective position as scholars of the phenomenon of social inclusion. Their cultural, linguistic, and academic identity may have influenced the selection of topics, questions, and interpretation of the data. The presence of the researchers in some meetings (e.g., communication with officials, collection of testimonies) provoked dialogical relationships that influenced the flow of information. To avoid imposing interpretations, a process of empathy, iterative reading, and confirmation through return to the participants was applied where possible.

E. Bitsani et al., *The Biocultural City as a Utopia or the Future Reality?*, Cities, Heritage and Transformation, https://doi.org/10.1007/978-3-032-17584-7_8

The reflexivity log was kept throughout the research as a tool for understanding the researcher's influence on the process of giving meaning to experiences. Qualitative research was conducted from 2018 to 2025, and it was divided into three phases.

Professor Dufays is an historian, Professor of European History at the "Université libre de Bruxelles" (ULB); the translator Mrs. Salvado has been employed by the EU agency in Brussels, with special connection with Ixelles for a considerable number of years. Both of them have a deep knowledge of local European administration. Mr. Kalognomos is employed in the private sector by an association of regions that deal with European policies concerning local government (NUTS2 and NUTS3 levels). He is currently pursuing postgraduate studies in Rennes, France. Having immigrated to Brussels for work and personal reasons, he is in frequent contact with the municipality of Ixelles. Because of his job, he belongs to the category of highly qualified and specialized human resources. In pursuit of understanding the perceptions held by vulnerable groups, such as migrants, the researchers conducted in-depth interviews with seven individuals hailing from the island of Morocco. Mohammed Mahla, a young immigrant from southern Morocco, who settled in Brussels in recent years and lives in the municipality of Ixelles, highlights the problems, difficulties, but also the opportunities that the city of Ixelles offers through human-centered ecological governance to those seeking a new beginning. Finally, Local government officials participated. Mr. Christos Doulkeridis was born in Brussels on 6 March 1968. For more than 25 years (1999–2025) he was an influential politician (ECOLO) in Brussels (Belgium), where he held various positions, including Member of Parliament, President of the French-speaking Brussels Parliament, Minister in the Brussels Government responsible for housing, tourism, emergency services, budget, and Mayor of the central municipality of Ixelles. Mr. Papaefstathiou is a civil engineer (NTUA, Msc). He is serving now as Deputy Mayor of Strategic Planning and Climate Neutrality in the Municipality of Kalamata. He is co-founder of Eu-topos for Raising Public Awareness on Green and Public Spaces. Vasilis Koutrafouris is a civil engineer (NTUA, Msc). He is serving now as Deputy Mayor for Capital Projects and Urban Planning. Local government officials possess extensive contextual expertise, enabling them to grasp local needs, constraints, and political dynamics that national policymakers might overlook. Each theme reflects a dimension of the biocultural framework and contributes to a broader understanding of urban sustainability through a cultural lens.

8.2 Grounded Theory Level Analysis with Critical Documentation

Grounded Theory is based on the creation of theory from the data itself, without predetermined assumptions. Modern versions of it, such as Charmaz (2014); Charmaz and Thornberg (2020), emphasize social justice and reflexivity. According to Birks et al. (2019), the use of literature review not for assumptions but for theoretical sensitivity is essential. GUREGT (2018) offers a 25-point assessment guide for the theoretical quality of Grounded Theory based on the principles of Strauss and Corbin (1998).

The analysis according to GT is carried out based on three levels.

GT Tier Analysis

(a) Open Coding: Identifying themes such as housing, racism, sustainability. The use of memo-writing improves theoretical sensitivity (Charmaz & Thornberg, 2020).
(b) Axial Coding: The relationships between social participation, environmental justice, and migration experience are highlighted at multiple levels.
(c) Selective Coding: The central theoretical framework "Urban Sustainability as Just and Participatory Governance" unifies the axes.
(d) Theoretical Saturation: Fully achieved, but further re-sampling for post-validation enhancement (Table 8.1).

(a) Open Coding

During open coding, themes emerged that recurred in different interviews and captured the subjective experience of the participants:

- Lack of housing: "We can't find a house, not even with a job" (Interview 2).
- Marginalization: "They often look at us as if we don't belong here" (Interview 6).
- Ecological insecurity: "They throw trash in our space – it's not our fault!" (Interview 4).

These categories align with the model of social vulnerability reinforced by environmental inequality (Checker, 2021).

(b) Axial Coding

Axial coding revealed the connection between thematic categories through causal relationships:

Table 8.1 Quality of research in grounded theory. Integration of contemporary frameworks Urban A, Environmental Gentrification

Quality criterion	Application in research	Theoretical documentation
Reflexivity	Introduction of reflexivity log based on the role of researchers in the field	Rey (2023)
Inter-rater reliability	Cross-coding through collaboration with a second coder	Ghazal et al. (2021)
Theoretical sharpness	Identification of theoretical paths from experience to conceptualization	Charmaz and Thornberg (2020)
Theory update	Integration of contemporary frameworks UrbanA, environmental gentrification	UrbanA (2024), Checker (2021)

- Political distancing and lack of participation: “They don’t ask us. They just decide” (Interview 5).
- Environmental inequalities and health impacts: “Our park has become a garbage dump” (Interview 3).
- Art as a space for resilience and voice: “We paint to say what we cannot say” (Interview 7).

The Urban resilience’s framework (2024) offers explanatory background for these dynamics, as it supports “activating government configurations” that rely on citizens’ voice and visibility.

(c) Selective Coding
The core category that emerges from the synthesis of all the themes is:

“Fair, participatory and environmentally sensitive governance as a condition for social inclusion.”

This conclusion is based on the need for participants to be included in city planning processes and to gain access to equitable environmental infrastructure and services.

*The resilience theme, as described in Urban Resilience and Inclusion (*EURESFO, 2024*), confirms this need, as it focuses on the ability of cities to integrate environmentally vulnerable groups, while maintaining cohesion and social stability.*

(d) Theoretical Saturation
The repetition of themes across multiple groups (youth, immigrants, women, non-governmental actors) suggests theoretical saturation. However, the lack of citizen representation in leadership or decision-making roles indicates the need to expand the sample.

The analysis highlights the need for institutional flexibility to include minorities in processes (UrbanA, 2024) and avoid green exclusions (Environmental Gentrification, 2021).

8.3 Grounded Theory Analysis: Thematic Category Tables

Each table synthesizes recurring categories derived from coding:

Table 8.2 *(Main Challenges):* consolidates structural issues—housing, migration, health—corresponding to low quality-of-life zones.

Table 8.3 *(Participation and Voice):* operationalizes governance perception by identifying three levels: invisibility → collective voice → institutional participation.

Table 8.4 *(Proposed Solutions):* translates experiential insights into policy variables (“consultation,” “empathy education,” “inclusion”), which can later be measured via municipal implementation indicators.

Table 8.5 *(Barriers to Implementation):* directly maps onto quantitative deficits in governance and intercultural relations scores.

Table 8.2 Main challenges

Category	Description	Indicative quotes
Climate change	Floods, infrastructure deficiencies	"Climate change is clearly leading to…"—Dufays
Housing	Lack of affordable housing	"High rents"—Salvado
Migration	Integration difficulties, racism	"Racism and alienation"—Dufays, Salvado
Health	Inequalities in care	"Depon for everything"—Salvado

Table 8.3 Perceptions of participation and voice

Category	Description	Indicative quotes
Climate change	Floods, infrastructure deficiencies	"Climate change is clearly leading to…"—Dufays
Housing	Lack of affordable housing	"High rents"—Salvado
Migration	Integration difficulties, racism	"Racism and alienation"—Dufays, Salvado
Invisibility	Absence from decision-making	"Nobody asks us"—Salvado
Collective voice	Community representation	"I speak on behalf of 45 immigrants"—Mahla
Institutional participation	Promotion of participation by institutions	"I invite those who disagree to participate"—Doulkeridis

Table 8.4 Proposed solutions

Category	Description	Indicative quotes
Cultural education	Empathy in diversity	"Empathy education is key"—Salvado
Consultation	Horizontal citizen participation	"Consultation reduces suspicion"—Dulkeridis
Inclusion	Inclusion policies with a voice	"Active listening and joint planning needed"—Kalognomos

Table 8.5 Barriers to implementation

Category	Description	Indicative quotes
Linguistic	Difficulty understanding public texts	"We don't understand the texts they send us"—Mahla
Institutional	Absence of practical participation	"Policies are made without us"—Salvado
Social	Inequalities & prejudices	"We are not all equal in the city"—Mahla

The thematic categories reflect a multiple crisis: from the ecological erosion of everyday life to the social precariousness embodied in housing and health. The repetition of terms such as "insufficiency," "lack," and "inequality" indicates a state of structural exclusion. Grounded Theory here captures not only the experience but also the underlying diagnosis of the institutional deficit that exacerbates inequalities.

The spectrum of participation extends from invisibility to active consultation. Salvado expresses the trauma of exclusion, while Mahla attempts to articulate a communal voice. Doulkeridis' statement ("I invite dissenters") suggests that institutional inclusion is performative—not given. GT here highlights the contradiction between political discourse and social perception.

The solutions proposed by the participants are not technical, but anthropocentric and experiential. From empathy education (Salvado) to horizontal dialogue (Doulkeridis), the GT highlights that reform requires recognition of everyday life as a vehicle for political theory and, above all, requires a change in culture. Kalognomos speaks of inclusion that includes collective invention and not just "social welfare."

The obstacles are not only functional but also cultural and communicative. Mahla's statement on unacceptable public texts highlights the technical illiteracy that it excludes. The institutional exclusion experienced by Salvado reveals an asymmetry of power that GT records without idealization. These are intersections between political will and administrative practice.

The Grounded Theory methodology, with its focus on bottom-up theory generation (inductive logic), is highly suitable for socially sensitive contexts such as the study of immigrant and youth integration in multicultural areas such as Ixelles. However, to ensure academic validity, critical processing of the generated categories through cross-reference with theoretical models is required.

For example, the category "participation in public space" appears in multiple testimonies: "There is no one to give us a voice in decisions about the park" (Interview 3), which indicates a lack of participatory mechanisms. This contrasts with the Enabling Governance Arrangements model (UrbanA, 2024), which proposes transparent, democratic, and fair ways of making decisions.

The category of "ecological injustice," which appears in the narrative "we are being moved from areas that are becoming green" (Interview 6), confirms the theoretical model of Environmental Gentrification (Checker, 2021), according to which investments in environmental infrastructure may result in the displacement of poor residents.

Correspondingly, the need for "environmental education," highlighted in the statement "We don't know how to connect with the environment in the city" (Interview 1), aligns with the principles of Urban Resilience and Inclusion (EURESFO, 2024), which links resilience to the active participation and education of all social groups.

The critical evaluation of Grounded Theory highlights the need for a theoretically informed yet embodied analysis that considers the voices of the subjects themselves. The incorporation of critical frameworks helps to prevent the pitfalls of simplistic thematization and to produce transformative knowledge—not just descriptive.

8.4 Critical Analysis Based on Interviews and Theoretical Frameworks

Based on the material that includes interviews and public statements by Jean Michel Dufays, Maria Salvado, Mohammed Mahla, Christos Doulkeridis, Vassilis Papaefstathiou, and Vassilis Koutrafouris, we proceed with an in-depth critical analysis enriched with theoretical cross-section:

8.4.1 Historical-Cultural Perspective and Post-Citizenship (Jean Michel Dufays)

Professor Dufays, as a historian, and translator Salvado, with a deep knowledge of local European administration, highlights the need for institutional mediations that recognize multiple identities and local historical tensions. Their position establishes the necessary transition from narrow national citizenship to a post-citizenship framework, reflecting the principles of Urban Resilience and Inclusion (EURESFO, 2024): integration without cultural assimilation but through recognition of cultural differences.

Dufays accurately identifies the core of the environmental and social crisis in contemporary urban environments. The key points of his analysis:

- Climate crisis and structural deficiencies

> Climate change is the most pressing threat we face... Floods, water shortages, outdated sewage systems

*His view is in line with the Urban Resilience and Inclusion framework (*EURESFO, 2024*), which points out that urban infrastructure must become not only "greener" but also more equitable.*

- Social cohesion and racism

> Immigration has caused integration challenges and, unfortunately, an increase in racist backlash.

*His statement highlights the social rift that often accompanies environmental strategies. This is central to the Environmental Gentrification theory (*Checker, 2021*), which explains how "green" policies can exacerbate social inequalities if they are not accompanied by integration policies.*

8.4.2 *Material, Symbolic, Institutional Exclusion Maria Salvado*

Maria Salvado, translator and active participant in the life of Ixelles, presents a deeply experiential and politically conscious position on structural inequalities, poverty, housing, and cultural tensions. Her interpretive analysis can be decomposed into the following three dimensions:

- Daily struggle and invisible poverty

> I've moved four houses because I can't afford the expenses anymore. I don't have enough money. And neither do my friends.
>
> No matter how much you love the city, it doesn't love your back.

These statements are not simple personal narratives; they constitute experiential data that reflect the institutionally normalized exclusion that embodies the phenomenon of "invisible poverty." Salvado here gives face to her theory of Environmental Gentrification (Checker, 2021), which explains how green policies (redevelopment, tourist development, aesthetic interventions) increase the cost of living and push the most vulnerable out of the centers.

- Cultural diversity and empathy deficit

> Most people don't want equality – they tolerate it, they don't seek it.
>
> The problem is not diversity, but the lack of education to understand it.

Here Salvado enters the level of society's cultural unconscious. She describes the deep cultural rift that makes diversity a threat rather than an opportunity. This finding aligns with the theory Urban Resilience and Inclusion (EURESFO, 2024), which focuses not only on the need for institutional inclusion, but also on changing social representations and educating in empathy. Salvado proposes critical cultural literacy as a prerequisite for political inclusion.

- Structural absence of voice and participation

> No one asks us. Policies consider us expendable.

This formulation reveals the absence of a meaningful institutional voice for socially marginalized urbanites. It is evidence that "participation" is superficial, often communicative. This confirms the principles of Enabling Governance Arrangements (UrbanA, 2024): only through active, co-led decision-making can the cycle of exclusion be broken.

In conclusion, Salvado embodies the "everyday theorist": she is a social subject, but also a vehicle of analysis. Through her narrative, a triple level emerges:

- Material exclusion (housing insecurity)
- Symbolic exclusion (lack of recognition of diversity)
- Institutional exclusion (absence of participation in governance)

When these themes intersect with GT principles and contemporary theoretical frameworks, the interpretive power of the analysis is dramatically enhanced: her voice is not simply reproduced but is established as a producer of social knowledge and transformation.

8.4.3 Political Empowerment and Social Representation Participation and Political Integration (Mohammed Mahla)

Mahla spoke on behalf of 45 migrants: "*We are never involved in planning. They represent us without us.*" This intervention is a clear argument in favor of Enabling Governance Arrangements (UrbanA, 2024) that propose social justice alliances and recognize the representation of "excluded others" not only as beneficiaries, but also as policy co-formulators. He also emphasized that "no one asks us, they just decide for us"—a phrase that reflects a lack of institutional participation and is in line with the interpretative framework of Enabling Governance Arrangements (UrbanA, 2024). This view proposes inclusive planning processes, especially at the local government level, in order to establish social justice through participation. Mahla offers a purely experiential reading of the experience of migration and integration in Ixelles:

- Immigration with simultaneous cultural affirmation
- "*My traditional clothing is the material evidence of my identity and my culture.*" This element shows a resistance to assimilation and a desire to preserve cultural identity, an element that is linked to the Enabling Governance Arrangements (UrbanA, 2024) precepts of pluralistic participation.
- Collective voice of youth

  I speak on behalf of 45 young immigrant friends – we discussed what I would say

This is an example of active social voice (grassroots representation) and makes Mahla bearer for collective social theory. GT here operates with bottom-up concept creation (Fig. 8.1).

- Human-centered ecological governance

Mahla links the issue of integration to the broader ecological and social policies of Ixelles. His position is fully consistent with the framework of human-centered ecological governance described by UrbanA (2024) and supports the embodied participation of citizens, regardless of origin.

His analysis also reveals a fundamental issue: the voice of migrants is not only invisible but structurally absent from institutions. This confirms the necessity for reflexive governance.

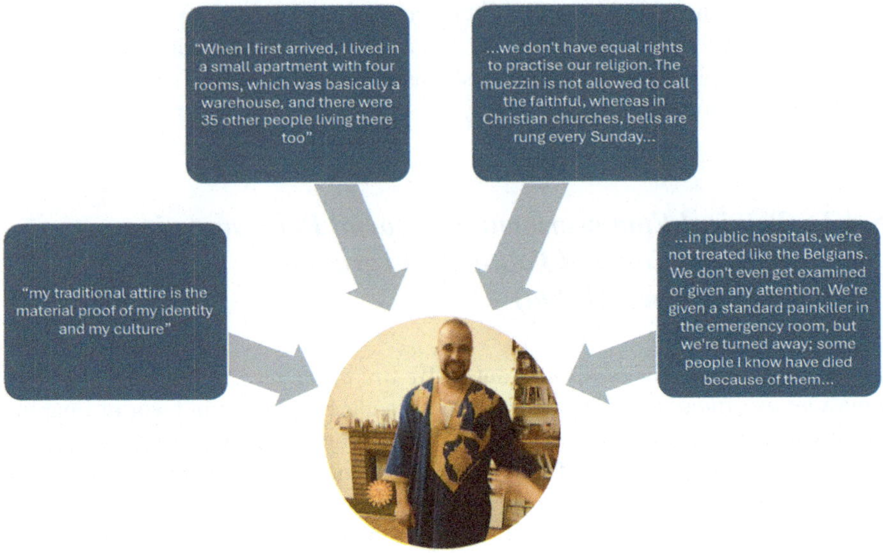

Fig. 8.1 Photo of Mohammed Mahla

8.4.4 *Critical Correlation and Reflection of Kolognomos*

Kalognomos, through his testimony, underlined the need for local coexistence, integration, and understanding of "others," emphasizing that "the future of public spaces is coexistence or dissolution"—a phrase that reinforces the moral foundation of participatory ecology and substantiates the moral basis of GT when approached with dispositive principles and reflexivity.

8.5 Forming Interpretive Axis

- Contrast between managerial and experiential discourse
- Dufays functions as a technocratic voice, describing the "crisis of infrastructure," while Salvado exposes the "crisis of daily survival" from below.
- Kalognomos offers an "intermediate discourse," combining institutional participation with cultural ecology.
- Loud and silent voices
- Mahla speaks on behalf of a silent majority of immigrants, introducing a form of "community theorizing" with collectivity.
- Salvado embodies invisible poverty and calls for substantive recognition instead of charitable surveillance.
- Cultural Integration vs. Political Assimilation

- Mahla carries the semiology of his cultural identity (traditional clothing) as an act of resistance to the decivilization of the immigrant experience.
- Kalognomos emphasizes the importance of public space as a cultural and ecological common ground.
- Urban governance as a point of crisis and opportunity
- Salvado and Mahla reveal the deficits of institutional participation.
- Dufays and Kalognomos propose alternative forms of participation, with the latter framing democracy as consultation rather than mere representation.

This interpretive framework shows how: (1) Grounded Theory is enriched when it incorporates a multiplicity of relational and acting subjects and (2) Political theory is transformed when it incorporates everyday experience as theory, and (3) Participatory governance requires a democratization of knowledge, not just institutions.

The critical value of these narratives is not descriptive. It is transformative (Table 8.6).

8.6 Interpretive Analysis of Focus Group and Semi-Structured Interviews of Immigrants

8.6.1 Methodological Approach

The focus group and semi-structured interviews with seven migrants from Morocco were conducted using the snowball method, ensuring a stratified and geographically diverse approach. The sample, although small, reflects distributed experiences of social exclusion, cultural isolation, and expectation of integration.

Table 8.6 Comparative interpretive analysis of interviews with stakeholders

Participant	Main topic	Sociological focus	Theoretical interconnection
Jean Michel Dufays	Climate threat and social instability	Structural infrastructural deficiencies and increasing racism	Urban Resilience (EURESFO, 2024), Environmental Gentrification (Checker, 2021)
Maria Salvado	Experience of invisible poverty and cultural exclusion	Symbolic exclusion, institutional absence	UrbanA (2024), Inclusion Frameworks
Mohammed Mahla	Resistance to Cultural Assimilation & Collective Representation	Intercultural voice, radical inclusion	Enabling Governance Arrangements, UrbanA (2024)
Stavros Kalognomos	Deliberative ecological governance	Participatory decision-making, intergenerational trust	Reflexive Governance UrbanA (2024)

8.6.2 *Interpretive Patterns and Codification*

The following main themes emerged from the participants' narratives:

(a) Normalization of unequal treatment
Quotes such as: "In public hospitals, we're not treated like the Belgians... some people I know have died because of them." and "We don't have equal rights to practice our religion... the muezzin is not allowed..." demonstrate an internalization of exclusion and at the same time testify to structural conditions that reinforce marginalization.

(b) Policy of voice and silence
Mahla's phrase: "I speak on behalf of at least 45 young immigrant friends..." reveals a community voice but also the absence of formal channels of representation. Political 'silence' here becomes an act of resistance and identity formation.

(c) Discrimination within migration
Interviewees underlined that: "There is a clear difference in treatment between European and non-European migrants" intensifying the analysis of the stratification of migration, not only as a legal or geographical category but also as a cultural category.

8.6.2.1 Crossing with Modern Theoretical Frameworks

UrbanA (2024): contrary to the principle of "coalition governance" and empowerment through participation, the migrant experience shows institutional withdrawal and lack of inclusion.

Environmental Gentrification (2021): narratives reveal how "green policies" do not focus on the needs of vulnerable groups—a factor of ecological injustice.

Urban Resilience and Inclusion (EURESFO, 2024): the concept of "resilience" in the interviews becomes an improvised survival strategy rather than an institutional guarantee of inclusion.

8.6.2.2 Critical Commentary

The constitution of narratives is not only intended to record but functions as an act of political visibility. The research team must go beyond description—to systemic interpretation of injustice and the design of participatory practices of transformation (see UrbanA, 2024).

8.7 Interpretive Analysis of the Three Institutional Panels

8.7.1 Cultural Ecology and Urban Resilience/Green Strategies and Social Exclusion (Christos Doulkeridis)

The former mayor of Ixelles said: "The environment is not a luxury, it is survival." Although he offers a visionary discourse of ecological transition, his policy analysis reveals ambiguity in relation to the social dimension. Some groups have not benefited from "green reforms"—an interpretation that aligns with Environmental Gentrification (Checker, 2021), which demonstrates how "ecological solutions" can exclude those who do not meet economic and cultural inclusion criteria Christos Dulkeridis, showed that he understood the problem due to the national institutional framework but highlighted his strategy of "cultural social ecology"—a position that confirms the principles of the Urban Resilience and Inclusion framework (EURESFO, 2024). His attempt to integrate migrant populations into ecological transitions supports the argument for social cohesion through environmental policy.

The former mayor of Ixelles presents a coherent program of ecological and participatory governance, with explicit references to:

Inter-parliamentary democracy: "I always invite those who disagree with a policy to present their arguments." This stance is consistent with the Enabling Governance Arrangements framework (UrbanA, 2024).

Environmental sensitivity and social inclusion: His perspective captures the resilient municipality model, with proposals for open accountability and participation of vulnerable populations in policy making.

Intergenerational and Intercultural Consciousness: It emphasizes that the environment "is not a technical issue, but a social and cultural one," an element that approaches political ecology theory with an emphasis on experiential engagement.

8.7.2 Urban Planning and Sustainable Regionalization/ Environmental Justice and Neighborhoods in Transition (Vassilis Papaevstathiou and Vassilis Koutrafouris)

Vassilis Papaevstathiou and Vassilis Koutrafouris, with experience in urban planning, raised issues of "green projects that are not always accompanied by social filters" and warned of population displacement phenomena due to the upgrading of spaces. This observation is found in the theory of Environmental Gentrification (Checker, 2021) which argues that green interventions can lead to displacement of vulnerable groups if not accompanied by protective mechanisms.

Papaeftathiou puts forward a critical institutional interpretation of local governance, where: The notion of European heterogeneity is foregrounded. He recognizes that decision-making mechanisms often alienate residents from the real consequences of policies. It shows a clear orientation towards networking and horizontal governance, commenting on the need for cooperation between cities on environmental and social crises. This is embodied in the institutional framework of UrbanA (2024) and Reflexive Governance.

Koutrafouris offers an endogenous interpretation of public space: He focuses on the socio-cultural reconstruction of the city, where public space becomes a context for participation, imagination, and equality. In his interview, he advocates the importance of cultural democracy as a precondition for ecological inclusion, highlighting the relationship between culture and social resilience—fully aligned with the Urban Resilience and Inclusion model (EURESFO, 2024) (Fig. 8.2).

Their panel statements highlighted the issue of upgrading the space without social guarantees: "The park has improved, but its old users have been lost." This phrase sums up the critique of "horizontal" policies and confirms the importance of jurisdictional filters in urban planning—a concept fundamental to Enabling Governance Arrangements and "territorial justice" mechanisms in metropolitan centers (Tables 8.7, 8.8 and 8.9).

Interpretive Conclusion

The comparison of the discourses shows that:

- Institutional actors process participation as an intention or tool, while migrants experience it as an unfulfilled need.
- Migrants produce informal strategies of collectivity, which have not yet been integrated into institutional mechanisms.
- Dialogue is not non-existent—but remains asymmetrical, which reinforces the stratification of political voice.

Fig. 8.2 Governance expert

Table 8.7 Comparative interpretive axis—Doulkeridis, Papaefstathiou, Koutrafouris

Participant	Core of discourse	Central issue	Theoretical connection
Christos Dulkeridis	Deliberative, ecological governance	Accountability, solidarity	Enabling governance, Urban resilience
Vassilis Papaefstathiou	European political criticism	Institutional alienation	Reflexive governance, political ecology
Vassilis Koutrafouris	Critical cultural space	Cultural integration	Environmental gentrification, Urban commons

Table 8.8 Summary of the theoretical and empirical connections

Chapter	Core focus/theoretical function	Key concepts	Connection to empirical part (Ixelles Case Study)
Chapter 8—Integrative framework: Biocultural Governance and Urban complexity	Synthesizes the analytical dimensions and theoretical contributions of previous chapters, integrating cultural, environmental, and political factors.	Political ecology, biocultural governance, complexity, interdependence, sustainability, Urban Systems	Links theoretical synthesis with the empirical findings; explains how cultural and ecological dimensions interact within governance systems in Ixelles.

Table 8.9 Comparative interpretive analysis: immigrants and institutional persons

Thematic axis	Migrant perspective	Institutional perspective	Critical reading
Participation	"Nobody asks us"—Salvado	"I invite those who disagree to participate"—Doulkeridis	There is a clear participation gap: The rhetoric of participation in institutions does not translate into experiential access for migrants.
Invisibility vs. representation	"We have no voice in the municipalities"—Mahla	"We need open spaces for consultation"—Kalognomos	Symbolic representation is not enough when it is not accompanied by structural integration into institutional patterns.
Public space	"I speak on behalf of 45 migrants"—Mahla	"We must represent all groups"—Papaefstathiou	There is a difference in the meaning of public space: for migrants it is a marginal space of acceptance, while for institutionalists it is a regulatory policy tool.
Resilience	"They only see us when they are needed"	"Public space is a tool for inclusion and coexistence"—Koutrafouris	Migrants experience resilience as a forced adaptation, not as empowerment.
Linguistic and cultural access	"We only feel accepted in the bazaars"—focus group	"Resilience is the result of planning policies"—Kalognomos	The absence of an explicit positioning of institutionalism shows that cultural accessibility remains institutionally unprocessed.

8.8 Discussion: Critical Summary—GT as an Empowerment Tool

Grounded Theory emerges as a flexible tool that can generate theoretical knowledge from the ground up—especially when it incorporates reflection and availability. The inclusion of the above narratives, when related to interpretive frameworks, creates not just qualitative description, but critically enriched theory. This theory functions as both knowledge and praxis.

The incorporation of narratives from key speakers theoretically enriches the Grounded Theory analysis, providing a connection to the everyday practices and perceptions of participants:

- Jean Michel Dufays and Maria Salvado stressed the need for institutional mediations that respect the multiple identities of citizens. This approach reinforces the Urban Resilience and Inclusion framework (EURESFO, 2024) as it promotes participation without cultural assimilation.
- Mohammed Mahla, expressing 45 migrants, said: "We never participate in planning. We are represented without us." This testimony vindicates the Enabling Governance Arrangements model (UrbanA, 2024), which calls for institutional activation of socially excluded voices.
- Christos Dulkeridis put forward the need to see ecology as a social cultural and survival issue, yet in practice there was no horizontal social benefit from green interventions. Environmental gentrification theory (Checker, 2021) explains this phenomenon as structural displacement through green policies.
- Vassilis Papaevstathiou and Vassilis Koutrafouris highlighted the need for planning with social filters: "The park has improved, but its old users have been lost." The park's old park was improved but its old users were lost. Placements that document the need for the implementation of 'territorial justice' and participatory urban planning as proposed by UrbanA (Tables 8.10 and 8.11).

Chapters 7 and 8 jointly show that resident satisfaction is not merely an attitudinal measure but a multi-variable construct grounded in environmental, social, and governance performance.

Quantitative measurement establishes patterns; qualitative interpretation explains the underlying mechanisms.

The result is a robust triangulated model of Urban Well-Being = f(Quality of Life, Governance, Cultural Participation, Sustainability, Safety, Intercultural Relations) that can guide municipal policy.

8.9 Triangulation Method of Qualitative Research

In keeping with the culture-centered and biocultural orientation of this study, the adoption of a triangulated methodological design was essential to capture the complexity of urban sustainability as both a measurable and a lived phenomenon. The

Table 8.10 Comparative table of interviews and tables with interpretive comments

Interview/table	Main thematic patterns	Interpretive comment
Dufays	Climate inequality, political alienation	It highlights the green transition as politically and socially unequal, denounces the institutional exclusion of disadvantaged groups.
Salvado	Health inequality, institutional exclusion	It denounces the degradation of everyday life at all levels
Mahla	Community representation, language barriers	The experience of insecurity is transformed into a reason for empathy and political vigilance. It denounces the lack of cultural education, exclusion in practice, and invisible racism due to economic (material) exclusion
Kalognomos	Institutional intention of inclusion	It articulates a collective voice of immigrants with reference to the lack of accessibility, offers an alternative political presence.
Doulkeridis	Open consultation, green governance	It attempts to harmonize political and cultural identity based on inclusion and cooperation.
Papaefstathiou	European political heterogeneity	It exercises a rhetoric of participation, without however ensuring practical accessibility for all.
Koutrafouris	Public space, cultural democracy	It recognizes the gap between institutional intention and social practice, focuses on collective organization.
Table 1: Main challenges	Climate change, housing, migration, health	Public space is proposed as a policy field, but the question remains who has access to it.
Table 2: Participation and voice	Invisibility, community voice, institutional participation	The multiple crisis under conditions of invisibility and stratification is traced.
Table 3: Solutions	Empathy, consultation, inclusion	The discrepancies between the experience and the declaration of participation are highlighted.
Table 4: Obstacles	Linguistic, institutional, social	The proposed solutions incorporate ethical and social demands from the grassroots.

diverse social, cultural, and ecological processes observed in Ixelles could not be adequately represented through a single mode of enquiry. Triangulation therefore provided a framework through which multiple epistemological perspectives—quantitative, qualitative, and visual—could be interrelated to yield a more comprehensive and contextually grounded understanding of the case. This pluralistic strategy reflects the conviction that the interplay between objective indicators and subjective meanings constitutes the very essence of biocultural research.

The mixed-methods triangulation employed in this study establishes internal validity through the convergence of all strands of qualitative evidence, thereby ensuring that quantitative results are grounded in lived experience and interpretative understanding. This approach positions triangulation not merely as a technical procedure but as a critical means of integrating diverse perspectives within the biocultural framework of the research. Quantitative indicators are interpreted through

Table 8.11 Theoretical cross-check table

Theoretical framework	Correlation with interviews/tables	Interpretive commentary
Enabling Governance (UrbanA, 2024)	Doulkeridis, Kalognomos, Table 2	Encourages participatory processes but often does not reach excluded populations.
Environmental Gentrification (Checker, 2021)	Dufays, Table 1	Green policies do not integrate the needs of vulnerable groups, reinforcing exclusion.
Urban Resilience and Inclusion (EURESFO, 2024)	Salvado, Mahla, Table 3	Migrant resilience is not institutionally supported but relies on informal networks.
Reflexive Governance	All interviews, Table 4	Highlighting the role of the researcher and power structures within the field.
Inter-rater Reliability (Ghazal et al., 2021)	Coding of all interviews	Strengthens validity through polyphonic analysis and avoiding one-sided interpretations.

qualitative narratives, interviews, and observational data, producing a dialogue between measurement and meaning. In this way, statistical patterns acquire explanatory depth, while qualitative accounts gain empirical grounding. The convergence of these complementary strands strengthens the analytical coherence of the study and reflects its central premise that cultural experience provides the interpretative lens through which sustainability must be understood.

The mixed-methods triangulation establishes internal validity through convergence: statistical correlations are explained by qualitative causality. For example:

The low intercultural relations index (ANOVA, $p = 0.11$) matches migrant testimonies about limited intercultural contact.

High variance in governance perception is echoed by institutional critique in interviews.

Thus, the expanded explanation clarifies how numeric indicators and lived experiences interrelate.

8.9.1 Triangulation of Findings from Three Qualitative Data Sources

This section applies the technique of triangulation between three data sources:

1. Individual interviews of Jean Michel Dufays, Maria Salvado, Mohhamed Mahla, and Stavros Kalognomou.
2. Interviews of migrants in the focus group.
3. Positions of institutional figures: Christos Doulkeridis, Vassilis Papaefstathiou, and Vassilis Koutrafouris.

Triangulation helps to detect divergences, synergies, and interpretive gaps between empirical, community, and institutional narratives. Triangulation allows for the emergence of polyphony, enhancing the meta-analysis of experiences without smoothing over differences. As Korstjens and Moser (2022) argue, it is a central element for the foundation of critically informed qualitative research, especially in social fields with structural asymmetries, such as that of migrant integration, and polymorphic urban coexistence (Bitsani, 2013).

The process of triangulation was operationalized through the systematic comparison and integration of data across methodological layers. Quantitative results were first analyzed to identify significant patterns and anomalies within key thematic areas such as intercultural relations, governance perception, and environmental engagement. These statistical tendencies were then cross-referenced with qualitative materials—namely interview transcripts, focus group discussions, and field observations—to verify, refine, or challenge emerging interpretations. The visual and spatial analyses provided an additional dimension, enabling the recognition of symbolic correspondences and contradictions between policy discourse, civic identity, and lived experience. Through iterative coding, thematic clustering, and matrix mapping, convergences were identified and inconsistencies examined, allowing for nuanced interpretation rather than mechanical alignment. This systematic cross-validation ensured that findings were both empirically robust and conceptually coherent, thereby reinforcing the internal validity and interpretative depth of the research as a whole (Table 8.12).

Then, after all the preceding discussion, below we proceed to an overall, interpretative and critical conclusion, which incorporates all the individual analyses, tables and interviews within the Grounded Theory approach, cross-checked with contemporary theoretical and empirical findings:

8.9.2 Overall Results of the Research

This analysis, based on semi-structured interviews, focus groups, and institutional positioning within the DELPHI analysis, revealed a multifaceted map of inequalities, participatory exclusions, and institutional ambiguity that runs through contemporary urban policies of inclusion and resilience. Through Grounded Theory, theory was not imposed a priori, but emerged from the empirical data itself, highlighting distinct patterns and contradictions.

Multiple Crisis and Stratification of Inequalities

The migrants' testimonies highlighted systemic deficiencies in housing, health, and institutional recognition. These challenges are not only social, but also political and cultural, as they shape boundaries of belonging and voice in the city (see Table 1). The phrase "Depon for all" (Salvado) summarizes not only a situation of health inadequacy, but also a symbolic individuation of political abandonment.

Table 8.12 Thematic data sources

Thematic	lInterviews	Focus group	Institutional positions
Participation and representation	Mahla: speaks of a latent community voice and lack of institutional connectivity. Salvado: strong sense of exclusion.	Migrants speak of distance from the political system and lack of voice in the cities.	Doulkeridis: rhetoric about the integration of dissenters.
Inequalities and access	Dufays: highlights ecological injustice and the class nature of the transition. Salvado: denounces unequal health care.	References to housing shortages and bureaucratic barriers—"they only accept us at the bazaars."	Papaefstathiou: supports theoretical representation. Solutions remain abstract and technocratic.
Cultural access	Mahla: emphasizes linguistic alienation and the absence of Muslim religious identification.	Difficulties in understanding public texts, silent acceptance of exclusion.	Absence of an explicit strategy for intercultural policy. The issue is ignored or minimized institutionally.
Resilience and empowerment	Kalognomos: proposes institutional alliances, without however radically questioning the hierarchy.	Resilience is presented as survival without institutional support—"we endure because we have no other choice."	Institutional settings present resilience as a planning goal rather than a response to abandonment.

Asymmetry of Participation

While institutional actors such as Doulkeridis and Kalognomos stress the need for inclusiveness, migrants themselves express the absence of institutional channels of inclusion. The contradiction between the rhetoric of inclusion and experiential silence is confirmed by the inability to access even linguistic tools. This "participatory paradox" was confirmed through Table 2.

Invisible Solutions and Structural Barriers

The solutions proposed by the stakeholders themselves (see Mahla, Salvado, Kalognomos) are not technical, but context-critical: they are about the formation of voice, the horizontality of consultation, and the shift of power toward the everyday. However, as Table 4 demonstrates, institutional inertia and the lack of normative cultural accessibility act as barriers of a structural nature.

Theoretical Cross-Validation and Critical Validation

The comparative juxtaposition of the findings with contemporary theoretical frameworks (Checker, 2021; EURESFO, 2024; UrbanA, 2024) confirms that urban resilience, participatory governance, and environmental justice remain asymmetrically distributed. Institutional actors act as agents of political intent, while migrants act as agents of biopolitical experience and survival.

Perspectives for Transformation

The material shows that change cannot be only institutional or technical—it must be interpretive, linguistic, and cultural. The politics of listening and empathy, as articulated by Salvado and Kalognomo, is a core element of critically conscious governance (see Voß, Bauknecht, & Kemp, 2006; Voß, Smith, & Grin, 2006). Grounded Theory allows these unanticipated perspectives to emerge, without reductionisms and theoretical transgressions.

8.9.3 Synthetic Interpretive Commentary

GT, through this theoretical and experiential intersection, is not limited to descriptive thematization but functions as a method of dialectical knowledge production with transformative value.

This qualitative analysis, enriched with Grounded Theory techniques and triangulation, has highlighted an intensely layered social and institutional map in which the institutional narrative of inclusion collides with the experiential experience of invisibility.

The voices of migrants (individually and collectively), as expressed by Dufays, Salvado, Mahla, and the focus group, capture not just material precarity but ontological precarity: the experience of being "present-absent" in a city that institutionally recognizes you but socially excludes you.

In the face of this, institutional positions (Doulkeridis, Papaeustathiou, Koutrafouris, Kalognomos) appear symbolically disposed but practically inadequate: the proposals of inclusion do not challenge the terms of access to power, while the language of inclusion often proves to be empty of mediation.

The triangulation revealed that:

- Migrants retain voice but not access.
- Institutions have access, but not a reflective voice.
- Intermediate policies (e.g., for public space, participation, language) remain at best advisory rather than transformative; Environmental Gentrification theory (Checker, 2021) and the frameworks of UrbanA (2024) and Reflexive Governance (Voß, Bauknecht, & Kemp, 2006) offer not just support for the findings, but conceptual tools to interpret institutional selectivity and the moralization of social inaction.

Linking empirical voices and theoretical models confirms that inclusion policy, environmental justice, and urban resilience are not distinct axes, but convergent forms of survival and dignity for vulnerable communities, with culture and pedagogy as the connecting link. It also demonstrates the urgent need for a change in institutional policy and governance model by shifting the center of gravity from the elite economy to people-centered multi-stakeholder governance through culture and cultural environmental education (pedagogy).

It becomes clear that inclusion is not a technical issue, but a political act of sharing voice, visibility, and access. Any policy that fails to embody these conditions reinforces exclusion through participatory pretense.

8.9.4 Discussion of the Results

This field research offers a deeply interdisciplinary and multimodal exploration of urban sustainability, governance, and inequality in the Municipality of Ixelles. Through the triangulation of visual, quantitative, and qualitative methods, the study constructs a layered understanding of how different actors—municipal institutions, residents, grassroots organizations—perceive, experience, and represent sustainable urban life.

8.9.4.1 Fragmented Sustainability Narratives

A core finding revealed in the triangulation chapter (Chap. 6) is the disconnect between official sustainability discourse—as constructed through municipal websites—and the lived realities of marginalized residents. The digital platforms emphasize ecological transition, green spaces, and inclusive branding (Chap. 3), yet fail to address structural barriers related to housing insecurity, gentrification, and unequal access to services (Chap. 4).

This echoes critical urban studies literature, which warns that institutional sustainability narratives often mask deeper social inequalities (Anguelovski et al., 2016; Purcell, 2009). The visual grammar of municipal websites, as shown in Chap. 3, strategically omits signs of conflict or exclusion, reinforcing a curated and depoliticized image of the city. The lack of visibility given to local activism, minority-led initiatives, or alternative sustainability imaginaries is particularly striking.

8.9.4.2 Socio-Spatial Inequalities and Symbolic Violence

The quantitative analysis (Chap. 4) maps striking socio-spatial disparities across Ixelles: higher income areas show improved access to services, education, and housing, while lower-income neighborhoods—often with racialized populations—experience marginalization. This confirms the "dual city" thesis (Marcuse, 1997), where cities evolve into parallel realities shaped by uneven development.

Bourdieu's (1986) notion of symbolic violence becomes relevant here: municipal messaging valorizes certain lifestyles and aesthetics (e.g., green living, digital connectivity), subtly marginalizing those who do not conform. The qualitative interviews (Chap. 5) further support this: many residents express frustration at being unheard in planning processes, or excluded from sustainability efforts due to linguistic, economic, or cultural barriers.

8.9.4.3 Power Asymmetries in Governance and Participation

The research reveals that urban governance in Ixelles operates within constrained participatory frameworks. While public consultations and workshops are technically open, actual access is often limited to more educated, French- or Dutch-speaking, middle-class residents. This resonates with Arnstein's (1969) critique of tokenistic participation: processes may appear inclusive but functionally reinforce elite control. Chapter 5 highlights grassroots responses, where local associations, often led by women or migrants, build bottom-up sustainability practices—from urban gardening to community repair cafés. However, their contributions are rarely integrated into municipal planning, suggesting a lack of institutional permeability (Healey, 1997).

8.9.4.4 9 Methodological Innovation and Limitations

The mixed-methods approach is one of the study's key strengths. The triangulation of visual, statistical, and narrative data allows for cross-validation and a more holistic representation of urban life. This corresponds with Denzin's (1978) call for methodological triangulation in complex social research.

However, studying is not without limitations. Language barriers, the digital divide, and pandemic constraints may have skewed representation in data collection. Moreover, as the researcher operated across multiple roles—observer, participant, analyst—issues of positionality and reflexivity were central. The methodology chapter notes this carefully, but further analysis of the researcher's influence on findings would strengthen transparency.

8.10 Conclusion

This longitudinal, transdisciplinary research provides critical insights into the contradictions and possibilities of sustainable urban governance in Ixelles. The findings illustrate that, while municipal institutions project a forward-thinking and inclusive image of sustainability, this narrative often diverges from the grounded experiences of diverse communities. The study reveals deep fractures in the urban fabric—fractures related to access, representation, and power.

By integrating multidimensional analysis, statistical mapping, and community-based inquiry, the project uncovers how sustainability is not only a technical or environmental goal but also a contested socio-cultural process. Issues of race, class, language, and cultural capital shape who benefits from "sustainable development" and who remains excluded.

Importantly, the research demonstrates the creative and resilient practices of marginalized groups, who adapt to and resist exclusion through their own forms of sustainability. These practices often operate below the radar of official policy but carry vital knowledge and transformative potential. Recognizing and supporting these bottom-up approaches is essential for any meaningful transition towards just urban futures.

In summary, the triangulation of methods within this study functions as both an analytical instrument and a philosophical stance, reflecting the biocultural premise that multiple realities coexist within urban experience. By correlating quantitative data with qualitative and visual insights, the research attains a balanced synthesis between measurement and meaning. This integrative process not only validates the findings internally but also lays the foundation for the subsequent stage of analysis, in which the interpretative patterns emerging from this convergence are examined in detail.

Conflicts of Interest The authors have no conflicts of interest to declare that are relevant to the content of this chapter.

References

Anguelovski, I., Shi, L., Chu, E., Gallagher, D., Goh, K., Lamb, Z., Reeve, K., & Teicher, H. (2016). Equity impacts of urban land use planning for climate adaptation: Critical perspectives from the global North and South. *Journal of Planning Education and Research, 36*(3), 333–348. https://doi.org/10.1177/0739456X16645166

Birks, M., Hoare, K., & Mills, J. (2019). Grounded theory: The FAQs. *International Journal of Qualitative Methods, 18*. https://doi.org/10.1177/1609406919882535. (Original work published 2019).

Bitsani, E. (2013). Theoretical approaches to organisational culture and organisational climate: Exploratory research examples and best policies in health care services. *Health Science Journal, 7*(1), 79–100. https://doi.org/10.11648/j.jhrm.20130104.11

Bourdieu, P. (1986). The forms of capital. In J. Richardson (Ed.), *Handbook of theory and research for the sociology of education* (pp. 241–258). Greenwood. https://www.marxists.org/reference/subject/philosophy/works/fr/bourdieu-forms-capital.htm

Charmaz, K. (2014). *Constructing grounded theory*. Sage.

Charmaz, K., & Thornberg, R. (2020). The pursuit of quality in grounded theory. *Qualitative Research in Psychology, 18*(3), 305–327. https://doi.org/10.1080/14780887.2020.1780357

Checker, M. (2021). Environmental justice and gentrification in New York city. *Environment: Science and Policy for Sustainable Development, 63*(2), 16–27. https://doi.org/10.1080/00139157.2021.1871293

Denzin, N. K. (1978). Triangulation: A case for methodological evaluation and combination. *Sociological Methods*, 339–357.

EURESFO. (2024). European Urban Resilience Forum: Full report. https://urbanresilienceforum.eu/fileadmin/user_upload/Post-even_documents/EURESFO-2024-Full-Report.pdf

Ghazal, R., Ahmed, N., & Khan, S. (2021). Inter-rater reliability in qualitative research: Methods and implications. *Qualitative Research Journal, 21*(4), 375–390. https://doi.org/10.1108/QRJ-05-2020-0063

GUREGT. (2018). *Governance of urban and regional ecological transitions: European thematic report*. European Commission, Directorate-General for Research and Innovation.

Healey, P. (1997). *Collaborative planning: Shaping places in fragmented societies*. Macmillan Press Ltd.. https://doi.org/10.1007/978-1-349-25538-2

Korstjens, I., & Moser, A. (2022). Series: Practical guidance to qualitative research. Part 6: Longitudinal qualitative and mixed-methods approaches for longitudinal and complex health themes in primary care research. *European Journal of General Practice, 28*(1), 118–124. https://doi.org/10.1080/13814788.2022.2053106

Marcuse, P. (1997). The ghetto of exclusion and the fortified enclave: New patterns in the United States: New patterns in the United States. *American Behavioral Scientist, 41*(3), 311–326. https://doi.org/10.1177/0002764297041003004. (Original work published 1997).

Purcell, M. (2009). Resisting neoliberalization: Communicative planning or counter-hegemonic movements? *Planning Theory, 8*, 140–165. https://doi.org/10.1177/1473095209102232

Rey, M. (2023). *Political ecology and urban transitions: Reframing sustainability through culture and governance*. Routledge. https://doi.org/10.4324/9781003257642

Strauss, A., & Corbin, J. (1998). *Basics of qualitative research: Techniques and procedures for developing grounded theory* (2nd ed.). Sage Publications.

Urban, A. (2024). *City resilience framework* (2024 ed.). Urban A Publishing.

Voß, J.-P., Bauknecht, D., & Kemp, R. (Eds.). (2006). *Reflexive governance for sustainable development*. Edward Elgar Publishing. https://doi.org/10.14279/depositonce-4543

Voß, J.-P., Smith, A., & Grin, J. (2006). Designing long-term policy: Rethinking transition management. *Policy Sciences, 39*(4), 321–343. https://doi.org/10.1007/s11077-006-9024-9

Chapter 9
Integrative Modelling of Three Complementary Research Strands: The Model of Biocultural City

9.1 Integrated Triangulation of Digital, Quantitative, and Qualitative Findings

The triangular model illustrates the integrative relationship between the three methodological pillars of this study: Digital Research, Quantitative Analysis, and Qualitative Inquiry. Each vertex represents a distinct epistemic lens—digital research captures the institutional discourse and online representations of Ixelles; quantitative analysis provides statistical evidence of resident satisfaction and socio-environmental variation; and qualitative inquiry interprets the lived experience and narratives of residents. The arrows converging toward the center symbolize the synthesis of these perspectives into an "Integrated Understanding of Ixelles Governance and Urban Life." This intersection embodies the principle of methodological complementarity, whereby diverse modes of knowledge co-produce a reflexive and multidimensional interpretation of urban sustainability, governance, and inclusion.

Within this research, triangulation serves as both a methodological design and an epistemological stance. It unites digital research, quantitative survey data, and qualitative narrative inquiry to examine the Municipality of Ixelles through three

---

Introducing Note
From Qualitative Insights to Integrated Triangulation
The preceding chapter has illuminated the nuanced ways in which residents of Ixelles articulate their lived experiences of governance, sustainability, and intercultural coexistence. Through interviews, focus groups, and narrative testimonies, the qualitative analysis revealed the emotional, symbolic, and relational dimensions that statistical data alone could not capture. These insights exposed both the vitality and the fragility of the municipality's social fabric—highlighting tensions between institutional representation and personal perception, as well as between inclusion and invisibility. Building upon these interpretive findings, the following chapter advances toward an integrative synthesis, combining digital, quantitative, and qualitative strands through a process of triangulation. This methodological convergence seeks to transcend disciplinary and epistemological boundaries, offering a comprehensive and reflexive interpretation of how governance, sustainability, and urban identity intersect in the lived reality of Ixelles.

E. Bitsani et al., *The Biocultural City as a Utopia or the Future Reality?*, Cities, Heritage and Transformation, https://doi.org/10.1007/978-3-032-17584-7_9

complementary perspectives: (a) the digital discourse that constructs the municipality's public image and institutional identity; (b) the quantitative evidence that measures resident satisfaction and socio-environmental well-being; and (c) the qualitative narratives that articulate the lived experiences and perceptions of residents. Together, these layers form what Creswell and Plano Clark (2018) term an "integrative design logic," where each dataset offers distinct yet interrelated insight into the same social phenomenon.

A conceptual representation of the integrative relationship between digital, quantitative, and qualitative research strands in the study of the Municipality of Ixelles. The figure illustrates how these three methodological perspectives converge to produce a multidimensional and reflexive understanding of governance, sustainability, and urban life.

The triangular model illustrates the integrative relationship between the three methodological pillars of this study: Digital Research, Quantitative Analysis, and Qualitative Inquiry. Each vertex represents a distinct epistemic lens—digital research captures the institutional discourse and online representations of Ixelles; quantitative analysis provides statistical evidence of resident satisfaction and socio-environmental variation; and qualitative inquiry interprets the lived experience and narratives of residents. The arrows converging toward the center symbolize the synthesis of these perspectives into an "Integrated Understanding of Ixelles Governance and Urban Life." This intersection embodies the principle of methodological complementarity, whereby diverse modes of knowledge co-produce a reflexive and multidimensional interpretation of urban sustainability, governance, and inclusion.

The epistemological grounding of this triangulation aligns with a critical realist ontology (Bitsani, 2013a, 2013b). Reality is viewed as stratified—comprising observable events, underlying mechanisms, and structural conditions. Quantitative data capture the empirical regularities of resident satisfaction; qualitative accounts uncover mechanisms of meaning-making and agency; and digital representations reveal discursive and institutional structures that mediate perception. This ontological plurality supports a more reflexive, holistic understanding of urban life in Ixelles.

Figure 9.1 *visualizes the epistemological foundation of the study, demonstrating how methodological complementarity generates a multidimensional understanding of urban reality.*

Moreover, triangulation in this study is interpretive rather than confirmatory. Following Greene (2007) and Johnson and Onwuegbuzie (2004), the purpose is not merely to cross-check for consistency but to engage the contradictions and tensions that emerge between methods. As Flick (2018) argues, the strength of triangulation lies precisely in exposing divergence—revealing how different modes of knowing construct alternative yet legitimate understandings of the same phenomenon. This study thus adopts what O'Cathain et al. (2010) call an "expansion purpose" of triangulation, using multiple lenses to extend and enrich interpretation.

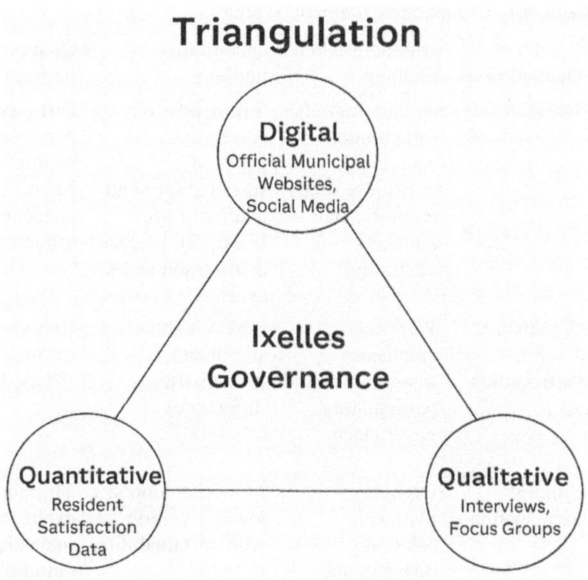

Fig. 9.1 Triangular model of triangulation

9.2 Comparative Thematic Synthesis

This section presents an integrative model derived from the synthesis of three complementary strands of research: the quantitative findings, the triangulated qualitative results, and the visual analysis conducted throughout the study. Each strand offers a distinct yet interconnected perspective on the biocultural dynamics of the urban context under investigation. The quantitative analysis provides measurable indicators of trends and relationships; the qualitative triangulation captures experiential and interpretative depth; and the visual analysis contributes an understanding of symbolic, spatial, and communicative dimensions. By correlating these outcomes conceptually, the study constructs a coherent analytical framework that reveals how cultural, social, and environmental factors interact to shape the lived experience of sustainability in Ixelles. This integrative modeling thereby translates diverse forms of evidence into a unified interpretation, reinforcing the study's culture-centered and biocultural orientation.

The integrated findings presented in Table 9.1 (retained from the previous version) summarize the principal convergences and contradictions across the three strands—digital, quantitative, and qualitative. The synthesis allows patterns of consistency and dissonance to emerge along six thematic axes: sustainability, governance, cultural participation, safety and cohesion, intercultural relations, and quality of life.

Table 9.1 *presents the principal thematic intersections, revealing how each methodological strand contributes distinct yet complementary insights into the urban dynamics of Ixelles.*

Table 9.1 Comparative thematic synthesis

Thematic axis	Digital research findings	Quantitative findings	Qualitative findings	Integrated interpretation
Sustainability	Online discourse foregrounds ecological transition, smart mobility, and climate neutrality.	Strong positive correlation between sustainability and quality of life ($r = 0.73$). Uneven distribution across districts.	Participants perceive "green inequality"—Improvements concentrate in affluent zones.	Sustainability operates simultaneously as environmental policy and social justice issue; digital optimism obscures territorial disparity.
Governance & Participation	Websites emphasize transparency and participatory governance.	ANOVA shows significant inter-district differences ($p < 0.001$).	Interviews express distrust: "They decide for us."	Governance is symbolically participatory but substantively uneven; co-decision remains limited.
Cultural participation	Digital platforms celebrate diversity and creative economy.	Moderate scores; low correlation with sustainability.	Migrants report exclusion from mainstream cultural venues.	Cultural participation framed as consumption rather than co-creation.
Safety & Social Cohesion	Online imagery depicts safe, lively spaces.	Perceptual gaps between high- and low-income districts ($F = 11.09$, $p < 0.001$).	Narratives from marginal areas describe fear and insecurity.	Safety becomes a symbolic asset online but a lived deficit offline.
Intercultural relations	Portrayed as harmonious multicultural coexistence.	Statistically insignificant variation ($p = 0.11$).	Accounts reveal subtle racism and linguistic barriers.	Intercultural harmony remains aspirational rather than actual.
Quality of life	Digitally represented through aesthetics of greenery and Well-being.	Wide range (2.0–4.7) between districts.	Residents link QoL to housing and recognition.	True quality of life derives from equity and governance, not imagery.

A cross-method analysis summarizes the principal patterns of convergence and divergence across the digital, quantitative, and qualitative strands of research. The table highlights the interrelations between sustainability, governance, cultural participation, safety, intercultural relations, and quality of life within the Municipality of Ixelles, offering an integrated empirical foundation for triangulated discussion.

The table illustrates that digital representations of Ixelles construct an image of inclusivity, ecological transition, and cultural vitality—echoing global "smart city" discourses (Evans et al., 2016). However, the quantitative and qualitative evidence nuance and at times contradict these portrayals. For instance, while sustainability is prominently branded online, survey data indicate unequal environmental satisfaction across districts, and qualitative narratives highlight "green inequality"—the concentration of ecological investments in already affluent areas (Anguelovski, 2015).

Similarly, governance is consistently presented in digital media as transparent and participatory. Yet, quantitative measures reveal significant inter-district variation in governance perception, and qualitative interviews expose feelings of exclusion and distrust among certain social groups. The synthesis thus highlights a fundamental tension between institutional performance and institutional perception gap often widened by the performative nature of municipal communication (Miller, 2007).

The comparative synthesis underscores three overarching dynamics:

Discursive idealization: Digital platforms amplify symbolic narratives of harmony and sustainability that exceed residents' lived realities.

1. **Empirical differentiation: Quantitative patterns expose spatial and social inequalities that contradict the homogeneity of digital discourse.**
2. **Experiential contestation: Qualitative narratives reclaim interpretive agency, revealing how residents negotiate, resist, or reinterpret institutional messages.**
3. **Together, these layers produce a composite understanding of Ixelles as a municipality of both cohesion and contradiction.**

9.3 Analytical Integration and Discussion

The integrated analysis reveals how the three strands of evidence interlock to depict a complex socio-political ecology. Ixelles emerges as a microcosm of contemporary European urbanism—progressive in discourse yet stratified in experience.

- **Governance as the Axis of Urban Legitimacy**

Across all data forms, governance emerges as the linchpin of urban satisfaction. While digital communication celebrates participatory democracy, the quantitative findings demonstrate stark variation in perceived transparency (ANOVA, $p < 0.001$), and qualitative data highlight persistent asymmetries of voice and influence. These contradictions illustrate what Healey (1997) terms the "paradox of collaborative planning": participatory frameworks exist discursively but are often underpinned by technocratic and centralized decision-making processes.

This suggests that governance in Ixelles operates simultaneously as a procedural ideal and a contested experience. The credibility of local governance is thus contingent upon reconciling procedural participation with substantive inclusion—ensuring that participation is not merely symbolic but redistributive in its effects (Voß et al., 2006).

- **Sustainability and the Politics of Environmental Justice**

Triangulation reveals sustainability to be the most visible yet most ambivalent domain. Quantitatively, it correlates strongly with quality of life ($r = 0.73$), indicating that environmental factors contribute significantly to residents' satisfaction. However, qualitative narratives complicate this correlation, pointing to uneven access to green amenities and the socio-economic exclusion produced by ecological regeneration projects.

This reflects the phenomenon of green gentrification, whereby environmental improvements disproportionately benefit higher-income residents while displacing lower-income populations (Anguelovski, 2015). Digital communication amplifies this tension by framing sustainability as universally beneficial, masking its classified and spatial dimensions. Consequently, the sustainability discourse functions ideologically, legitimizing municipal action while obscuring the uneven geography of its benefits (Harvey, 1996).

- **Intercultural Relations and the Symbolic Politics of Diversity**

Digital representations of Ixelles celebrate multicultural coexistence through imagery of conviviality and diversity. However, both quantitative and qualitative findings reveal that intercultural relations remain superficial. The statistical neutrality ($p = 0.11$) indicates limited differentiation across districts, but qualitative testimonies expose subtle forms of racism, linguistic segregation, and unequal access to cultural capital.

Drawing on Bourdieu's (1986) theory of cultural capital, this can be interpreted as the symbolic commodification of diversity: cultural representation becomes a marker of prestige rather than an instrument of equity. True intercultural inclusion, as suggested by Sandercock (2003), requires intercultural dialogue and shared urban imaginaries conditions that remain only partially realized in Ixelles.

- **Quality of Life and Relational Well-Being**

The triangulated data reveals that perceptions of quality of life are deeply relational. Statistical measures quantify satisfaction with housing, services, and safety, but qualitative data enrich this by demonstrating how residents define well-being through belonging, respect, and recognition. This resonates with Honneth's notion of the struggle for recognition—the idea that social esteem and mutual respect are essential dimensions of justice. Thus, the quantitative variables of quality of life and governance should be interpreted not merely as administrative indicators but as proxies for moral and affective relations within the urban community.

Integrating the Do No Harm principle (Anderson, 1999) into biocultural urban governance provides a normative and ethical compass for policy design. It suggests that sustainable urban development must be guided not only by efficiency and innovation but also by a conscious commitment to prevent harm, foster equity, and promote social repair in culturally diverse settings.

9.4 Reflexivity and Methodological Limits

The process of triangulation invites critical reflection on the epistemological boundaries of knowledge production. Following Flick (2018) and Silverman (2021), the legitimacy of mixed-method integration depends not on homogenizing differences but on embracing them as epistemic resources.

Digital data offers insight into institutional discourse but are constrained by curation and intentionality; quantitative surveys provide measurable reliability but risk reductionism; qualitative interviews uncover depth and nuance but are situated within specific interpretive contexts. The triangulation thus acknowledges that each method illuminates certain dimensions while obscuring others.

Researcher positionality is central to this reflexivity. The dual role of the researcher—as both analyst of institutional communication and interpreter of resident narratives—necessitates an awareness of interpretive bias and the power asymmetries embedded in representation. Ethical reflexivity, therefore, entails recognizing that data are co-constructed through social interaction rather than discovered as neutral facts (Finlay, 2002).

Methodologically, this study's integration across asynchronous datasets (digital 2022–2024; quantitative 2023–2024; qualitative 2024–2025) introduces temporal fluidity. Yet this temporal layering enriches rather than undermines validity, enabling a longitudinal perspective on evolving municipal discourse. Nonetheless, the limited survey sample constrains generalizability; the study's strength lies not in representativeness but in analytical depth and interpretive triangulation.

9.5 Synthesis and Implications for Urban Governance

Triangulation provides a comprehensive lens through which to interpret the interplay of discourse, structure, and experience in Ixelles. The convergence of digital idealization, quantitative differentiation, and qualitative contestation reveals an urban system negotiating between legitimacy and inequality.

From a governance perspective, this integration highlights the need for reflexive governance—an adaptive model that continuously interrogates the alignment between institutional representation and lived reality (Voß et al., 2006). In practice, this entails embedding participatory mechanisms that go beyond consultation, fostering co-production of policy design and evaluation.

Theoretically, triangulation substantiates the argument that sustainability and inclusion cannot be separated: environmental policies must be assessed in relation to their distributive and symbolic impacts. Quantitative monitoring should therefore be complemented by qualitative dialogue and digital transparency to ensure that data-driven governance remains accountable to citizens' narratives.

As evidenced by the patterns delineated in Table 9.1 and visually reinforced in Fig. 9.1, the triangulation of digital, quantitative, and qualitative perspectives reveals the multidimensional interplay of legitimacy, participation, and socio-spatial equity within Ixelles.

Finally, from a methodological standpoint, this study contributes to the evolving field of integrated urban inquiry. It demonstrates that the synthesis of digital, quantitative, and qualitative data can move beyond validation toward interpretation, producing what Creswell calls "meta-inference"—a higher-order understanding derived from the interaction of multiple data forms.

The model developed in this section demonstrates that the convergence of quantitative evidence, qualitative insights, and visual interpretation produces a more nuanced understanding of the biocultural mechanisms operating within the urban fabric of Ixelles. It highlights the interdependence of empirical measurement, human perception, and symbolic representation in shaping sustainable urban identities. By aligning these dimensions, the model substantiates the study's central claim that culture functions as the integrative medium through which ecological and social processes acquire coherence and direction. The findings presented here therefore provide the conceptual bridge to the concluding discussion, where the broader theoretical and policy implications of this integrative approach are examined in relation to the future governance of biocultural cities.

9.6 The Model of the Biocultural City

The biocultural approach has gained increasing relevance in contemporary discourses of sustainability, particularly in the context of urban development. Rooted in interdisciplinary fields such as anthropology, ecology, and political theory, this approach emphasizes the inseparability of cultural and biological diversity.

Biocultural diversity—defined as the interconnectedness of cultural and biological diversity—is increasingly viewed as a policy priority. The EU-funded projects like "Territorial and Urban Potentials Connected to Culture" (TURaS) and "ROCK" (Regeneration and Optimisation of Cultural heritage in creative and Knowledge cities) highlight the role of heritage-led urban regeneration in enhancing sustainability.

Biocultural perspectives recognize that human cultures are not isolated from nature but are co-constitutive with ecological systems.

This entanglement underscores the need for policies that integrate both biodiversity and cultural knowledge as foundational to resilient urban futures.

Culture, in this framework, is not merely a set of symbolic expressions but a vital system of knowledge production, land management, and social organization that interacts dynamically with ecological systems.

From traditional agricultural practices to spiritual engagements with landscapes, cultural expressions shape and are shaped by the environments in which they emerge.

As such, cultural diversity becomes a resource for ecological resilience, while ecological integrity is essential to the survival of many cultural identities.

The urban context presents both challenges and opportunities for biocultural integration.

On one hand, cities are often seen as loci of ecological degradation and cultural homogenization, driven by consumerism, gentrification, and standardized planning models. On the other hand, urban environments offer spaces for hybrid ecological practices and intercultural engagement. Migrant communities, for example, bring with them distinct agricultural knowledge, culinary traditions, and ecological ethics that enrich urban biodiversity. Critically, the biocultural approach reveals how dominant models of urban development often marginalize local cultures and indigenous ecological knowledge in favor of technocratic and market-driven strategies.

This critique aligns with postcolonial and post-development theories that challenge the epistemic violence embedded in Western development paradigms. By privileging universal metrics of growth and efficiency, urban planning frequently neglects place-based knowledge and participatory practices.

In contrast, a biocultural paradigm advocates for urban sustainability that is culturally grounded and ecologically informed. This involves participatory governance structures that allow diverse communities to shape environmental policies, design public spaces, and manage urban ecosystems. Initiatives such as community gardens, ethnobotanical parks, and intercultural environmental education programs embody this vision by integrating cultural expression with biodiversity conservation.

Moreover, the biocultural approach advances the notion of "epistemic plurality" a recognition that multiple ways of knowing the world, including indigenous, diasporic, and local perspectives, hold value in addressing ecological crises. This shift challenges the dominance of scientific rationalism and opens space for ethical and relational understandings of sustainability.

In this regard, biocultural frameworks resonate with the call for decolonizing sustainability science and creating knowledge systems that are inclusive, democratic, and grounded in lived experience.

To conclude, the biocultural approach offers a critical and holistic lens through which to reimagine urban sustainability.

By foregrounding the interdependence of culture and ecology, it pushes beyond technical solutions toward transformative practices rooted in social justice, ecological balance, and cultural respect.

The integration of biocultural values into urban policy is not merely a matter of inclusion, it is foundational to creating cities that are resilient, equitable, and truly sustainable.

Indicators for evaluating a city's sustainable development may include air pollution levels, public transportation availability, employment rates, educational attainment, green space ratio, energy consumption, and access to drinking water. The term resilience has been applied across various scientific disciplines and has acquired multiple meanings, making its precise definition complex. Despite its polysemous nature, resilience originally referred to a system's ability to respond to disruptions and return to its initial state.

9.6.1 *Definition and Essence of the Biocultural City*

The concept of the biocultural city represents an emergent paradigm within urban theory and practice that seeks to integrate ecological, cultural, and social dimensions of sustainability into a cohesive framework for urban governance and development. The modeling of this concept is not limited to a technical or spatial exercise; rather, it constitutes an interpretive and reflexive process that captures the dynamic interplay between nature, culture, and human agency. Drawing on the principles of constructivism and grounded theory (Charmaz, 2014), the model operates as a conceptual bridge between theoretical understanding and empirical observation, allowing culture to be examined not as an accessory to urban systems but as a structuring force within them.

At its core, the biocultural city model posits that the urban environment functions as a living organism—a relational system shaped by material infrastructures, social practices, and symbolic values. This perspective echoes Bourdieu's (1986) notion of habitus and social capital, wherein everyday practices and shared meanings collectively generate and reproduce urban space. Through this lens, the city becomes both a site and a product of biocultural interdependence: ecosystems support human livelihoods, while cultural narratives and governance practices determine how those ecosystems are perceived, managed, and sustained. As Healey (1997) argues, spatial planning must be understood as a communicative process shaped by values, discourse, and institutional negotiation—principles that also underpin the biocultural approach.

In this framework, the biocultural city is a living, adaptive ecosystem where biological diversity and cultural diversity co-evolve in mutual support.

It views the city not as a mechanical or economic structure but as a biocultural organism—a space where human creativity, social justice, and ecological integrity form one continuous system.

"Culture must be at the heart of all policy areas if sustainability goals are to be achieved, as biocultural lenses are the key tool for urban development." (Table 9.2).

Table 9.2 Biocultural city foundational paradigm

Core concept	Description
Biocultural approach	Integrates nature and culture as interdependent forces of sustainability.
Biopolitical awareness	Recognizes how governance, knowledge, and power shape human–environment relations.
Cultural ecology	Understands cities as evolving cultural ecosystems influenced by technology and environment.
Holistic sustainability	Merges economic, social, environmental, and cultural dimensions.

Core Principles of the Biocultural City

1. **Culture–Ecology Integration**

   - Urban policy, planning, and technology must reflect ecological consciousness and cultural meaning.
   - Both tangible (land, heritage) and intangible (memory, identity) assets are equally valuable.

2. **Cultural Sustainability as Urban Backbone**
   Culture is not a "sector" but a structural principle that guides all sustainable policies.
3. **Interdependence and Coexistence**
   Encourages intercultural dialogue, mutual respect, and social equity across communities.
4. **Participatory Governance**

   - Promotes multi-actor collaboration: citizens, local authorities, universities, NGOs, and the private sector.

5. **Ethics of Care and Responsibility**

   - Based on *biocultural ethics*: respect for both human and non-human life.

6. **Right to the City (Lefebvre)**

   - Ensures all inhabitants—not just elites—shape ecological and cultural future (Table 9.3).

The biocultural city model operates through four interconnected dimensions—environmental, cultural, social, and governance-based—each reinforcing the others in the pursuit of equitable sustainability.

1. The environmental dimension focuses on ecological stewardship, sustainable mobility, and climate adaptation. It interprets environmental policy as a cultural practice, shaped by values, narratives, and perceptions of risk and responsibility.

Table 9.3 Interconnected dimensions of the biocultural model

Dimension	Key components	Biocultural goal
Ecological	Green infrastructure, biodiversity, circular economy, climate adaptation	Regenerate the natural fabric of the city
Cultural	Heritage, creative expression, collective memory, local knowledge	Strengthen identity and continuity
Social	Inclusion, diversity, solidarity, equality	Foster coexistence and cohesion
Technological	Digital inclusion, smart governance, AI ethics	Support sustainability through responsible innovation
Economic	Green jobs, local economies, fair distribution	Create equitable prosperity
Governance	Participatory planning, transparent institutions	Enable shared ownership of decisions

2. The cultural dimension addresses the symbolic and affective meanings of place, heritage, and identity. Drawing from Healey's (1997) communicative planning and Bitsani's (2013a, 2013b) work on cultural climate, it highlights how cultural participation fosters social cohesion and inclusion.
3. The social dimension explores issues of diversity, gender equity, and migrant inclusion, aligning with the United Nations Sustainable Development Goals (Goals 5, 10, and 11). It regards social sustainability as the capacity to live together in difference, sustained by intercultural dialogue and collective well-being.
4. The governance dimension centers on transparency, participation, and accountability. Following Voß et al. (2006), governance is conceptualized as reflexive and adaptive, evolving through feedback between institutions and citizens.

The interplay of these dimensions produces a dynamic field of interaction in which policies, perceptions, and everyday practices coalesce. Modeling these interdependencies allows researchers to map not only structural inequalities but also the latent capacities for transformation embedded within urban systems.

1. **Resilience and Adaptation**
   The city anticipates, absorbs, and recovers from crises (climate, health, social).
2. **Equity and Inclusion**
   Addresses systemic inequalities (class, race, gender, migration) in access to resources.
3. **Knowledge and Innovation**
   Blends scientific, indigenous, and community knowledge systems.
4. **Cultural-Ecological Memory**
   Heritage and tradition are used as guides for future urban resilience.
5. **Participatory co-Governance**

9.6.2 The Biocultural Urban Ecosystem: The Model of Biocultural City

The model also adopts a multi-scalar perspective, situating Ixelles as both a local and translocal context within European urban systems. Through the study's mixed-methods design, the modelling process integrates statistical indicators (e.g., quality of life, governance perception, sustainability) with ethnographic insight into how residents, particularly those from migrant and minority backgrounds, experience the city. This layered analysis allows for the emergence of what Charmaz and Thornberg (2020) describe as theoretical integration—a synthesis between grounded data and interpretive abstraction. In this sense, the biocultural model becomes both an analytical framework and a reflexive practice of co-creation.

A **biocultural city** operates as an **interactive system** of:

- **Human capital** (education, creativity, citizenship).
- **Natural capital** (biodiversity, water, air, landscape).
- **Cultural capital** (heritage, values, identity).
- **Social capital** (trust, cooperation, empathy).
- **Technological capital** (digital innovation, data ethics).

Each dimension reinforces the others, creating a **self-renewing urban metabolism** based on solidarity and ecological justice. Citizens co-design urban spaces and sustainability indicators.

Case Study Reference: Ixelles (Brussels)

- The research uses Ixelles as a living laboratory:

- Multicultural urban district balancing historic heritage and social diversity.
- Field studies (2018–2025) combining qualitative and quantitative data on sustainability perception. Mixed methods research (2018–2025): ethnography, surveys, participatory workshops.
- Demonstrates *how cultural diversity can drive sustainability* when coupled with inclusive governance.
- Reveals contradictions—such as *green gentrification*—showing that real biocultural transformation requires redistribution of power and resources (Table 9.4).

The strategic goals of the biocultural city emphasize the integration of cultural, ecological, and social values as fundamental components of urban development. They aim to foster resilient, inclusive, and participatory cities where sustainability is achieved through cooperation, creativity, and respect for diversity.

In practical terms, the biocultural model integrates concepts of cultural capital (Bourdieu, 1986), social trust, and participatory governance (Arnstein, 1969) to articulate how citizens engage with sustainability agendas. Culture is conceived not merely as heritage or artistic production, but as a living infrastructure of meaning that enables communities to interpret, contest, and shape their environments. Bitsani (2013a, 2013b) emphasizes that organizational and cultural climates within public institutions directly influence their capacity to adapt and respond to complex

Table 9.4 Biocultural city strategic goals

Goal	Action direction
1. Embed culture in sustainability frameworks	Integrate culture as the fourth pillar of sustainability (UN 2030 Agenda).
2. Co-create local biocultural policies	Develop city-specific cultural–ecological strategies with community participation.
3. Ensure environmental justice	Redistribute access to green and public spaces equitably.
4. Promote biocultural rights	Recognize community rights to cultural and natural resources.
5. Foster circular and regenerative economies	Prioritize reuse, creativity, and low-carbon production.
6. Educate for biocultural citizenship	Empower citizens to act as ecological stewards and cultural innovators.

challenges—a principle that applies equally to municipalities navigating transitions toward sustainability. Within this framework, the Human Intercultural City Model (Bitsani, 2013a, 2013b) provides a complementary lens, conceptualizing the city as a dynamic ecosystem of intercultural dialogue, participatory governance, and shared values. This model underscores that urban resilience and social cohesion emerge from the capacity of diverse communities to co-create cultural and institutional meanings through continuous interaction. The incorporation of this intercultural dimension strengthens the biocultural approach by embedding diversity and dialogue as central drivers of sustainable urban transformation (Fig. 9.2).

The diagram conceptualizes the *biocultural city* as a dynamic system formed through the intersection of **Culture**, **Ecology**, and **Technology**, illustrating their mutual influence in shaping sustainable urban development. At the convergence of these dimensions lie the core principles of **Participation**, **Equity**, **Innovation**, **Resilience**, and **Justice**, which collectively constitute the ethical and operational foundation of biocultural urbanism. The model emphasizes that true sustainability emerges from the integration of cultural meaning, ecological balance, and technological responsibility within participatory governance frameworks.

The model also adopts a multi-scalar perspective, situating Ixelles as both a local and translocal context within European urban systems. Through the study's mixed-methods design, the modelling process integrates statistical indicators (e.g., quality of life, governance perception, sustainability) with ethnographic insight into how residents, particularly those from migrant and minority backgrounds, experience the city. This layered analysis allows for the emergence of what Charmaz and Thornberg (2020) describe as theoretical integration—a synthesis between grounded data and interpretive abstraction. In this sense, the biocultural model becomes both an analytical framework and a reflexive practice of co-creation, uniting intercultural understanding, cultural participation, and ecological awareness within a coherent structure of urban sustainability. The experience of Ixelles exemplifies this integrative vision: its cultural diversity, participatory initiatives, and evolving governance structures mirror the principles of the Human Intercultural City Model, demonstrating how interculturalism can serve as a foundation for biocultural resilience and inclusive urban regeneration.

9.6.2.1 Critical Reflections on the Model's Application

While the biocultural model offers a holistic framework, it is not without limitations. Its interdisciplinary nature challenges conventional academic and policy silos, requiring new forms of collaboration between planners, cultural practitioners, and ecologists. Furthermore, its implementation depends on the willingness of institutions to engage in participatory processes that may destabilize established hierarchies. As Denzin (1978) argue, methodological triangulation and longitudinal engagement are vital for capturing the complexity of human experience within

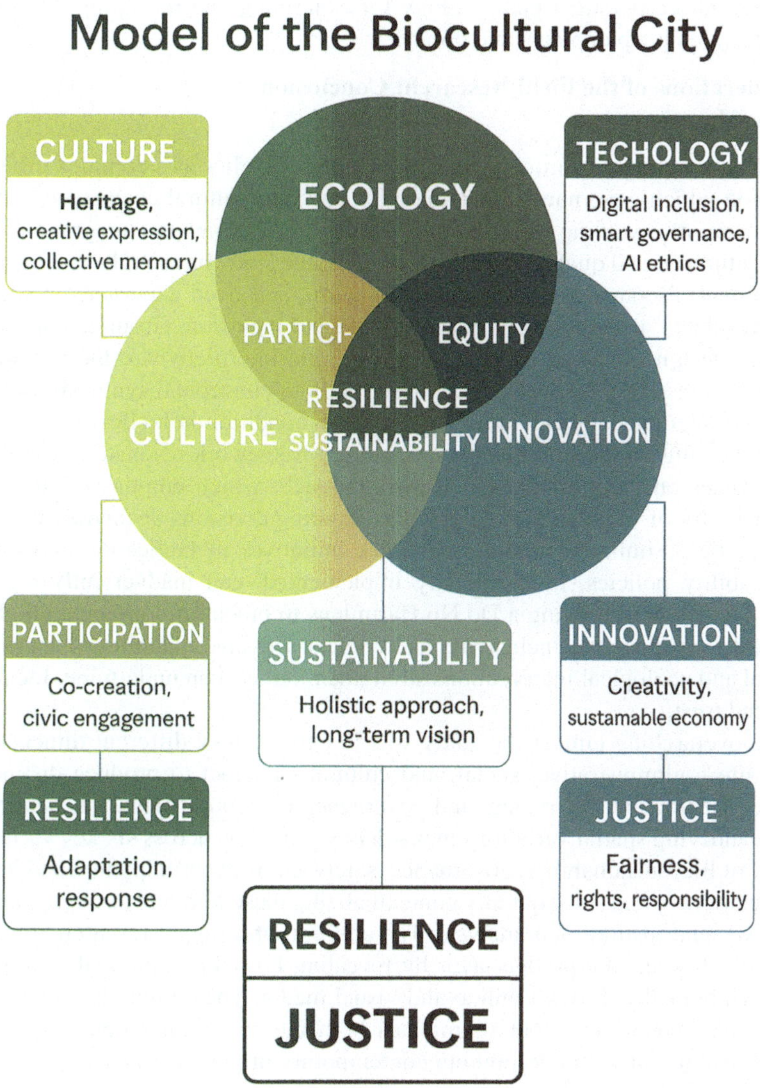

Fig. 9.2 Model of the biocultural city

evolving contexts. This reinforces the importance of sustained reflexivity in both research and governance.

In applying this model to Ixelles, the research demonstrates how culture operates as both a diagnostic and transformative element in urban life. Cultural narratives reveal how residents make sense of environmental challenges, governance structures, and social inequalities, while also articulating visions of belonging and resilience. The model thereby translates the abstract principles of the biocultural paradigm into a practical methodology for examining and reshaping the relationships between people, place, and policy.

Considerations of the Field Research: Conclusion

I. **Considerations**

This research has examined the Municipality of Ixelles as a complex urban ecosystem in which governance, sustainability, and intercultural coexistence intersect within a rapidly evolving socio-political landscape. Through the integration of digital, quantitative, and qualitative methodologies, the study has sought to unravel how institutional discourse, statistical representation, and lived experience coalesce to shape residents' perceptions of quality of life and inclusion. Grounded in a mixed-methods design informed by critical realism and interpretivism, the research has moved progressively from empirical description to conceptual synthesis, culminating in a triangulated understanding of the urban condition in Ixelles.

In reflecting on the findings, the biocultural framework resonates strongly with Mary Anderson's (1999) Do No Harm approach, which emphasizes the ethical responsibility of interventions to avoid reinforcing divisions or inequalities within communities. Similarly, urban governance initiatives in Ixelles demonstrate that sustainability policies, if uncritically implemented, can inadvertently reproduce social exclusions. Applying a Do No Harm lens to biocultural urbanism highlights the need for reflexivity, inclusivity, and context-sensitive planning—ensuring that cultural and ecological interventions strengthen, rather than undermine, local cohesion, and trust.

The overarching aim of the study was to explore how different dimensions of urban life—administrative, social, and cultural—interact to produce distinct patterns of satisfaction, belonging, and governance perception. This was achieved by first quantifying spatial variations in resident satisfaction across six key variables—quality of life, sustainability, governance, safety, cultural participation, and intercultural relations—and subsequently contextualizing these findings through qualitative narratives and institutional analysis. The addition of a digital research component enabled a further interpretive layer by revealing how the municipality represents itself symbolically through online and visual media. The cumulative outcome is a multi-scalar perspective that illuminates both the structural inequalities and the aspirational narratives underpinning contemporary urban governance.

- **Key Findings and Theoretical Contributions**

The findings presented throughout the preceding chapters converge on a central insight: that urban governance in Ixelles, while discursively inclusive, operates through layered inequalities that are spatial, symbolic, and participatory. The quantitative analysis revealed statistically significant disparities between districts, indicating that material well-being and institutional trust are unevenly distributed across the municipality. Districts characterized by higher environmental quality and economic capital demonstrated stronger perceptions of governance efficiency and social safety, whereas others—particularly those with limited access to green space or cultural infrastructure—registered lower satisfaction and higher levels of perceived exclusion.

The qualitative and digital analyses deepened this picture by exposing the contrast between the municipality's public-facing discourse of sustainability and diversity, and the residents' lived experiences of fragmentation and selective participation. The triangulation process revealed that governance functions simultaneously as an administrative practice and as a symbolic performance. By situating these dynamics within frameworks of collaborative planning (Healey, 1997), environmental justice (Harvey, 1996; Anguelovski, 2015), and reflexive governance (Voß et al., 2006), the study contributes to an understanding of how local governments mediate legitimacy through communicative and representational strategies.

Methodologically, the study advances the application of interpretive triangulation as a model for mixed-methods urban research. Rather than using multiple datasets solely for corroboration, the integration pursued here demonstrates how epistemic tensions between digital, quantitative, and qualitative perspectives can become generative of new knowledge. This approach aligns with recent calls for methodological pluralism (Flick, 2018; Greene, 2007), acknowledging that cities are not static objects of study but dynamic sites of overlapping narratives and power relations.

- **Towards a Reflexive Understanding of Urban Knowledge**

The synthesis achieved through triangulation offers more than empirical insight—it foregrounds the reflexive processes through which knowledge about cities is produced, contested, and mobilized. The juxtaposition of digital representation, statistical regularity, and experiential depth highlights that urban knowledge is always partial, situated, and negotiated. By embracing this multiplicity, the study challenges positivist assumptions about objectivity and instead promotes a relational epistemology grounded in dialogue between data and experience.

In the context of Ixelles, this reflexive stance underscores the importance of aligning institutional communication with lived realities, ensuring that participatory governance transcends symbolic performance. The research thus positions triangulation not only as a methodological framework but also as a democratic principle—one that insists on the coexistence of diverse voices, forms of evidence, and interpretive standpoints in the pursuit of sustainable and inclusive urban futures.

- **Policy and Governance Implications**

The triangulated findings of this research bear significant implications for local governance, urban policy, and community engagement. The Municipality of Ixelles stands at a crucial juncture where communicative transparency and socio-spatial justice must converge if sustainability and inclusion are to be more than rhetorical commitments. The results reveal a dual reality: on one hand, a municipality rhetorically committed to participatory democracy and environmental innovation; on the other, uneven patterns of satisfaction, access, and representation that reflect deep-seated socio-economic divisions.

First, the evidence indicates that governance reforms should move beyond procedural participation toward substantive co-production of urban policy. Participatory mechanisms such as public consultations and online forums, while symbolically inclusive, must evolve into structures that redistribute decision-making power and ensure diverse representation particularly of migrants, youth, and lower-income residents.

Second, sustainability initiatives require spatially targeted strategies. While ecological projects have enhanced liveability in affluent districts, the uneven distribution of green spaces and mobility infrastructure risks entrenching environmental privilege. Urban planning should adopt the principles of environmental justice, ensuring that ecological regeneration contributes to social equity rather than displacement.

Third, cultural participation must be reframed as a process of co-creation rather than consumption. The data highlight that cultural policy in Ixelles tends to privilege institutional venues and established artistic communities, often marginalizing grassroots intercultural expression. Encouraging participatory cultural production across linguistic and ethnic lines would strengthen intercultural trust and civic solidarity.

Finally, digital governance—particularly the municipality's online communication and data management—should embrace reflexive transparency. Beyond showcasing achievements, digital platforms can serve as interactive tools for accountability, citizen science, and feedback loops. Such integration would align municipal communication with the relational realities uncovered through this study, fostering a more dialogical and trust-based relationship between government and residents.

- **Methodological Reflections and Limitations**

Methodologically, this research exemplifies the strengths and challenges of interpretive triangulation in urban studies. The integration of digital, quantitative, and qualitative data provided a multi-perspectival view of Ixelles, enabling a richer and more nuanced interpretation of governance and sustainability. However, several methodological reflections and limitations deserve consideration.

Firstly, while triangulation enhanced depth and validity, the heterogeneity of data sources required constant negotiation of scale, temporality, and epistemic stance. The digital materials (2022–2024) reflected institutional representation, whereas the quantitative survey (2023–2024) and qualitative fieldwork (2024–2025) captured evolving perceptions and lived realities. Although this temporal layering enriched the analysis, it also introduced interpretive complexity that necessitated careful synthesizing.

Secondly, sample representation in the quantitative component was limited by both response rate and geographic concentration. While statistical tests provided meaningful patterns, they cannot claim exhaustive synthesizing. Similarly, the qualitative findings, though analytically deep, were shaped by the researcher's interpretive lens and the subjectivity inherent in narrative data. Reflexivity and transparency were therefore essential methodological principles throughout the analysis.

Thirdly, the process of digital discourse analysis brought its own challenges. Online materials are curated, selective, and often politically performative. This required an awareness that digital narratives do not mirror reality but construct it discursively. Integrating these sources with empirical data allowed the research to balance representation with experience—a methodological achievement but also an interpretive responsibility.

Ultimately, these reflections affirm that the strength of triangulation lies not in achieving methodological homogeneity but in negotiating epistemic diversity. The dialogical tension between datasets is what renders this approach analytically powerful and theoretically innovative.

- **Recommendations for Future Research**

Building on the theoretical and methodological insights of this study, several avenues for future research emerge.

Longitudinal Triangulation: Future studies could replicate this multi-method approach over time to examine how perceptions of governance, sustainability, and intercultural relations evolve. Such longitudinal triangulation would provide a dynamic understanding of policy outcomes and urban change.

Comparative Urban Analysis: Applying this framework to other Brussels municipalities or to cities with similar socio-cultural diversity could reveal whether the patterns identified in Ixelles reflect broader metropolitan dynamics or are context-specific. Comparative triangulation would advance theory-building on multi-level governance and urban equity.

Expanded Digital Ethnography: The digital dimension could be developed through network analysis, citizen engagement metrics, or sentiment analysis of online discourse. This would allow for a more granular understanding of how institutional narratives circulate and influence public perception.

Participatory Action Research (PAR): Involving residents directly in data collection and interpretation could strengthen the democratic dimension of triangulated inquiry. PAR would transform participants from subjects into co-researchers, embodying the reflexive governance model advocated in this study.

Interdisciplinary Integration: Future research should continue to blur the boundaries between social science, digital humanities, and environmental studies. The urban experience is inherently interdisciplinary, and triangulated designs offer a methodological bridge across disciplines and epistemologies.

II. **Conclusion**

This chapter has demonstrated that triangulation, when applied reflexively, transforms from a methodological technique into an epistemological practice. In Ixelles, the digital, quantitative, and qualitative findings converge to reveal an urban fabric that is at once cohesive and fractured—progressive in rhetoric but stratified in experience.

By integrating diverse evidentiary modes, this chapter constructs a multidimensional portrait of Ixelles an evolving, negotiated, and contested urban assemblage. The findings suggest that for municipalities seeking genuine sustainability and inclusiveness, triangulation should not end with research but inform governance itself. A reflexive, multi-voiced approach where digital communication, statistical performance, and citizen narrative continuously intersect offers a pathway toward more equitable and authentic urban future.

The analysis identifies governance as the central mediator of urban satisfaction, sustainability as both opportunity and exclusion, and interculturality as a symbolic yet synthesizing aspiration. The contradictions between online representation and lived experience underscore the political nature of data and the performance of discourse.

The integrated interpretation articulated through Table 9.1 and Fig. 9.1 reinforces the argument that triangulation is not merely a methodological strategy but an epistemological framework for apprehending the layered realities of contemporary urban governance.

In conclusion, the interrelations captured in Table 9.1 and represented graphically in Fig. 9.1 encapsulate the study's central contribution: that the synthesis of diverse evidentiary forms yields a ref. In synthesizing the digital, quantitative, and qualitative strands of inquiry, this chapter has demonstrated how triangulation can transcend methodological boundaries to generate a more holistic understanding of urban reality. The Municipality of Ixelles emerges from this integrated perspective not as a uniform or static entity, but as a dynamic assemblage of discourses, data, and lived experiences that continuously shape and reconfigure one another. By engaging with these intersecting layers, the research moves beyond the mere validation of findings toward a critical reflection on the processes through which knowledge about cities is constructed, represented, and acted upon. This integrative approach thus establishes the conceptual and empirical foundation for the concluding reflections that follow, where the implications of this study for theory, policy, and future urban research are further explored.

Ultimately, the modeling of the biocultural city provides a framework for understanding urban sustainability as a process of continual negotiation between cultural meaning and material conditions. It foregrounds culture as a transformative agent capable of bridging ecological and social divides, fostering solidarity across differences, and enabling cities to adapt ethically to future challenges. As Healey (1997) observes, planning in fragmented societies requires dialogical spaces where diverse voices can co-produce shared visions. The biocultural city model, grounded in this ethos, aspires to such inclusivity—an adaptive, participatory, and justice-oriented approach that synthesizes the interdependence of life, culture, and environment.

Drawing on constructivist and grounded theory perspectives (Charmaz, 2014; Strauss & Corbin, 1998), and informed by frameworks such as the Human Intercultural City Model (Bitsani, 2013a, 2013b) and Reflexive Governance (Voß et al., 2006), the study develops a multidimensional model encompassing environmental, cultural, social, and governance components. Through the case study of Ixelles, the model synthesizes quantitative indicators with ethnographic insight to reveal how urban resilience and inclusion emerge from intercultural dialogue, participatory governance, and shared cultural values. The chapter contributes to contemporary urban theory by demonstrating that sustainable development is inherently biocultural—dependent on both ecological adaptation and the cultural meanings through which communities negotiate change. The biocultural model thus provides a reflexive and participatory framework for guiding equitable, inclusive, and ecologically grounded urban transformation and comprehensive understanding of Ixelles' governance and urban life.

Drawing on constructivist and grounded theory perspectives (Charmaz, 2014; Strauss & Corbin, 1998) and informed by frameworks such as the Human Intercultural City Model (Bitsani, 2013a, 2013b) and Reflexive Governance (Voß et al., 2006), the study develops a multidimensional model encompassing environmental, cultural, social, and governance components. Through the case study of Ixelles, the model synthesizes quantitative indicators with ethnographic insight to reveal how urban resilience and inclusion emerge from intercultural dialogue, participatory governance, and shared cultural values. The chapter contributes to contemporary urban theory by demonstrating that sustainable development is inherently biocultural—dependent on both ecological adaptation and the cultural meanings through which communities negotiate change. The biocultural model thus provides a reflexive and participatory framework for guiding equitable, inclusive, and ecologically grounded urban transformation.

In synthesizing the digital, quantitative, and qualitative strands of inquiry, this chapter has demonstrated how triangulation can transcend methodological boundaries to generate a more holistic understanding of urban reality. The Municipality of Ixelles emerges from this integrated perspective not as a uniform or static entity, but as a dynamic assemblage of discourses, data, and lived experiences that continuously shape and reconfigure one another. By engaging with these intersecting layers, the research moves beyond the mere validation of findings toward a critical reflection on the processes through which knowledge about cities is constructed, represented, and acted upon. This integrative approach thus establishes the conceptual and empirical foundation for the concluding reflections that follow, where the implications of this study for theory, policy, and future urban research are further explored.

Ultimately, the modeling of the biocultural city provides a framework for understanding urban sustainability as a process of continual negotiation between cultural meaning and material conditions. It foregrounds culture as a transformative agent capable of bridging ecological and social divides, fostering solidarity across differences, and enabling cities to adapt ethically to future challenges. As Healey (1997) observes, planning in fragmented societies requires dialogical spaces where diverse voices can co-produce shared visions.

The biocultural city model, grounded in this ethos, aspires to such inclusivity—an adaptive, participatory, and justice-oriented approach that synthesizes the interdependence of life, culture, and environment. Drawing on constructivist and grounded theory perspectives (Charmaz, 2014; Strauss & Corbin, 1998), and informed by frameworks such as the Human Intercultural City Model (Bitsani, 2013a, 2013b) and Reflexive Governance (Voß et al., 2006), the study develops a multidimensional model encompassing environmental, cultural, social, and governance components. Through the case study of Ixelles, the model synthesizes quantitative indicators with ethnographic insight to reveal how urban resilience and inclusion emerge from intercultural dialogue, participatory governance, and shared cultural values. The chapter contributes to contemporary urban theory by demonstrating that sustainable development is inherently biocultural—dependent on both ecological adaptation and the cultural meanings through which communities negotiate change. The biocultural model thus provides a reflexive and participatory framework for guiding equitable, inclusive, and ecologically grounded urban transformation.

- **Concluding Remarks**

This chapter has consolidated the empirical and interpretative strands of the research through the integrative modelling of quantitative, qualitative, and visual findings. By correlating these complementary forms of evidence, the analysis has revealed how cultural, social, and environmental dimensions interact within the biocultural dynamics of Ixelles. The resulting model demonstrates that culture operates not as a peripheral variable but as the connective medium through which sustainability acquires meaning, coherence, and civic legitimacy. In synthesizing these diverse insights, the chapter has translated empirical observation into c onceptual understanding, thereby preparing the ground for the final discussion in Chap. 10, where the theoretical and policy implications of this biocultural framework are examined within a broader urban and European context.

Conflicts of Interest The authors have no conflicts of interest to declare that are relevant to the content of this chapter.

References

Anderson, M. B. (1999). *Do no harm: How aid can support peace—Or war*. Lynne Rienner Publishers.

Anguelovski, I. (2015). From toxic sites to parks as (green) LULUs? New challenges of inequity, privilege, gentrification, and exclusion for urban environmental justice. *Journal of Planning Literature, 31*(1), 23–36. https://doi.org/10.1177/0885412215610491. (Original work published 2016).

Arnstein, S. R. (1969). A ladder of citizen participation. *Journal of the American Institute of Planners, 35*(4), 216–224. https://doi.org/10.1080/01944366908977225

Bitsani, E. (2013a). Theoretical approaches to organisational culture and organisational climate: Exploratory research examples and best policies in health care services. *Health Science Journal, 7*(1), 79–100. https://doi.org/10.11648/j.jhrm.20130104.11

Bitsani, E. (2013b). *"Intercultural City identity" and "human intercultural cities" (H.I.C). An ontological/dynamic model for the social cohesion of modern and post-modern cities. The research example of Trieste Italy*. Nova Publishers.

Bourdieu, P. (1986). The forms of capital. In J. Richardson (Ed.), *Handbook of theory and research for the sociology of education* (pp. 241–258). Greenwood.

Charmaz, K., & Thornberg, R. (2020). The pursuit of quality in grounded theory. *Qualitative Research in Psychology, 18*(3), 305–327. https://doi.org/10.1080/14780887.2020.1780357

Denzin, N. K. (1978). Triangulation: A case for methodological evaluation and combination. *Sociological Methods*, 339–357. https://www.scirp.org/reference/referencespapers?referenceid=2398870

Evans, J., Karvonen, A., & Raven, R. (2016). *The experimental city*. Routledge.

Finlay, L. (2002). "Outing" the researcher: The provenance, process, and practice of reflexivity. *Qualitative Health Research, 12*(4), 531–545. https://doi.org/10.1177/104973202129120052

Flick, U. (2018). *An introduction to qualitative research* (6th ed.). Sage.

Greene, J. C. (2007). *Mixed methods in social inquiry*. Jossey-Bass.

Harvey, D. (1996). *Justice, nature, and the geography of difference*. Blackwell.

Healey, P. (1997). *Collaborative planning: Shaping places in fragmented societies*. Macmillan.

Johnson, R. B., & Onwuegbuzie, A. J. (2004). Mixed methods research: A research paradigm whose time has come. *Educational Researcher, 33*(7), 14–26. https://doi.org/10.3102/0013189X033007014

Miller, P. (2007). Governing by numbers: Why calculative practices matter. *Social Research, 68*(2), 379–396.

O'Cathain, A., Murphy, E., & Nicholl, J. (2010). Three techniques for integrating data in mixed methods studies. *BMJ (Clinical Research Ed.), 341*, c4587. https://doi.org/10.1136/bmj.c4587

Sandercock, L. (2003). *Cosmopolis II: Mongrel cities of the 21st century*. Continuum.

Silverman, D. (2021). *Interpreting qualitative data* (6th ed.). Sage.

Voß, J.-P., Bauknecht, D., & Kemp, R. (Eds.). (2006). *Reflexive governance for sustainable development*. Edward Elgar Publishing. https://doi.org/10.14279/depositonce-4543

Chapter 10
Final Overall Conclusions

10.1 Overview

This concluding chapter synthesizes the principal findings and contributions of the research, situating them within the broader theoretical and policy discourse on urban sustainability and cultural development. It reflects upon the journey undertaken throughout the study—from conceptual formulation to empirical exploration—and demonstrates how the biocultural city paradigm constitutes both an analytical framework and a transformative vision for contemporary urban governance. The chapter brings together the theoretical, methodological, and empirical dimensions of the research, showing how they converge in the case of Ixelles to articulate a comprehensive understanding of sustainability as a biocultural process. It also delineates the study's significance for future urban research, governance practice, and the advancement of participatory and intercultural models of sustainable development.

The study's central premise has been that sustainability cannot be adequately understood or achieved through environmental policy alone. Rather, it must be conceptualized as a living and evolving relationship between ecological systems, cultural practices, and social structures. This view foregrounds the biocultural interdependence of urban life—a recognition that ecological resilience and social justice are inseparable from cultural expression, shared meaning, and collective participation. Within this framework, culture is understood not as a decorative adjunct to sustainability, but as its structuring and transformative force—an active medium through which people interpret and shape their environment. The study's interdisciplinary approach—drawing from urban sociology, cultural studies, political ecology, and participatory planning—reflects this understanding by integrating multiple dimensions of human experience into a unified theoretical model.

By developing and modeling the biocultural city, this research contributes to the redefinition of urban sustainability as a dialogical and participatory process that

E. Bitsani et al., *The Biocultural City as a Utopia or the Future Reality?*, Cities, Heritage and Transformation, https://doi.org/10.1007/978-3-032-17584-7_10

emerges through the interplay of values, practices, and institutions. The model challenges the compartmentalization of ecological, economic, and social domains by highlighting the cultural mechanisms that link them. In doing so, it extends the communicative and reflexive planning traditions articulated by Healey (1997) and Voß et al. (2006), while also building upon Bitsani's (2013) Human Intercultural City Model, which frames intercultural dialogue as a driver of resilience and cohesion. The following sections elaborate upon these contributions by addressing, respectively, the theoretical, methodological, and empirical dimensions of the research, followed by a synthesis of its policy implications, limitations, and future directions.

10.2 Theoretical Contributions

The research offers a substantial theoretical contribution by positioning the biocultural city as an integrative paradigm for understanding the relational dynamics between culture, ecology, and governance in urban sustainability. It bridges multiple strands of thought—including cultural ecology, social capital theory, reflexive governance, and intercultural dialogue—to propose a model in which culture serves as the connective tissue of sustainable development. This reconceptualization draws upon Bourdieu's (1986) notion of cultural and social capital, which frames culture as a form of collective resource capable of generating cooperation and trust, and upon Putnam et al. (1993) analysis of civic traditions, which underscores the social embeddedness of institutional performance. Together, these perspectives inform a model that situates sustainability within the lived cultural fabric of the city, emphasizing that ecological transformation depends upon social cohesion, participatory governance, and the ethical imagination of communities.

It rejects the dichotomy between nature and culture, proposing instead that both dimensions coexist within a dynamic continuum. This paradigm views human societies as integral components of ecosystems, whose survival depends on balance, reciprocity, and the creative adaptation of cultural practices to environmental conditions (Berkes, 2008).

In rethinking sustainability as a biocultural rather than a purely ecological or economic construct, the study contributes to the ongoing critique of technocratic urbanism, which often isolates environmental policy from its human and cultural dimensions. The biocultural city paradigm instead conceives of urban sustainability as an evolving cultural project, wherein governance, creativity, and environmental awareness converge. This theoretical innovation responds directly to the limitations of modernist planning paradigms by offering a framework grounded in interpretive pluralism—a recognition that sustainability is negotiated through diverse meanings and contested practices rather than imposed as a universal blueprint. The model thus aligns with the constructivist epistemology of Guba and Lincoln (1994), while also advancing the reflexive governance principles identified by Voß et al. (2006), which stress adaptive learning and co-creation among stakeholders.

Furthermore, the study enhances theoretical understanding of interculturalism as an urban asset. Drawing from Bitsani's (2013) Human Intercultural City Model, the research demonstrates that diversity is not a challenge to sustainability but a precondition for it. Intercultural dialogue generates forms of social capital that strengthen resilience by fostering empathy, mutual learning, and shared responsibility for the urban commons. The biocultural model therefore provides a theoretical bridge between environmental ethics and cultural pluralism, suggesting that sustainable cities must be conceived as ecosystems of meaning in which difference, creativity, and cooperation are integral to ecological and social balance. This theoretical synthesis represents the study's core contribution to urban studies—reframing the sustainable city not as a technical system but as a living, reflexive organism animated by its cultural and ethical lifeworlds.

10.3 Methodological Reflections

From a methodological perspective, the research contributes to the evolution of reflexive and mixed-methods inquiry within urban and cultural studies. The methodological design combined abductive reasoning, longitudinal observation, and multi-scalar analysis, thereby capturing both the structural and experiential dimensions of sustainability in Ixelles. This integrative design—informed by the "research onion" model (Saunders et al., 2007)—enabled an iterative dialogue between theory and data, in which conceptual understanding evolved alongside empirical evidence. The abductive approach was particularly significant: it facilitated the continuous reconfiguration of theoretical assumptions in response to observed realities, ensuring that the model of the biocultural city remained open, dynamic, and context-sensitive.

The research employed triangulation at multiple levels—across methods (quantitative and qualitative), sources (primary and secondary data), and investigators (researchers from diverse academic and cultural backgrounds). This comprehensive triangulation enhanced validity and interpretive richness, reflecting Denzin's (1978) principle that methodological diversity generates deeper understanding of complex social phenomena. The inclusion of ethnographic and participatory components, including interviews, workshops, and digital observation, grounded the analysis in lived experience and local knowledge, consistent with the interpretivist orientation of grounded theory (Charmaz & Thornberg, 2020). By combining these approaches with quantitative analysis—such as surveys and spatial mapping—the study exemplified a methodological pluralism capable of addressing both the measurable and the meaningful dimensions of sustainability.

In addition, the methodological framework was pragmatic in its orientation, recognizing that research on sustainability must engage both theoretical reflection and applied problem-solving. The dual time horizon—longitudinal (2018–2025) and cross-sectional—allowed the study to trace continuity and change across social, cultural, and ecological variables. This design revealed how perceptions, governance practices, and cultural participation evolved in response to shifting urban conditions. Importantly, the research also incorporated digital ethnography, acknowledging that contemporary urban life increasingly unfolds in hybrid spaces where physical and virtual interactions intersect. This innovation expands methodological practice in urban sustainability research, demonstrating how digital platforms can function as spaces of intercultural communication, identity formation, and civic participation.

10.4 Empirical Insights from the Ixelles Case

Empirically, the research demonstrates how the biocultural city framework manifests within the lived realities of Ixelles, revealing a municipality marked by cultural diversity, environmental ambition, and social asymmetry. The quantitative analysis identified statistically significant disparities between neighborhoods in terms of quality of life, governance perception, and sustainability engagement ($p < 0.001$), indicating that urban well-being in Ixelles is spatially differentiated. Districts 5 and 8 formed a "satisfied–resilient" cluster characterized by strong cultural participation, environmental awareness, and institutional trust. In contrast, District 4 exhibited low scores across most variables, signaling socio-environmental vulnerability and limited inclusion. These patterns reflect broader dynamics of urban segregation in Brussels and confirm the interdependence of material, social, and cultural capital (Bourdieu, 1986).

The qualitative findings complemented these patterns by uncovering how residents interpret sustainability through everyday narratives of belonging, fairness, and adaptation. Interviews and focus groups revealed that many participants viewed culture as both a bridge and a boundary—a resource for community cohesion yet also a marker of inequality when access to participation was uneven. The data thus illustrated that sustainability is experienced not as a uniform condition but as a process negotiated through cultural meaning and institutional interaction. Importantly, the integration of Bitsani's (2013) Human Intercultural City Model within the analysis clarified how intercultural dialogue and shared symbolic spaces foster social resilience. These insights underscore that sustainability must be understood as a relational construct—emerging at the intersection of identity, governance, and environment.

By linking quantitative measures to qualitative interpretation, the study confirmed that perceived sustainability is not determined solely by infrastructure or policy, but by the degree of biocultural coherence—the alignment between environmental quality, cultural vitality, and social inclusion. This empirical validation grounds the biocultural model in observable realities, showing how theoretical constructs translate into patterns of urban life. The case of Ixelles therefore serves not merely as an example but as an analytical lens through which broader processes of urban transformation in Europe can be understood.

10.5 Policy and Practical Implications

The findings carry significant implications for urban policy, planning, and governance. First, they affirm that culture must be embedded as a strategic dimension of sustainability frameworks rather than treated as a peripheral concern. Municipalities should adopt biocultural planning approaches that integrate cultural values, artistic expression, and intercultural dialogue into environmental and social policy. This would align with global agendas such as the UN Sustainable Development Goals (particularly Goals 10, 11, and 13) while grounding them in locally meaningful practices.

Second, the study highlights the need for participatory governance as a cornerstone of sustainable urbanism. Policies fostering co-creation—such as participatory budgeting, community-led cultural initiatives, and intercultural forums—can enhance transparency and trust between institutions and citizens. Drawing on Arnstein's ladder of participation, the research suggests that cities like Ixelles can strengthen their democratic legitimacy by advancing from consultation to co-decision-making models, thus embedding governance within everyday civic life.

Third, the results reveal that symbolic representation and communication are integral to sustainability. The municipality's digital platforms and visual narratives play a key role in shaping perceptions of identity and belonging. Policymakers should therefore treat digital communication not merely as information dissemination but as a form of cultural governance capable of fostering inclusion and visibility for marginalized groups. The biocultural city model thus offers a roadmap for linking environmental policy, cultural participation, and digital transformation within a unified vision of just and inclusive urban development.

10.6 Limitations and Future Research

While this study provides a robust theoretical and empirical foundation, several limitations warrant consideration. The focus on Ixelles, while allowing for depth and contextual specificity, limits the generalizability of findings across diverse urban contexts. Comparative research across European cities with varying governance systems, demographic profiles, and environmental challenges could test and refine the biocultural model's applicability. Similarly, extending the longitudinal dimension beyond 2025 would allow researchers to observe how socio-cultural transformations respond to evolving climate realities, migration patterns, and digitalization.

Another limitation lies in the scope of quantitative indicators used to operationalize the biocultural framework. While they captured key variables such as quality of life and governance perception, additional dimensions—such as informal cultural practices, mental well-being, or ecological literacy—could provide a fuller representation of urban sustainability. Future studies could also employ participatory mapping, cultural analytics, or digital ethnography to visualize how networks of meaning and belonging intersect with spatial and policy dynamics. Interdisciplinary collaboration between sociologists, environmental scientists, artists, and data analysts would further enhance the model's explanatory power.

Ultimately, the biocultural city remains an evolving research agenda. Its full potential lies in the development of comparative, cross-cultural, and transdisciplinary studies capable of addressing the global challenges of inequality, migration, and climate change through a cultural lens. Future research should continue to explore how biocultural processes can inform adaptive, ethical, and inclusive urban governance models capable of sustaining the complex life of contemporary cities.

10.7 Final Reflections

This chapter bridges theory and practice by advancing a biocultural perspective on urban sustainability, integrating ethical governance, cultural resilience, and participatory approaches to foster socially equitable and ecologically grounded cities.

The findings of this research affirm that sustainability is not merely a technical or managerial endeavor but a cultural and ethical project—a continuous process of dialogue between humans, environments, and institutions. The biocultural city offers a way of seeing and shaping urban life that foregrounds interdependence, reflexivity, and creativity as the foundations of resilience. In Ixelles, this vision is expressed through the coexistence of multiple identities, narratives, and practices that together form the living texture of the city. Despite disparities, these interactions reveal an underlying capacity for cooperation and renewal, testifying to the strength of cultural life as a resource for sustainability.

In the context of urban sustainability, the biocultural approach highlights how cultural expressions—traditions, architectural forms, artistic production, and collective rituals—are embedded in ecological processes. These elements not only reflect a community's environmental awareness but also shape its capacity for resilience and innovation. As such, culture becomes a key driver of sustainable transformation, In conclusion, the biocultural city stands not only as a theoretical construct but as a vision for the future of urban life. It invites scholars, policymakers, and citizens to embrace a deeper understanding of sustainability as a shared cultural commitment—one that binds ecological responsibility with social justice and creative expression. The experience of Ixelles demonstrates that when culture, ecology, and governance are aligned within a participatory and reflexive framework, the city becomes more than a physical territory: it becomes a living organism capable of sustaining both the planet and the plural worlds of human meaning that inhabit it.

The concept of **Biocultural City** based on biocultural approach and biocultural development synthesizes Foucault's notion of biopolitics with the cultural dimension of human society, producing a framework that acknowledges the entanglement of power, knowledge, identity, and environmental sustainability. Biopolitics, as conceptualized by Michel Foucault, refers to the way in which political power exerts control over human life through scientific knowledge, surveillance mechanisms, and institutional strategies (Foucault, 1978). In this context, life is not simply biological but becomes a subject of political governance. Foucault's distinction between sovereign power and biopower is key to understanding how societies shifted from disciplining individuals to managing populations. In works such as Discipline and Punish (1975) and The History of Sexuality (1978), Foucault explores how power operates not only through direct oppression but through normalization, regulation, and medicalization. These insights allow us to trace how governmental systems structure life from molecular (health, reproduction, genes) to the societal (education, labor, social behavior), using science and institutions to shape populations' conduct. Biocultural development expands biopolitics by incorporating cultural and environmental sustainability into the management of human life. In this framework, power is not only biological and institutional, but also symbolic and epistemological.

Policies in areas such as education, public health, and environmental regulation shape social practices while often marginalizing indigenous identities and traditional ecological knowledge. For instance, modern agricultural reforms may displace local farming practices that embody sustainable, culturally embedded forms of land stewardship.

Critically, the model of the biocultural city calls for a reconsideration of sustainability that is not solely technocratic or economized but rooted in local knowledge systems and cultural continuity.

It promotes coexistence between scientific modernity and traditional lifeways, aiming to preserve both ecological integrity and cultural diversity (Escobar, 2008).

This approach addresses criticisms leveled at sustainable development paradigms that neglect local agency and cultural specificity in favor of universal, top-down models.

Theoretically, the biocultural approach and the biocultural model rest on a post-structuralist critique of power. By integrating Foucault's theories, it reveals how control is exercised not just through institutions but through norms, sciences, and discourses that define the boundaries of legitimate knowledge and behavior. Practically, it offers a model for inclusive policy-making that respects cultural autonomy and ecological interdependence.

In fields like eco-tourism, this means privileging community narratives, protecting biodiversity, and resisting mass consumerist approaches that erode local identities. Furthermore, biocultural development aligns with calls for epistemic justice in global development discourse. It challenges hegemonic Western scientific paradigms and values indigenous perspectives as equally valid sources of knowledge. It thus represents a paradigmatic shift toward pluralism and relational sustainability, moving beyond anthropocentrism and toward multispecies and place-based ethics.

Biocultural approaches facilitate synthesis in action and encourage composite programs, aligning with contemporary policy methods stemming from polycrisis and wicked problems theories that advocate for interdisciplinarity and cross-sector policy cooperation.

As global societies grapple with ecological crises and cultural erosion, such integrated models of development are essential for fostering a just and resilient future.

Sustainable development in the twenty-first century has evolved into a multidimensional concept that extends beyond environmental concerns to encompass cultural identity, social inclusion, and governance. The European Union's Agenda 2021 and the global Agenda 2030 provide guiding frameworks for sustainability that highlight the importance of cultural and ecological integrity as essential to long-term prosperity.

In this framework, the recent decades, cultural policy has evolved from a sector-specific concern into a multidimensional field that intersects with economic, social, and environmental agendas. This transformation reflects a growing recognition of culture's pivotal role in shaping collective identities, influencing behavior, and fostering social cohesion (Throsby, 2010; UNESCO, 2013). Culture is no longer perceived merely as heritage or artistic production but as a dynamic force that drives innovation and contributes directly to sustainable development.

Urban governance has become a central concept in the analysis of contemporary cities. It encompasses the processes through which local authorities, private actors, civil society, and citizens interact to manage urban change and shape collective futures (Pierre, 2011; UN-Habitat, 2020). Within this framework, governance is no longer viewed merely as top-down administration but as a participatory and multi-scalar system of negotiation, coordination, and shared responsibility.

The implications extend beyond Ixelles. As cities across Europe grapple with climate crises, migration, and social fragmentation, this study offers both a methodological model and a political call to action for sustainability to be effective, it must be rooted in equity, co-production, and the recognition of diverse urban voices.

These insights allow us to trace how governmental systems structure life from the molecular (health, reproduction, genes) to societal (education, labor, social behavior), using science and institutions to shape the population's conduct.

Overall, the true innovation of this study resides in its reconfiguration of the prevailing paradigm—and, by extension, of the dominant narrative—by positioning culture at the very heart of perception, decision-making, policy, and governance. In this re-envisioned framework, culture functions as the connective tissue of the urban system, traversing and integrating all policy domains to ensure coherence, inclusivity, and transformative capacity.

In addition, this study opens new areas of questioning:

- ***Who defines the "center" of political integration?***
- ***Who always remains in the "shadow" of the consultation?***
- ***And, most importantly, who has the right to interpret the needs of others without them?***

In line with Mary B. Anderson's (1999) Do No Harm framework, the findings of this study underscore the ethical imperative for urban governance to remain aware of the unintended consequences of sustainability interventions. Just as Anderson demonstrated that humanitarian and development assistance can either exacerbate existing divisions or strengthen social cohesion, biocultural urban policies may likewise reproduce inequalities if they are detached from local realities and cultural sensitivities. The Do No Harm principle therefore complements the biocultural approach by foregrounding reflexivity, inclusivity, and ethical awareness as essential conditions for sustainable transformation. Applied to the governance of Ixelles and other urban contexts, it calls for policies that not only avoid harm but actively cultivate trust, dialogue, and cooperation across cultural and ecological divides.

In this context, the model of a biocultural city is not a visionary future proposal but an urgent necessity for the survival of our world. It encapsulates a paradigm of coexistence grounded in care, justice, and interdependence—values that guide societies toward a more sustainable and humane urban condition. The biocultural city stands as both an analytical lens and a call to action, reminding us that the ethics of "doing no harm" must extend beyond humanitarian aid to encompass the everyday governance of life itself. Only by aligning cultural, ecological, and political systems within a reflexive and participatory framework can cities truly become agents of regeneration and custodians of the shared planetary future.

Anti-Epilogue

Our Vision Statement: *"The Biocultural City is not an ideal utopia but a living reality in formation—a city where humans, cultures, and ecosystems co-create a shared future of justice, creativity, and resilience."*

European Cities of the North and South: An Interplay of Dystopia and Eutopia

> "There's one left that you never talk about."
> Marco Polo bowed his head.
> "Venice," said Khan.
> Marco smiled. "And what else did you think I was talking to you about?"
> The emperor didn't even bat an eyelid. "And yet, I never heard you say her name."
> And Polo: "Every time I describe a city, I say something about Venice." (Calvino 1972, p. 112)

Calvino's Invisible Cities functions as a magnet, drawing together utopia and memory, dream and architecture. Among his imagined cities, Zenobia (pp. 57–58) and Valdrada (pp. 75–76) appear as small reflections of Venice—each lacking something essential in its charm. One stands on stilts without water; the other floats upon water, haunted by mirrored lives. Thus, the enigmatic Marco Polo seems to recount many cities, but he is in fact describing one. As Kublai Khan gradually realizes:

> Kublai Khan had realised that Marco Polo's cities resembled each other, as if passing from one to another did not imply a journey but simply a change of their elements. (Calvino, 1972, p. 65)

How distinct, then, is urban life in the Mediterranean compared with that of Northern Europe? This question invites reflection not only on urban form but also on the multiplicity of cultures that animate city life. The contrast becomes vivid when we juxtapose Calvino's luminous, utopian cities with the often-bleak urban landscapes portrayed by James Joyce in Dubliners and Ulysses, alongside other Northern European literary depictions. Although neither author explicitly situates his work in the Mediterranean, Calvino evokes its spirit and sensuality, whereas Joyce's cityscapes resonate with the moral austerity and psychological confinement associated with Northern urbanism.

Here lies the essential divergence: the Mediterranean city as a cultural magnet and site of conviviality, versus the densely populated, morally burdened, and environmentally degraded metropolizes of the North. Joyce's anti-urban portrayals align with a wider Anglo-American literary tradition that idealizes rural life while casting cities as dystopian spaces. Scholars such as Raymond Williams (1972) and Carl Schorske (1963) have traced this sentiment through English and Irish literature from the eighteenth century onwards, in authors like Thomas Hardy and D. H. Lawrence, for whom the countryside represents moral authenticity against the corruption of the city. The creation of suburbs, garden cities, and public parks in Northern Europe—and the social distancing of residents from industrial urban cores—may be understood as physical manifestations of this persistent anti-urban impulse.

In contrast, the Mediterranean urban tradition subverts this dystopian imaginary. Southern European cities have historically functioned as cradles of culture, innovation, and civic life. Their vitality stems from the enduring legacy of the polis and the city-state—urban cultures that flourished from antiquity through the Renaissance and continue to illuminate the collective memory of the region (Bitsani, 2013, 2016). The Greek conception of urbanism (astikotita), unlike its often pejorative Western usage, carries positive moral and intellectual connotations. Southern cities maintained their dynamism well into the twentieth century, even as many Northern cities underwent de-urbanization. Plato perhaps expressed this ideal most eloquently when he described the city as "the highest expression of human intellect and of community with others" (Plato, 1974, p. 11). Calvino stands firmly within this lineage.

From Marco Polo's narratives emerge places that are multifaceted, luminous, and contradictory—much like the Mediterranean cities themselves. Venice and Barcelona, Naples and Athens, Rome and Marseille: all are kaleidoscopic spaces at the crossroads of worlds between North and South, East and West. These are urban eutopias—centers of culture, imagination, and exchange—yet, like Calvino's cities, they contain within them their own shadows, their own latent dystopias.

Ultimately, Invisible Cities challenges urban scholars, planners, and citizens to think differently about the city—not as a fixed entity to be managed, but as a living process to be interpreted, co-created, and felt. In this light, literature becomes a crucial dimension of critical urban theory, capable of revealing the contradictions, aspirations, and spectral memories that inhabit every street and skyline.

Ixelles and the Invisible Mediterranean Cities

Through years of observation, research, and writing about Ixelles, we gradually came to realize that we were, in essence, writing about our own Mediterranean cities—our own invisible cities, to borrow Calvino's phrase. Beneath the surface of Ixelles, we encountered the spirit of Bari and Venice, Thessaloniki and Barcelona, Rhodes and Marseille. These southern cities were not absent; they were hidden within the urban fabric of Ixelles, shaping its cultural rhythms, its forms of conviviality, and its subtle resistance to alienation.

In our view, this synthesis reflects the profound success of the municipality under **Mayor Christos Doulkeridis**, whose vision of ecological and intercultural urbanism has reimagined Ixelles as a village within the metropolis—a Mediterranean enclave in the heart of industrial Brussels. Doulkeridis, of Greek origin, sought to transplant something of the warmth and solidarity of his childhood summers on the island of Rhodes into the civic life of Ixelles, and he succeeded. As he stated in an early interview, his aim was to "build a village within the city"—a statement that, when read through the lens of this research, encapsulates the essence of the biocultural city as a living synthesis of nature, community, and culture.

In intercultural Ixelles we thus rediscovered the invisible Mediterranean cities that dwell within the heart of Europe: Bari and Venice, Marrakech and Fez, Thessaloniki and Granada, Chania and Corfu, Rhodes and Constantinople. These symbolic geographies inhabit the imagination of the biocultural city, offering a vision of sustainability grounded not in domination but in coexistence—with humanity not as a sovereign over nature, but as a companion and guardian of the landscape.

Final Reflection

> I think I have written something like a final love poem for cities, at a time when it is becoming increasingly difficult to live in them. Perhaps we are approaching a time of crisis in urban life, and Invisible Cities is a dream born from the heart of invisible cities. (Calvino, 1972, p. 16)

This study, too, is our own final love poem to the city—to the cities in which we were born, raised, and continue to live. Like Calvino, we write in a time of crisis for urban life, when questions of sustainability, belonging, and justice demand new ways of imagining the urban future. Through the lens of the biocultural city, we have sought to illuminate the invisible connections that sustain urban existence: between ecology and culture, between governance and imagination, between the visible structures of policy and the unseen textures of everyday life.

In this sense, the anti-epilogue is both an ending and a beginning — a reflection on the living cities that inspire our research and the invisible cities that dwell within them. Together, they form a single vision: the biocultural city as a space of shared creation, where humanity, culture, and nature co-evolve in pursuit of a more just and resilient future.

So, this book is our last love poem for our cities, in which we were born, raised, and live, and is dedicated to them, as Calvin would say.

Conflicts of Interest The authors have no conflicts of interest to declare that are relevant to the content of this chapter.

References

Anderson, M. B. (1999). *Do no harm: How aid can support peace—Or war*. Lynne Rienner Publishers.

Berkes, F. (2008). *Sacred ecology* (2nd ed.). Routledge.

Bitsani, E. (2013). *"Intercultural City identity" and "human intercultural cities" (H.I.C). An ontological/dynamic model for the social cohesion of modern and post-modern cities. The research example of Trieste Italy*. Nova Publishers.

Bitsani, E. (2016). Migration memory, cultural heritage: A vehicle of the city' intercultural identity. The case study of the Eastern Greek Community and the Greek museums in Trieste Italy. *Cogents Arts and Humanities, 3*(1), 2–15.

Bourdieu, P. (1986). The forms of capital. In J. Richardson (Ed.), *Handbook of theory and research for the sociology of education* (pp. 241–258). Greenwood. https://www.marxists.org/reference/subject/philosophy/works/fr/bourdieu-forms-capital.htm

Calvino, I. (1972). *Le città invisibili*. Einaudi.

Charmaz, K., & Thornberg, R. (2020). The pursuit of quality in grounded theory. *Qualitative Research in Psychology, 18*(3), 305–327. https://doi.org/10.1080/14780887.2020.1780357

Denzin, N. K. (1978). Triangulation: A case for methodological evaluation and combination. *Sociological Methods*, 339–357. https://www.scirp.org/reference/referencespapers?referenceid=2398870

Escobar, A. (2008). *Territories of difference: Place, movements, life, Redes*. Duke University Press.

Foucault, M. (1978). *The history of sexuality, Volume 1: An introduction*. Vintage Books.

Guba, E. G., & Lincoln, Y. S. (1994). Competing paradigms in qualitative research. In N. K. Denzin & Y. S. Lincoln (Eds.), *Handbook of qualitative research* (pp. 105–117). Sage.

Healey, P. (1997). *Collaborative planning: Shaping places in fragmented societies*. Macmillan Press. https://doi.org/10.1007/978-1-349-25538-2

Pierre, J. (2011). *The politics of urban governance*. Palgrave Macmillan.

Plato. (1974). *The republic* (2nd ed.). Penguin Books.

Putnam, R. D., Leonardi, R., & Nanetti, R. Y. (1993). *Making democracy work: Civic traditions in modern Italy*. Princeton University Press.

Saunders, M., Lewis, P., & Thornhill, A. (2007). *Research methods for business students* (4th ed.). Financial Times Prentice Hall.

Schorske, C. E. (1963). *The idea of the city in European thought: Voltaire to Spengler*. Princeton University Press.

Throsby, D. (2010). *The economics of cultural policy*. Cambridge University Press.

UNESCO. (2013). *The Hangzhou declaration: Placing culture at the heart of sustainable development policies*. United Nations Educational, Scientific and Cultural Organization. https://unesdoc.unesco.org/ark:/48223/pf0000221238

UN-Habitat. (2020). *World cities report 2020: The value of sustainable urbanization*. United Nations Human Settlements Programme. https://unhabitat.org/sites/default/files/2020/10/wcr_2020_report.pdf

Voß, J.-P., Bauknecht, D., & Kemp, R. (2006). *Reflexive governance for sustainable development*. Edward Elgar. https://doi.org/10.14279/depositonce-4543

Williams, R. (1972). *Thomas hardy and rural England*. Chatto and Windus.

Index

E. Bitsani et al., *The Biocultural City as a Utopia or the Future Reality?*, Cities, Heritage and Transformation, https://doi.org/10.1007/978-3-032-17584-7

The manufacturer's authorised representative in the EU is Springer Nature Customer Service Centre GmbH, Europaplatz 3, 69115 Heidelberg, Germany. If you have any concerns regarding our products, please contact ProductSafety@springernature.com

Printed and bound by CPI Group (UK) Ltd, Croydon, CR0 4YY
16/09/2026
02217815-0006